Artificial Plasmonics for VLSI Interconnects

Artificial Plasmonics for VLSI Interconnects

Bridging the Gap between Electronics and Optics

Soumitra R. Joy

University of North Carolina at Charlotte
United States

Pinaki Mazumder

University of Michigan
United States

IEEE Press
WILEY

Published by John Wiley & Sons, Inc., Hoboken, New Jersey.

Library of Congress Cataloging-in-Publication Data has been applied for:

Hardback ISBN: 9781394289950
ePDF ISBN: 9781394289981
ePub ISBN: 9781394289967
oBook ISBN: 9781394290000

Cover Design: Wiley
Cover Image: Imagery by the Authors, © MF3d/Getty Images

To our children and grandchildren

Contents

Preface *xiii*
Acknowledgments *xvii*
About the Companion Website *xix*

1 Prospects and Pitfalls of Modern Interconnect Technologies *1*
1.1 Overview and Motivation *1*
1.1.1 Problem Specifics *2*
1.2 Communications Challenges: Human-Level vs Machine-Level *4*
1.3 Modes of Interconnects: A Technology Gap *5*
1.4 Innovations in Interconnect Frontier *6*
1.4.1 Interconnect Research at Material Level *6*
1.4.2 Interconnect Research at the Network Level *8*
1.4.3 Interconnect Research at Waveguide Level *10*
1.4.4 Chip-scale Interconnect Technologies: Major Industrial Steppingstones *14*
1.5 Scaling Issue of System Level Interconnect *16*
1.5.1 Chip Package Signaling *18*
1.5.2 Issue of Crosstalk *20*
1.5.3 High Power Consumption in System-level Interconnect *22*
1.6 Optical Interconnect: Evolution Toward Chip-scale Communication *23*
1.6.1 Integrated Photonic Circuits on Silicon *24*
1.6.2 Is Optical Interconnect Viable at Short Range? *27*
1.6.3 Wireless Network on Chip *29*
1.6.4 Carbon Nanotube Interconnect *32*

1.7 Complexity and Dilemma in Data Transfer *33*
1.7.1 The Last Centimeter Barrier: The Most Critical Distance in Data Transfer *34*
1.7.2 Time Variation in Data Traffic: A Dilemma in Selecting Interconnect Technology *35*
1.7.3 Do We Have Any Alternative Interconnect Technology in Hand? *35*
1.7.4 Spoof Plasmon Interconnect: A New Paradigm in Communication Technology *36*
1.8 Research on Spoof Plasmon Wave: Toward CMOS Compatibility *42*
1.8.1 Leading Researchers on Spoof Surface Plasmon Technology *43*
1.9 Summary of the Chapter *47*
References *47*

2 Spoof Plasmonics: Origin and State-of-the-Art Development *57*
2.1 Slow Wave Structure: A Historical Perspective *57*
2.1.1 Variants on Slow Wave Structures *59*
2.2 Surface Plasmon Polariton in Metal *62*
2.2.1 Extraordinary Transmission by Surface Plasmonics *64*
2.3 Surface Plasmon Polariton: Explanation Through Drude's Model *66*
2.4 SSPP in Planar Geometry *69*
2.4.1 1D Groove Pattern *69*
2.5 SSPP-based THz Circuits: Research in Mazumder Laboratory *76*
2.5.1 Electromagnetic Analysis of the SSPP Mode *78*
2.5.2 THz SSPP Switch Design *79*
2.5.3 THz SSPP Circuit Component Design *82*
2.5.4 THz Biosensor Design *91*
2.6 Particle-Motion Control by SSPP Waveguide *93*
2.7 Recent Advances in Spoof Plasmonics *94*
2.8 Conclusion *97*
References *98*

3 Fundamental Electrodynamics of Spoof Plasmonic Mode *103*
3.1 Baleen Whales: What They Teach Us on Novel Communication *103*
3.2 Plasmonics Aided High Speed VLSI Communication *104*
3.2.1 Innovative Spice Simulation Tool Development for Plasmonics *106*
3.3 A Universal Theoretical Framework for Spoof Plasmonics *109*
3.3.1 Dispersion Law of SSPP Waveguides *111*
3.4 Electrodynamics of Spoof Plasmon in Finite Structure *116*
3.5 Modal Analysis of Spoof Plasmon *117*

3.5.1 Dispersion Relation in Wide Structures *117*
3.5.2 Impact of Dielectric Half-space on Dispersion *119*
3.5.3 Dispersion Relation Correction in Structures of Finite Thickness *120*
3.5.4 Effect of Substrate *123*
3.5.5 Effective Refractive Index of a Planar SSPP *124*
3.6 Thin-film of SSPP *124*
3.6.1 The Concept of Effective Thickness *124*
3.6.2 Capacitance Between Edges of Neighboring Conductor Plates *126*
3.6.3 Effect of a Substrate on Thin SSPP *127*
3.7 Properties of Confined Modes *129*
3.7.1 Degree of Confinement *129*
3.7.2 Bandwidth Modulation of Confined Mode *131*
3.8 Summary of the Chapter *132*
References *133*

4 Information Capacity of Spoof Plasmonic Interconnect *137*
4.1 The 1858 Transatlantic Telegraph: Lessons from a Failed Project *137*
4.2 Data Transfer Through Noisy Channel: Condition the Signal, Don't "Brute-Force" *138*
4.3 Challenges in Millimeter-scale Communication: A Call for Innovation Beyond Shannon's Paradigm *140*
4.4 Millimeters-scale Communication: Its Growing Relevance in Data-driven World *141*
4.5 A Growing Industry Investment in Millimeter-scale Chip Packaging *142*
4.6 Quest for a Fundamentally Different Propagation Mode for Millimeter-scale Packaging *144*
4.7 Limitations of Standard Interconnect Technologies *145*
4.8 Authors' Contribution to the Field of Interconnect Design *145*
4.9 Bandwidth in Crosstalk-mediated SSPP Channels *147*
4.10 Traveling Length of SSPP Mode in Lossy Metal *152*
4.11 Information Capacity in the Limit of Thermal Noise *154*
4.12 SSPP Interconnect in Comparison with Others: The Benefit of Minimized Interference *157*
4.12.1 Crosstalk in Optical Interconnect *158*
4.12.2 Crosstalk in Electrical Interconnects *160*
4.12.3 Crosstalk in Spoof Plasmon Interconnects *162*
4.13 Dual Mode in Spoof Plasmon Waveguide *165*
4.14 Summary *166*
References *167*

5 Augmented Bandwidth by Spoof Plasmonics *171*
5.1 Introduction *171*
5.2 Background Studies: Severity of Crosstalk *171*
5.3 Conventional Strategies for Crosstalk Reduction *173*
5.4 Authors' Contribution: Dealing with Crosstalk in Data Bus *175*
5.4.1 Advantages of Proposed Interconnect *175*
5.4.2 Crosstalk Limited Bandwidth in Lossless Electrical Bus *177*
5.4.3 Crosstalk Limited Bandwidth in Lossless SSPP Bus *179*
5.5 Hybrid-SSPP Mode: Theory and Property Analysis *183*
5.5.1 Coupling Between Two Channels of Hybrid-SSPP Mode *188*
5.5.2 Coupling Among a Large Number of Parallel Hybrid-SSPP Channels *190*
5.5.3 Design Technique of Hybrid-SSPP for High Frequency Modulated Data Transmission *192*
5.6 Optimal Design Technique for Hybrid-SSPP Waveguide for Baseband Communication *193*
5.7 Experimental Characterizations of SSPP Data Bus *194*
5.7.1 Verification of Electronic-SSPP Mode *195*
5.7.2 Verification of Optical-SSPP Mode *196*
5.7.3 Microfabrication Process *198*
5.8 Mechanism for Bandwidth Augmentation *198*
5.8.1 Experimental Validation of Bandwidth Enhancement *208*
5.9 How Spoof Plasmon Advances the Engineering of Interconnect *210*
5.10 Summary *212*
References *212*

6 Signal Modulation by Spoof Plasmonics *215*
6.1 Introduction *215*
6.2 Background Studies: Design of Modulator *215*
6.3 Authors' Contribution in the Field of Controlling Spoof Plasmon *217*
6.4 Transmission Spectra of Homogeneous and Heterogeneous Structures *218*
6.5 SSPP Scattering in Heterogeneous Structures *221*
6.6 *Q*-factor and Enhanced Radiation Rate *225*
6.7 Dynamic Switching of SSPP Transmission Property *226*
6.7.1 SSPP Dispersion in the Presence of Modulator: Theoretical Framework *227*
6.7.2 Prediction of Dispersion-limited Modulation Speed *232*
6.7.3 Analysis of Energy Efficiency *234*
6.7.4 Numerical Analysis *235*
6.7.5 Trade-off Between Modulation Speed and Energy-efficiency *237*

6.7.6 Extension of the Theory for a Broad Class of SSPP Modulators *238*
6.8 Experimental Considerations and Signal Modulation *240*
6.8.1 Design Considerations *240*
6.8.2 Transmission Characteristics *241*
6.8.3 Modulation Characteristics *244*
6.9 Summary *248*
References *248*

7 Process Variation Effect on Spoof Plasmonic Interconnect: Compensations *253*
7.1 Introduction *253*
7.2 Background Studies: Process Variation in Interconnects *253*
7.3 Author's Contribution in the Field of Spoof Plasmon Signal Restoration *254*
7.4 Frequency Response of SSPP Channel *254*
7.4.1 SSPP Channel with Ideal Pattern *254*
7.4.2 SSPP with Pattern Irregularity *255*
7.5 Performance Loss for Structural Imperfections *257*
7.5.1 Bandwidth Degradations *257*
7.5.2 Loss of Signal Integrity *258*
7.6 Mitigation of Performance Degradation *260*
7.6.1 Mathematical Functions to Compensate for Signal Loss *260*
7.6.2 Nonlinear Circuit Design for Real-time Compensations *261*
7.6.3 Dynamic Tunability of Compensation Circuit *262*
7.7 Summary *263*
References *264*

8 Future Research Avenues for Spoof Plasmonic Interconnects *265*
8.1 Introduction *265*
8.2 New Research Frontiers for Spoof Plasmonic Interconnect *266*
8.2.1 New Material Integration with Spoof Plasmonics *266*
8.2.2 Spoof Plasmonics for Higher Device Integration *268*
8.2.3 Nanoscale Realization of Spoof Plasmonic Interconnect *269*
8.2.4 Development of Multi-level Computation Model for THz Surface Wave Network *270*
8.3 Summary of the Chapter *273*

Index *275*

Preface

The accelerating demand for data driven by advancements in AI and large-scale model training has made data centers consume substantial amounts of energy. This computational enhancement, while being transformative, raises significant concerns about energy sustainability. As power requirements become increasingly critical, the need for continuing innovations in energy-efficient computing has become imperative.

At the heart of modern computing are transistors and interconnects. For decades, transistor miniaturization fueled performance scaling. However, as transistor limits become apparent, the focus has shifted to interconnects, now crucial to integrating multiple dies/chiplets and extending computational performance in advanced VLSI design.

Despite their newfound prominence, interconnects face a paradox: unlike transistors, their performance degrades with downscaling. Reduced wire cross sections increase resistance, and higher density in advanced nodes elevates capacitance. These challenges worsen signal integrity, limit the bandwidth, and contribute significantly to power dissipation, bottlenecking terabit-per-second data transfer. Existing solutions offer only incremental improvements, leaving the core problem unresolved.

This book *Artificial Plasmonics for VLSI Interconnects: Bridging the Gap Between Electronics and Optics* introduces a fundamentally novel approach. It presents a pioneering concept that manipulates electromagnetic waves to enhance interconnect performance. By ingeniously engineering the electromagnetic environment, we demonstrate a viable path to overcome traditional interconnect limitations, enabling unprecedented energy efficiency, bandwidth, and signal integrity for future computing architectures.

Much of the new discoveries on spoof plasmon interconnects in this book originated from the Mazumder Lab at the University of Michigan, led by Prof. Pinaki

Mazumder, who has researched surface plasmonics and spoof surface plasmonics for over a decade in the timeframe of 2008 to 2020. Dr Soumitra Joy, the lead author, is the latest PhD graduate from his laboratory to contribute to this subfield. After working at Intel, Dr Joy returned to academia in 2024, joining the University of North Carolina at Charlotte with a vision to advance high-performance, reliable interconnects for next-generation computing.

For graduate students and researchers specializing in VLSI design, the inherent challenges of interconnects are acutely recognized. While optical interconnects offer compelling advantages for long-distance data transmission, and electrical interconnects suffice for micron-scale distance, limited by their RC delay and parasitic coupling, a critical technological gap persists for centimeter-scale communication, particularly concerning terascale data transfer among electronic chiplets. This book attempts to address this significant gap by introducing an emergent technology: the spoof surface plasmon polariton (SSPP) waveguide. We elucidate its profound potential when deployed especially in on-package and on-board integration scenarios. Owing to the unique electromagnetic interaction, the resultant guided modes are robustly confined to the surface of the structure. This strong confinement ensures that adjacent interconnects maintain substantial electrical isolation even within highly dense designs. This considerable reduction in cross-coupling among conductors not only enhances signal quality but also augments bandwidth, substantially decreases the bit-error-rate, and ultimately conserves both power consumption and the silicon area that would otherwise be allocated for signal compensation circuitry and buffer designs.

Artificial Plasmonics for VLSI Interconnects: Bridging the Gap Between Electronics and Optics systematically delineates the operational principles and design methodologies of this novel interconnect technology across multiple hierarchical levels. In Chapter 1, we commence with a comparative analysis of various state-of-the-art technologies in interconnect design, thereby identifying the existing technological gap and articulating the compelling necessity for a new class of interconnects based on the artificial plasmonic mode, more formally known as spoof surface plasmon polariton. Following this, in Chapter 2, we discuss the origination of the concept of spoof surface plasmon polariton from other branches of microwave engineering and optical science, and its state-of-the-art development in various applications ranging from sensing, computing, and signal processing. We also discuss the contribution of Mazumder Laboratory in this emergent field of research on spoof plasmonics. In Chapter 3, we discuss the fundamental physics of electrodynamics governing the propagation of this special mode through periodic structures. We also discuss the technique of generalizing the analysis to cover a wide range of variation of spoof plasmonic waveguides. In Chapter 4, we establish analytical solutions to rigorously analyze signal transmission, crosstalk, and signal integrity within these systems. Chapter 5 details the technique of augmenting the bandwidth of standard electrical interconnects with periodically structured, spoof plasmonic waveguide and compares the improvement in signal integrity. A theoretical framework for optimization of such interconnect designs and experimental verification of the same are provided. Chapter 6 discusses the details of modulation circuit design for spoof plasmonic mode based the data transfer system and estimates the relationship between energy efficiency and modulation speed. Chapter 7 examines practical considerations, such as imperfections inherent in fabrication technology, and their consequent impact on the performance of the newly introduced waveguides, concurrently developing circuits to compensate for any resulting losses. Finally, in Chapter 8, we explore prospective future directions that may leverage this novel technology and outlines avenues for further research to meet the material integration challenges, circuit component design, and system development, to fully exploit its extensive potential. This comprehensive exploration aims at furnishing the readers with requisite knowledge and tools to harness artificial plasmonics for the next-generation high-performance, energy-efficient VLSI systems.

Acknowledgments

This book is a culmination of many years of research into artificial plasmonics for development of new interconnects in VLSI technology. A significant portion of this work took place during Dr Joy's doctoral research in Dr Mazumder's laboratory at the University of Michigan. We extend our sincere gratitude to the current and former graduate students, post-docs, and research scholars in the Mazumder Lab, as well as to the collaborating research groups worldwide, for their significant contributions, insights, and support.

We would like to specifically thank the following individuals whose essential work forms the foundation of this book and resulted in them earning their PhD degree from the University of Michigan:

- K. Song, for his work on active terahertz switches based on spoof surface plasmon polariton waveguides
- X. Zhao, for his work on biosensor design utilizing double-sided, enclosed corrugated structures
- M. Aghadjani, for his work on spoof plasmon polariton-based THz circuit components, including beam steering, analog-to-digital converters, and filter designs

We also wish to acknowledge the early work of Jack East and Dr George Haddad, who provided important insights in the initial stages of this research.

In addition, we express our profound thanks to:

- Dr M. Erementchouk, who was instrumental in developing the theoretical framework for spoof plasmonic mode formation and propagation
- Dr F. Lan, a visiting scholar at the University of Michigan, who provided invaluable fabrication and measurement support

- Dr Z. Baten and his research laboratory in Bangladesh for advancing our understanding of spoof plasmonic wave propagation and its application in data transfer

This research was made possible by the generous and prolonged support of the National Science Foundation (NSF) and the Air Force Office of Scientific Research (AFOSR). We are especially grateful to Dr Gernot Pomerenke of the AFOSR, who was a steadfast champion of this work and provided critical funding from 2006 to 2020.

Finally, we thank our families for their unwavering support and sacrifice. Without them, this book could not have been written. We hope this book will be a valuable resource for researchers in the field of VLSI technology.

About the Companion Website

This book is accompanied by a companion website.

https://www.wiley.com/go/artificialplasmonics

This website includes:

- Images from the book rendered in color

1

Prospects and Pitfalls of Modern Interconnect Technologies

1.1 Overview and Motivation

In the brief time, it takes you to read this sentence—about 10 seconds—US data centers will consume 700 GWh of energy, enough to power the entire country of Nigeria for a minute. This massive energy demand, driven largely by the recent AI boom, is projected to triple by 2030, posing a risk of nationwide grid failures (see Figure 1.1).

This highlights the urgent need to examine the physics of power dissipation and innovate device materials to significantly cut energy consumption and ensure a sustainable computing future. A significant portion of this power consumption and heat generation comes from servers of data centers, which include processors, memory, and interconnects. About 20% of a server's power usage is due to interconnects and switches. High-density interconnects in package-level integrations are especially power-intensive due to high parasitics, causing excessive power loss, signal delay, distortion, and crosstalk. For instance, next-generation 6G chips, designed to handle 4–8 Tb/s bandwidth through wide-lane data buses, could have interconnect densities up to 50,000/cm^2. At these densities, crosstalk among short-reach interconnects might increase by 100 dB, and power loss could rise 1,000-fold, leading to severe power penalties and potential device malfunctions. The excessive device parasitics originate from the capacitance among the interconnect networks. Capacitances are proportional to the dielectric constant of the material. Traditionally, capacitance has been mitigated using materials with low dielectric constants (low-k materials) such as porous silicon dioxide, organosilicate glass, and polymeric dielectrics. However, even with advancements in material processing, no

Artificial Plasmonics for VLSI Interconnects: Bridging the Gap between Electronics and Optics, First Edition. Soumitra R. Joy and Pinaki Mazumder.
 Companion Website: https://www.wiley.com/go/artificialplasmonics

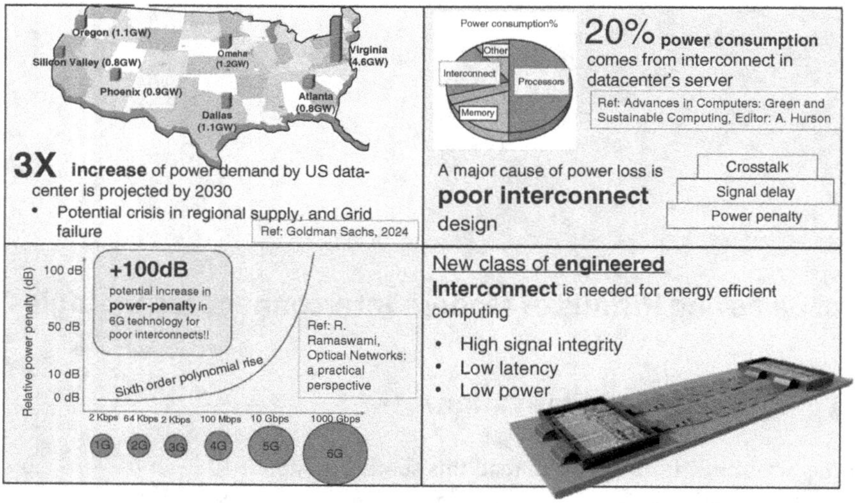

Figure 1.1 A dire power crisis has loomed in the advent of soaring demand of computing. A breakthrough technology for energy-efficient computing hardware is needed to address the issue. This proposal aims to introduce new technologies for chip-to-chip interconnects with potentially lower wire parasitic in order to realize energy-efficient, high performance computing.

natural material can achieve a dielectric constant less than 1, making it unlikely that current materials can reduce parasitic capacitances by more than an order of magnitude in silicon devices, which is nowhere near being sufficient. Therefore, a breakthrough technology for a new class of interconnect is essential to tackle the parasitic capacitance issue and thus mitigate the high-power penalties in microelectronic devices.

1.1.1 Problem Specifics

Design of high-performance interconnects is mission-critical for next-generation IC packaging to achieve silicon scaling and heterogeneous integration, while allowing optimization of all five major design variables of a technology IP, namely power, performance, area, cost, and time-to-market. Design of short-reach interconnects (≤1 cm) among chiplets in system-in-packages is particularly challenging due to high wire-parasitic among thousands of lanes, causing significant power loss and crosstalk.

Let us consider how catastrophic wire-parasitic effects can become in a dense interconnect fabric. In a state-of-the-art silicon die with 1,000 I/O pins packed into a 1 cm^2 area, mutual capacitance can create up to an astonishing 498,500 undesired parasitic paths through the connecting wires and silicon substrate. This vast number of parasitic paths can severely impair signal integrity and lead to significant energy loss.

There are two primary methods to mitigate parasitic capacitance: increasing the spacing between wires or reducing the dielectric constant of the substrate. Increasing spacing is impractical due to the loss of device density, while reducing the dielectric constant of the substrate is limited by the materials available. For example, replacing silicon ($\epsilon_r = 12$) with silicon dioxide ($\epsilon_r = 4$) only reduces capacitance by a factor of 3. Even using highly porous materials ($\epsilon_r \geq 1$) offers only a factor of 12 reduction, which is insufficient for the anticipated 100 dB increase in crosstalk in future technology nodes. This level of reduction falls short as we move from GHz to high-speed THz electronics (see Figure 1.2).

A potential solution could be the development of artificially engineered electromagnetic materials with an ultralow dielectric constant beyond what is achievable by natural materials. These engineered materials leverage deep subwavelength unit-cell spacing to create conjugate dipole moments that effectively cancel each other's electromagnetic induction, resulting in minimal parasitic capacitance. This approach depends more on geometric features than on the surrounding material properties and can be realized through advanced lithography.

This book seeks to discuss the development of a new technology for short-reach interconnects using standard complementary metal-oxide semiconductor (CMOS)-compatible materials and to engineer them in such a way as to behave as an artificial ultra-low-k material to minimize parasitics and improve power efficiency. The book will focus on modeling and system analysis of these interconnects to enable their integration into commercial design kits (PDKs), accelerating their adoption in microelectronics.

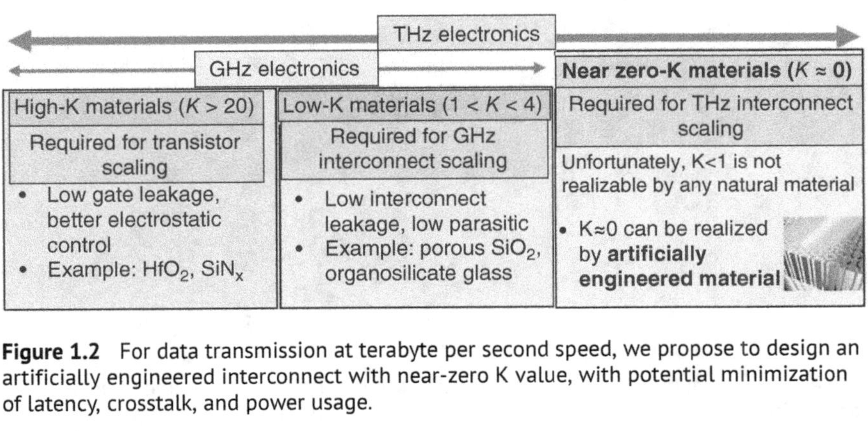

Figure 1.2 For data transmission at terabyte per second speed, we propose to design an artificially engineered interconnect with near-zero K value, with potential minimization of latency, crosstalk, and power usage.

1.2 Communications Challenges: Human-Level vs Machine-Level

When we deal with communication between person-to-person, typically the greater the distance of communication, the greater the complexity of the information transfer, owing to data loss, security and privacy loss, etc. But how about communication among machines? Does the information become safer from data leakage and crosstalk as we keep scaling down the length of communication distance? Well, quite counter-intuitively, no. In fact, the aggregate data density among machines increases exponentially with the scaling down of length of distance of interest. As the device area shrinks down, all pairs of point-to-point communication on the chip-scale begin to compete against each other, and signal integrity severely suffers from interference from neighboring channels, as illustrated in Figure 1.3. Keeping signal integrity unharmed from data leakage and interference is one of the Herculean challenges for modern chip design technology. For machines, especially machines at the scale of sub-micrometers and nanometers

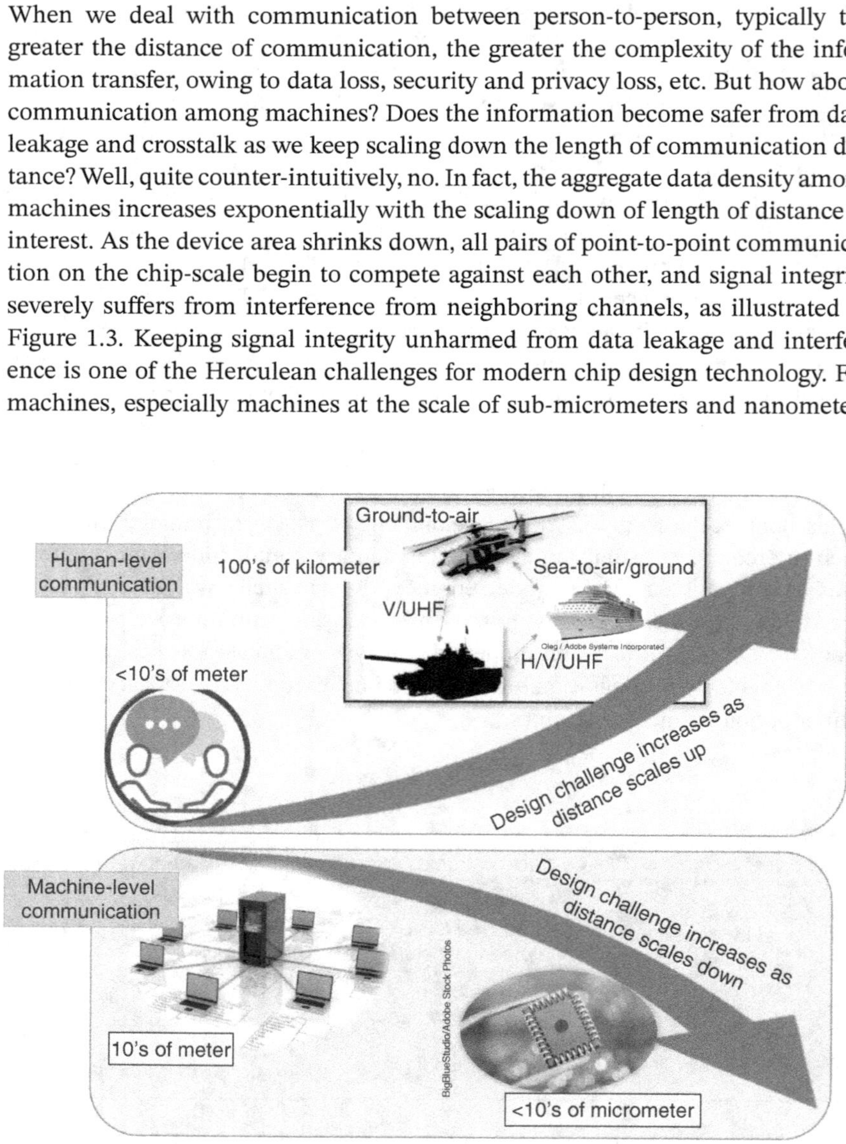

Figure 1.3 In human-level vs in machine-level communication, the degree of design complexity for preventing data leakage shows opposite trend with length scaling.

in a highly dense system, we need a very different kind of engineering strategy to prevent loss of signal integrity due to interference and crosstalk.

1.3 Modes of Interconnects: A Technology Gap

A modern computing system, by and large, is based on von Neumann architecture, and is made of three basic components: the processors, the memories, and a machine-to-human interface. The way all these critical units exchange information among themselves is called interconnects. Modern interconnect technologies used in digital systems have primarily emerged from two opposite ends of the electromagnetic spectrum [1–3]. On the electronic side of the spectrum, copper wires are employed to transfer data at the rate of nearly gigabits per second (Gbps). On the other hand, on the photonic side of the electromagnetic spectrum, optical interconnect technology with a wavelength of 1.55 μm (i.e., $\sim$ 200 THz) is used for communication between circuit boards and racks in large digital systems by deploying photons to transfer data at a much higher rate [4–7].

The two interconnect technologies, namely electrical and optical ones, rely on two different mechanisms of guiding electromagnetic modes through the channel. Electrical interconnects confine electromagnetic modes by setting a pair of conductors: each conductor carries equal current with opposite phase. As long as the spacing between the conductors is much smaller than the wavelength, the pair of wires guides electromagnetic waves along the longitudinal direction with negligible radiation to the surroundings. Optical interconnect, on the other hand, guides electromagnetic modes by using the difference in refractive index of the channel material with respect to its surroundings.

Figure 1.4 shows different types of interconnects, classified according to the distance they cover and the nominal data rates they are supposed to carry. As of its current standing, long distance, such as intercontinental or interdata center communication, is managed by optical interconnect, while the chip-level data transfer is supported by electrical interconnects. However, in the coming days, all the different levels of communication will keep shifting to their left in Figure 1.4, as the need for data rate will increase over time. At their current stature, it is clear that optical interconnects will remain the unchallenged candidate for long-haul communication. However, as the communication distance reduces down to chip scale, we will enter a gray area where neither optical nor electrical interconnect technology has a clear edge over the other. A great deal of research is going on in order to cover that coveted critical area of communication: some of the endeavors focus on pushing the limits of the present optical interconnect technology, some on pushing the limits of electrical ones. Meanwhile, a few researchers,

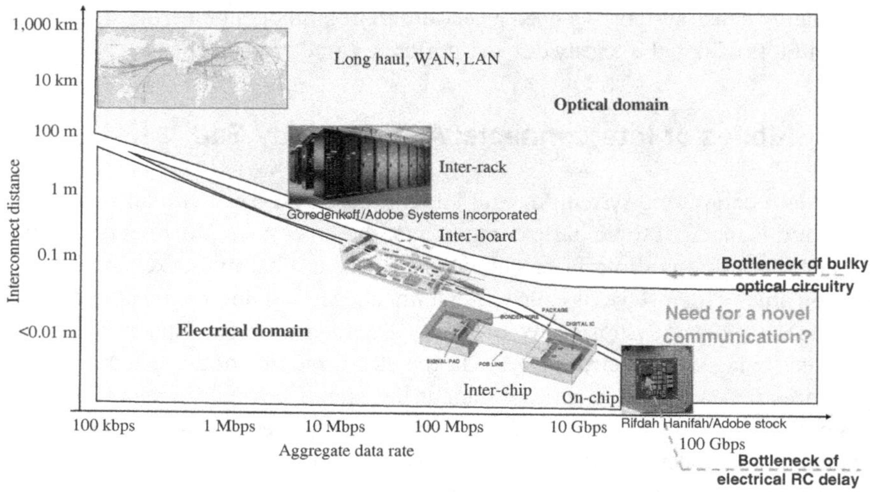

Figure 1.4 The illustration of how interconnects can be divided into multiple classes depending on what distance to cover and what data rate to support. While optical interconnects will remain unrivaled for long-haul communication for the foreseeable future, a technology gap between electrical and optical interconnect is emerging for centimeter-scale communication with data rate beyond 100 Gbps.

including the authors of the present book, have taken on venturing a novel, unforeseen interconnect technology with the aim to high-density information transfer while keeping energy consumption at a reasonable limit.

1.4 Innovations in Interconnect Frontier

From a bird's-eye view, we can categorize the approaches' innovation in interconnect domains into four different hierarchical levels, as shown in Figure 1.5.

- Innovation at the material level (cobalt, ruthenium are being considered for chip-scale communication; porous materials)
- Innovation at the architecture level (3D IC, reconfigurable differential pair channel)
- Innovation beyond von Neumann architecture (memory-in-processor)
- Innovation at the signal-guiding mechanism

1.4.1 Interconnect Research at Material Level

As chip scaling progresses below the 7 nm threshold, the interconnects at the chip level are undergoing significant transformations. The density of smaller transistors on a die is increasing, accompanied by a growing volume of data

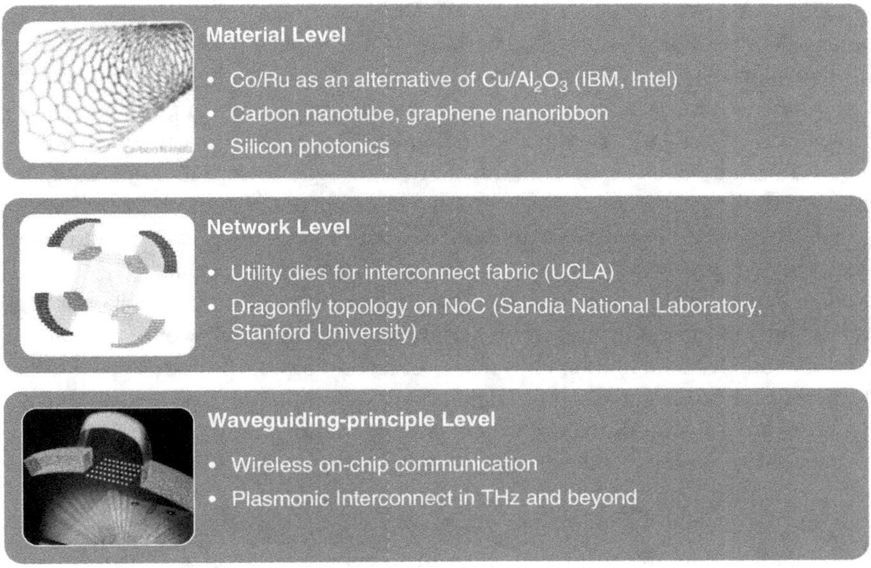

Figure 1.5 Different hierarchies of research venues on emerging interconnect technologies.

processing and transfer within and beyond chips or packages, as shown in Figure 1.26. Consequently, there are ongoing changes in the materials used for interconnects. Establishing reliable connections between different layers poses the primary challenge in interconnect fabrication. Copper, the long-standing material of choice for interconnects since the 130 nm node, has reached its limitations. One of the issues with copper is its tendency to diffuse into adjacent materials at advanced nodes. This necessitates the use of barrier layers, which must become thinner as scaling continues to 5 nm and beyond 3 nm while retaining conductivity. Additionally, copper encounters other challenges, including: (i) reduced reliability as interconnect dimensions decrease due to increased current density; (ii) increased resistivity at lower dimensions caused by grain-boundary scattering and surface scattering; (iii) rapidly rising resistivity due to Joule heating; and (iv) decreased conductivity at high operating frequencies resulting from the skin effect. Consequently, copper falls short of meeting the requirements for future technology demands in interconnect materials. The growing disparity between the needs of future on-chip interconnect materials and the current reliance on copper has spurred researchers and designers to seek innovative material solutions. Intel has already initiated the use of cobalt instead of copper for the local interconnect layers M0 and M1 starting from the 10 nm technology node. Cobalt is also being employed for liners in the back-end-of-line (BEOL) process, although ruthenium is gaining momentum as an alternative. Furthermore, numerous other materials are being explored as potential options. At a more advanced stage of research, carbon nanotubes (CNTs) and graphene nanoribbons are being considered as potential alternatives to copper. CNTs consist of single layers of graphene sheets rolled into cylindrical structures with diameters ranging from 1 to 5 nm. Metallic CNTs exhibit ballistic electron transport, allowing electrons to move along the nanotube axis without scattering, resulting in a long mean free path (MFP) spanning micrometers. In contrast, the MFP of electrons in copper is limited to a few nanometers. Due to their large MFP and small diameter, electrons in CNTs experience fewer scattering events, leading to lower resistance. In the field of optical interconnect innovation, silicon photonics is on the verge of revolutionizing the datacom and telecom market for transceiver technology. Its primary appeal stems from its significantly lower cost compared to other alternatives, especially InP-based transceivers. Additionally, silicon photonics brings considerable advantages in terms of reduced upfront expenses, as numerous companies can take advantage of the established silicon photonics fabrication ecosystem.

Traditionally, silicon photonics has heavily relied on the silicon-on-insulator (SOI) platform as the primary means of integration, enabling the incorporation of various passive and active devices. However, certain passive devices, like wavelength-division multiplexing (WDM), present challenges related to process control and phase error due to the high index contrast of silicon material, resulting in stricter fabrication tolerances. To tackle these issues, a silicon nitride (SiN) platform has been proposed, utilizing silicon test wafers. While the SiN platform offers unique benefits such as extremely low propagation loss and wider fabrication tolerances, it lacks inherent active capabilities. As a solution, a SiN-on-SOI platform has been developed to leverage the advantages of both Si and SiN platforms. Figure 1.6 illustrates the cross-sectional schematic of a typical silicon photonics integration platform, encompassing SiN and SOI devices. The SiN waveguide layer is responsible for functions such as large-scale light routing, while all other passive and active devices can be manufactured within the SiN waveguide layers. However, it is important to note that silicon photonics does not fulfill all the requirements for a complete photonics integrated circuit, primarily due to the absence of a laser. Consequently, a conventional silicon transceiver experiences additional losses in coupling from the laser to the silicon photonics chip.

1.4.2 Interconnect Research at the Network Level

The growing computational demands of modern applications and services have prompted researchers to focus on enhancing semiconductor-based technology. As a result, there has been a significant evolution in on-chip networks, which are now receiving considerable attention due to their beneficial characteristics,

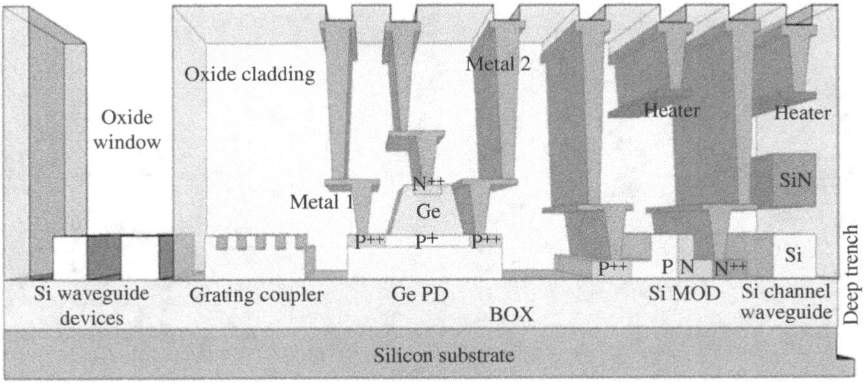

Figure 1.6 Different hierarchies of research venues on emerging interconnect technologies.

including small footprint and low power consumption. Noteworthy examples of on-chip architectures include System-on-Chip (SoC), Pasricha and Dutt [8], and multiprocessor-SoC (MPSoC), Wolf [9]. Additionally, various components of the on-chip network are interconnected using different interconnection networks, such as shared buses and regular buses. In a bus-based topology, devices primarily communicate through bus links, while wire collections are employed in shared bus topologies. Comparatively, the shared bus architecture offers a cost-effective solution with straightforward control features. These advantages make it the preferred choice for facilitating communication among the integrated processing units on the chip [10].

Topology, routing, and switching are prominent and promising research areas within on-chip design [11], see Figure 1.7. Active research is also focused on developing diverse application mapping algorithms to enhance on-chip performance by minimizing network latency. Thanks to advancements in integration technology, it is now possible to accommodate a significant number of heterogeneous cores on a single chip [12]. Consequently, when designing such chips, the interconnections between multiple cores play a vital role in system performance and communication. In addition to managing memory transactions, the on-chip communication architecture also handles I/O traffic and establishes a reliable channel for data sharing between processors. Therefore, the design of high-performance and scalable on-chip communications is critical for SoC design [13]. However, achieving efficient communication among on-chip components presents significant challenges. The network-on-chip (NoC) topology defines the physical organization of the architecture and serves as a key design criterion. It determines the interconnection scheme of the NoC elements. Traditionally, regular tile-based topologies have

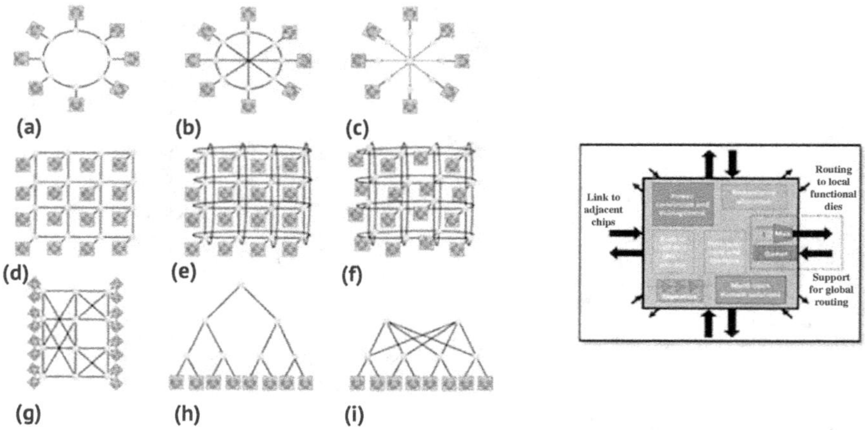

Figure 1.7 Network-on-chip (NoC) topologies, i.e. (a–i) ring, octagon, star, 4×4 mesh, 4×4 torus, 4×4 folded torus, butterfly, binary tree, and fat tree. The right-hand side figure shows the template for NoC chiplets organization. *Source*: Alimi et al. [11]/IntechOpen/CC BY 3.0.

been utilized to connect homogeneous cores. However, there is a growing focus on custom-based, domain-specific irregular topologies that can support heterogeneous cores with varying sizes, functionalities, and communication requirements. Figure 1.7 showcases a range of interconnect topologies and routing techniques currently being explored and experimented with for high-density chip-scale communication [11].

1.4.3 Interconnect Research at Waveguide Level

1.4.3.1 Wireless Communication at Chip Scale

The landscape of communication infrastructure within a chip is undergoing a transformative shift, transitioning from traditional bus-based systems to NoC architectures in response to the proliferation of many-core chips [15]. Simultaneously, interchip communication is rapidly evolving to meet the escalating bandwidth requirements while adhering to stringent power constraints. The conventional interchip interconnections encompass various methods, ranging from solder bumps or C4 interconnects in multichip modules within a System-in-Package (SiP), to more contemporary designs such as arrays of solder balls and silicon interposers [16]. In recent years, there has been a growing interest in leveraging wireless communication as a solution to address the limitations posed by wired metallic interconnections. The feasibility of incorporating on-chip antennas into a silicon die has been explored, leading to the proposition of wireless NoC (WiNoC) architectures [17], as illustrated in Figure 1.8. These WiNoCs

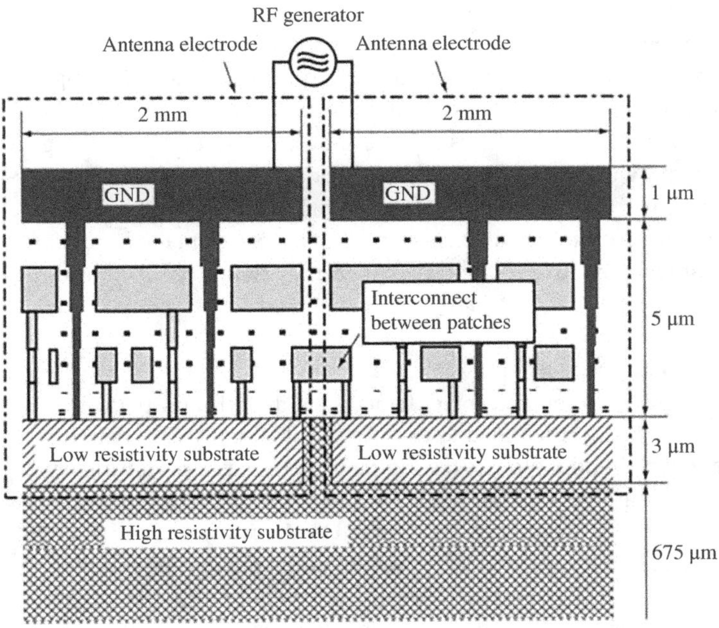

Figure 1.8 Schematic illustration of wireless network on chip. *Source*: Yordanov and Russer [14].

have demonstrated enhancements in energy efficiency and bandwidth for on-chip data communication within many-core chips, leveraging intra-chip wireless links [18]. Although extensive research has been conducted on WiNoCs, the potential of wireless interconnects for interchip communication has not been adequately explored, despite the potential advantages gained by eliminating chip-to-chip metallic interconnections. Researchers are currently investigating several emerging interconnect technologies, including silicon photonics and three-dimensional (3D) integration, to mitigate the challenges associated with metal interconnects. Wireless interconnects utilizing on-chip antennas tuned to millimeter-wave bands are considered the most mature interconnect technology, as these antennas can be fabricated using CMOS processes. Numerous on-chip antennas, such as meander dipoles, zigzag dipoles, slots, loops, inverted F, bowtie, and Yagi, have been designed and studied [19, 20]. Compactness is a critical requirement for on-chip antennas, as they need to fit within the limited footprint of the core, typically a few square millimeters, in current and future technology nodes. The concept of subwavelength interfacing between optics and electronics holds promise as a means to reduce the size of communication devices while maintaining the required speed and bandwidth [21]. However, this presents a formidable challenge, as existing electronic technology struggles to respond to the rapid oscillating fields associated with optical frequencies, and the size of photonic components remains significantly larger than their electronic counterparts. A. Alù and N. Engheta proposed, in theory, an optical wireless channel between two nanoscale antennas [22]. Partial demonstrations of this idea have been achieved by directing an optical signal toward a distant luminescent receiver or by utilizing surface plasmons. Recent progress has showcased a proof-of-principle wireless link, where gold nano-antennas serve as transmitters, directing an out-of-plane laser signal toward an optical rectenna, where the transmitted radiation is rectified as direct current (DC).

1.4.3.2 Plasmonic Interconnects

To meet the ever-growing demand for enhanced bandwidth and performance, extensive research is being conducted to explore novel technologies. Among the numerous contenders, nanophotonics technology emerges as a promising candidate for revolutionizing integrated devices. However, the efficient guidance and modulation of light in these devices are constrained by the diffraction limit of light, which imposes a size limitation. Even with the utilization of high-index materials, achieving a minimum size of approximately 200 nm for telecommunication wavelengths remains challenging. Nevertheless, a subfield known as nanophotonics or plasmonics has emerged as a potential solution. By exploiting the momentum of free electrons at a metal-dielectric interface, referred to as surface plasmon polaritons (SPPs), metallic nanophotonics facilitates the confinement and guidance of light at the nanoscale [24], see Figure 1.9. Despite the notable advancement in integration offered by this approach, it is not without drawbacks. Metal nanophotonics inherently suffer from unavoidable losses, which are contingent on the level of modal confinement within the structure [25]. Consequently, although metal nanophotonics can confine light to dimensions as small as tens of nanometers, the associated losses pose significant challenges in the integration of these structures into current devices. However, researchers are actively engaged in studying potential compensatory measures, such as gain, to mitigate these losses.

The pursuit of low-loss metal nanophotonic structures initially focused on utilizing gold and silver as the metallic core. Through continuous refinements in fabrication techniques and optimization, metal nanophotonics has achieved sub dB/cm propagation, indicating a substantial reduction in losses [26]. Nevertheless, the larger size of the mode in comparison to corresponding photonic structures, approximately 10 μm^2, diminishes the Figure-of-Merit (FoM) associated with these structures. Recent efforts have explored various geometries aiming to reduce the mode size while maintaining low propagation loss. As a result, hybrid and dielectric-loaded waveguide structures have elevated the Figure-of-Merit (FoM)—defined as the inverse of the mode size multiplied by the attenuation—to values surpassing 1000 for mode sizes around 1 μm^2. Like the advancements

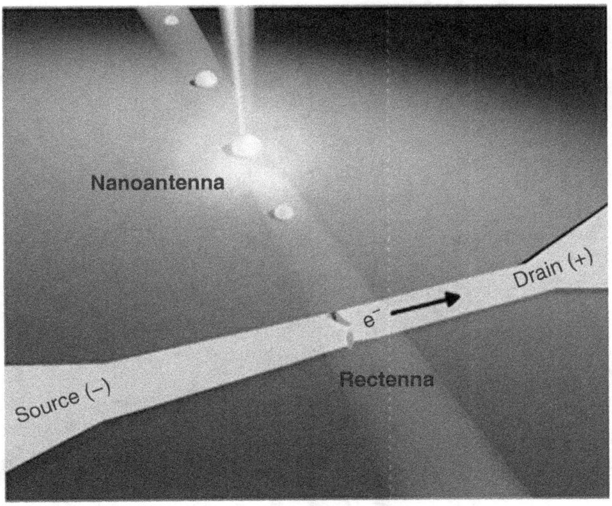

Figure 1.9 Wireless channel between nanoscale antenna. *Source*: Dasgupta et al. [23].

witnessed in photonic waveguides, further improvements in materials, fabrication techniques, and structural design are anticipated to drive the progress of metal nanophotonic waveguides in the future.

1.4.3.3 Terahertz Interconnects

The THz spectral region presents a highly promising opportunity to meet the ever-increasing demand for higher data transfer rates (adapted from Sun et al. [28]). Its extensive bandwidth is of paramount importance in the development of 6G mobile communication networks, where the wireless transmission of data at terabits per second over long distances is a prerequisite. Furthermore, this spectral band facilitates the establishment of an expansive "internet of things" network, connecting billions of devices. Additionally, the THz spectral band holds great potential in addressing the persistent challenge of achieving high-speed, energy-efficient, and cost-effective intrachip/interchip communication links within short distances, thereby enabling the realization of massive multicore processors, NoC architectures, and system-in-package solutions [29]. To realize these ambitious objectives, it is crucial to develop integrated and economically viable solutions for on-chip THz waveguiding and manipulation. Traditional approaches to THz waveguiding, including hollow metallic waveguides, metallic transmission lines, photonic crystals, metal wires, and THz fibers, suffer from inherent susceptibility to defects such as fabrication imperfections and significant losses at sharp corners due to bending. However, recent breakthroughs in the field of the topological phase of light have opened new possibilities for addressing these challenges

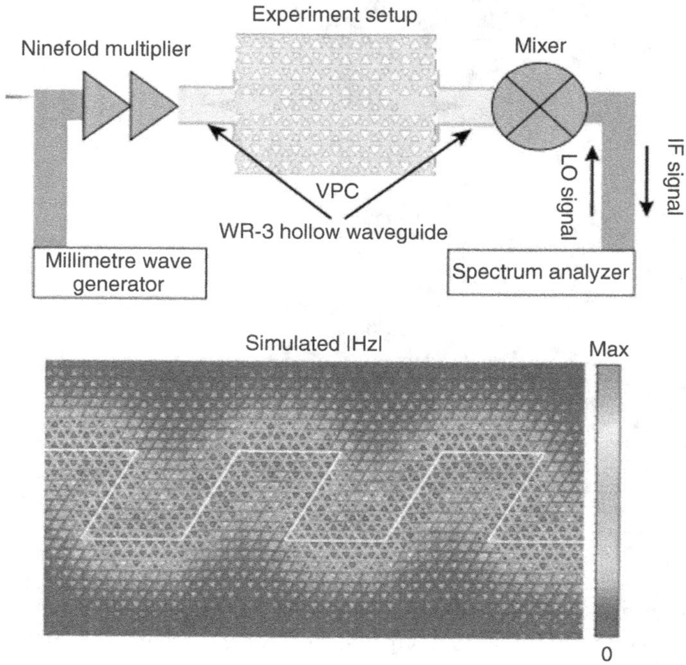

Figure 1.10 Terahertz signal propagation utilizing photonic topological insulators. *Source*: Yang et al. [27].

[27]. One promising solution lies in the utilization of photonic topological insulators (PTIs), which possess an "insulating" bulk while exhibiting "conducting" characteristics at their edges. These PTIs demonstrate robust edge transport, effectively suppressing backscattering even in the presence of disorder and sharp bends. Building upon the principles of the topological phase of light, Yang et al. successfully conducted an experimental demonstration of robust THz topological valley transport using an all-silicon chip; see Figure 1.10.

1.4.4 Chip-scale Interconnect Technologies: Major Industrial Steppingstones

The emergence of on-chip planar interconnect technology can be attributed to the innovative thinking of Robert Noyce, one of the co-founders of Fairchild Semiconductor. In 1959, Noyce introduced the concept of interconnecting silicon circuit components using aluminum metal lines placed on top of a protective oxide coating; see Figure 1.11. This advancement replaced the cumbersome practice of manually connecting wires, leading to the realization of Jack Kilby's solid circuit

Intel Free Press/ Wikimedia Commons / CC BY SA 2.0

Figure 1.11 Robert Noyce, a pioneer in integrated interconnect on chip.

through integrated metallic interconnects. Since its inception, interconnect technology has experienced a significant shift, marking a paradigmatic transformation.

1.4.4.1 Electrical Interconnects: Evolution Toward Nanoscopic, High Density Network

During the period spanning from 1970s to 1990s, transistors had relatively large gate lengths, exceeding 1 μm. At that time, interconnect delays were minimal in comparison to gate delays. However, in the latter half of the 1990s, aggressive miniaturization of transistors significantly reduced gate delays, resulting in interconnect delays becoming comparable to gate delays. This shift sparked an urgent need for research and development in high-speed interconnect technologies to propel the advancement of Moore's law and enable scaling. By the late 1990s, at the 0.25 μm technology node, the delays introduced by the aluminum/silicon dioxide (Al/SiO_2) interconnects became comparable to the intrinsic gate delay of transistors. Over time, as technology progressed, interconnect delays started dominating the total delays within a chip. As the slowest component of the circuit dictates the overall speed, interconnect delays emerged as a fundamental bottleneck in achieving high-performance devices.

In 1997, IBM introduced copper interconnects with low-k dielectric materials, allowing processors to achieve unprecedented speeds. Copper exhibited 40% lower electrical resistance than aluminum, resulting in a 15% boost in chip performance. However, copper presented challenges due to its tendency to electromigrate into the silicon substrate. IBM's research focused on developing barrier materials, particularly tungsten, and implementing a unique metal deposition technique to isolate copper from silicon and enhance its performance. To reduce parasitic RC delay in interconnect routes, low-k materials such as SiOF (with $k \leq 2.2$) were paired with copper interconnects. In 1998, GlobalFoundries Inc. filed a patent for a semiconductor interconnect structure that incorporated air gaps to reduce intralayer capacitance in metal layers. Various research groups pursued the idea of including air or porosity in the silicon substrate to address the mechanical fragility

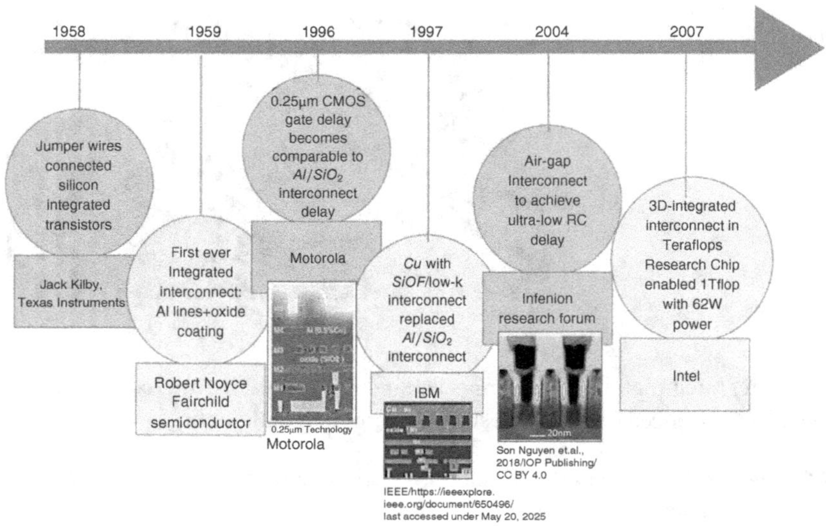

Figure 1.12 Major milestones over the time toward ultrafast chip-scale interconnect technologies.

of these structures. For instance, Infineon Research Forum demonstrated the use of air-gap interconnects in 2004, while D. Kioussis et al. discussed optimization techniques for processing ultra-low-k films to reduce RC delay and power consumption in 2011. The advent of 3D integration using copper interconnect technology opened up new possibilities for high-performance chips. Examples include Intel's Teraflop Research chip, unveiled in 2007, which achieved 1TFLOP operation while consuming significantly less than 100 W of power. The 3D integration of wiring systems, along with proper placement within the circuit, promises higher clock rates, lower power dissipation, and increased integration density. Through-silicon vias (TSVs) have already been implemented in the production of CMOS image sensors, where the active silicon area is bonded into glass and accessed from the backside. Figure 1.12 illustrates some major milestones in realizing high-performance commercial interconnects.

1.5 Scaling Issue of System Level Interconnect

Over the past decade, we have seen an exponential growth of IC speed and integration levels, as the IC fabrication technology, coupled with aggressive circuit design, has advanced significantly. However, as the gate delay reduced with the shrinkage of transistor size at newer process nodes, the interconnect delay, unfortunately, increased for two reasons: as interconnects are getting narrower, it causes metallic resistance to rise, and higher interconnect density causes parasitic capacitance to rise. The effect of the interconnect delay bottleneck is presented in Figure 1.13. In order to get the benefit of the improvement of the chip-level speed at the overall system performance, the communication bandwidth among different ICs and sub-systems must scale accordingly. Currently, communication links in various applications approach Gbps data rates. These applications include computer-to-peripheral connections, local area networks, memory buses, and multiprocessor interconnection networks. Figure 1.14 shows the cross-section of Integrated Circuit (IC) modules on board to reveal the wire pitch of different levels of interconnects.

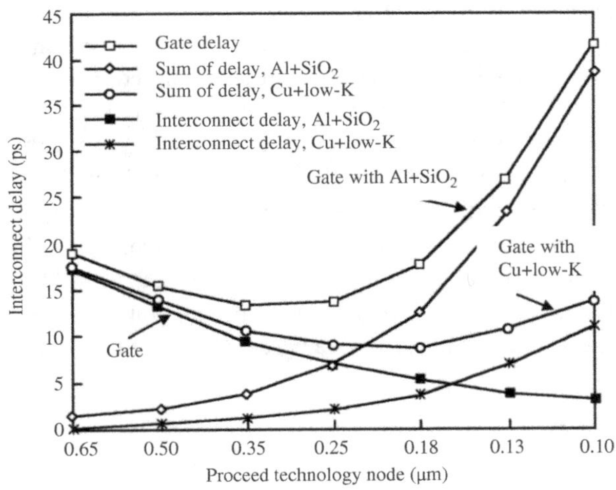

Figure 1.13 Trend of interconnect delay as it evolves through every process node.

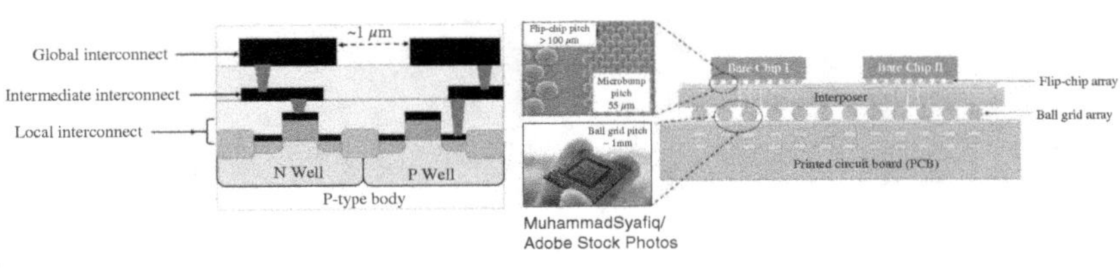

MuhammadSyafiq/
Adobe Stock Photos

Figure 1.14 Cross-sectional schematic of a very large scale Integration chip, revealing interconnect at various hierarchical levels.

1.5.1 Chip Package Signaling

Board-based interconnects have been used for relatively long-range interconnects (~ 30 – 100 cm) between modules. To keep pace with the chip functionality that doubles every 2 years or so, the off-chip bandwidth also multiplies by a factor of 2 every other year, thanks to the increases in signal pin count and per-pin bandwidth. According to the International Technology Roadmap for Semiconductors (ITRS) prediction, the chip package bandwidth would have to surpass 10 Tbps in the timeframe of 2030 and beyond, refer to Figure 1.15. This unprecedented bandwidth enhancement relies on two factors: an increase in pin-count per package, and the increase in data rate per pin. Figure 1.16 summarizes the growth in pin count and data rate per-pin over time. While pin count per package increases at a linear rate, the data rate per pin has been increasing at an exponential rate over the last 20 years. For example, a state-of-the-art chip can support several hundred differential I/O ports, where each of the channels can carry 10 Gbps data. Unfortunately, neither of these trends of growth can continue too far into the future, as they are subject to fundamental barriers such as power, area, and frequency wall. The increases in the bandwidth of system-level interconnects do not scale at the same rate as on-chip core frequencies [30]. This is primarily because of the inherent limit of the minimum channel pitch achievable on the motherboard. For instance, high-density differential stripline pair typically has a pitch around ~ 500 μm. Therefore, with a 10 Gbps channel data rate, the maximum bandwidth density that we can achieve on-board-based interconnect is in the ballpark of ~ 15 Gbps/mm. Figure 1.17, reproduced from Yamada et al. [31], shows the trend in the wire/pad pitch of the Very Large Scale Integration (VLSI) technology at different levels of interconnects, where wire pitch is defined as the

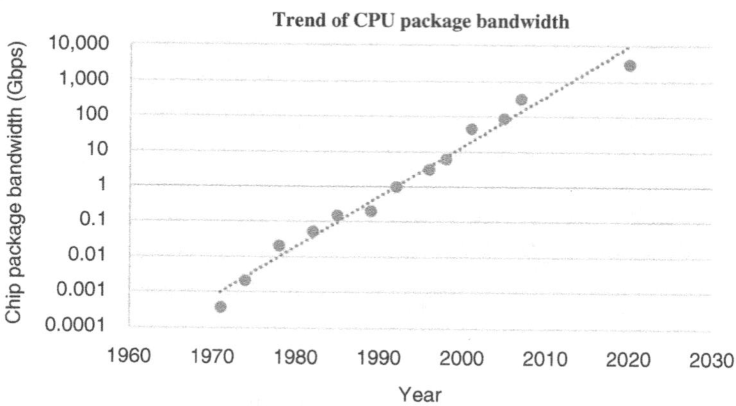

Figure 1.15 Chip package bandwidth over the course of the generation in microprocessor since its inception.

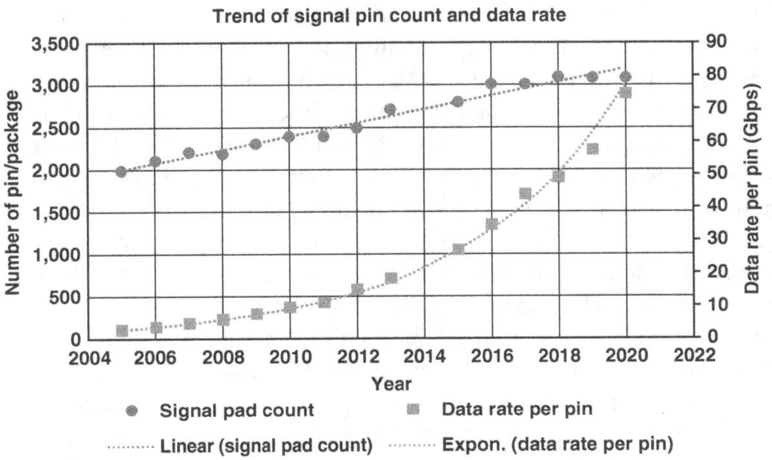

Figure 1.16 ITRS projections of signal pin count and per-pin bandwidth.

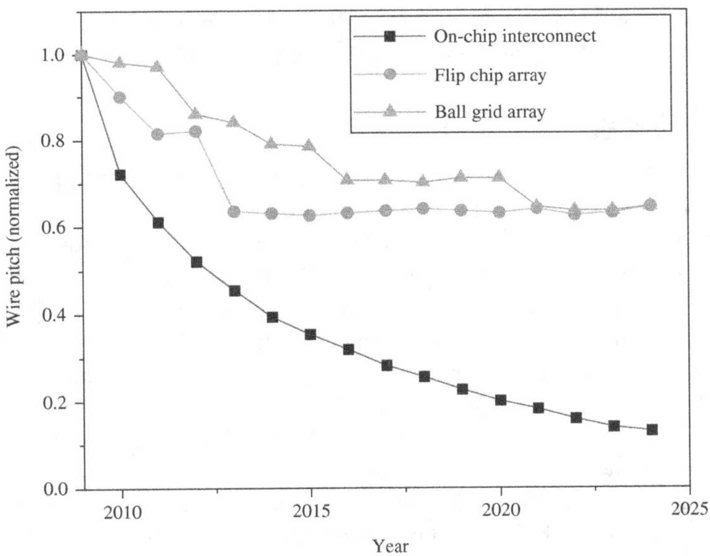

Figure 1.17 Trend of wire pitch/pad pitch at different levels of interconnect. *Source*: Adapted from Yamada et al. [31].

spacing between neighboring interconnects. It is evident that for on-chip interconnects, the wire pitch is being scaled down by a factor of 2 every 6 years, whereas the pad pitch on the PCB has barely scaled for the last few years.

For the best performance in data transfer, a bidirectional, differential channel would have to be designed to occupy eight bumps on the system-on-a-chip (SoC), where each directional link is associated with two signal lines, power, and ground. In addition, the pins will have to be designed as close as possible to the bump to reduce the parasitic capacitance to overcome the RC delay. The pad capacitance is related to the physical geometry, and unfortunately, the geometric size of the pad cannot be aggressively scaled down. Signal pad capacitance is made up of the capacitance of the physical landing pad as well as the gate capacitance of driver/receiver transistors of the interconnect, and the capacitance of the ESD circuitry. None of these capacitances are amenable to exponential scaling over generation. To elaborate the reason, the capacitance of the physical landing pad derives from the size of the landing pad, which is dictated by C4 bumping packaging technology and the separation from the ground plane. This separation between signal wire and ground plane is related to the technological metal stack-up. On the contrary, the geometric width of the off-chip interconnect cannot be scaled too much, as it must provide a low-impedance path to ESD circuitry that combats any issue with current surge. The reliability issue, such as electromigration, also sets a lower bound for the feature size of the interconnects. All these factors combine to raise the parasitic capacitance of the channel, which takes a toll on the data rate.

1.5.2 Issue of Crosstalk

A second category of interference is crosstalk, where signaling channels in proximity capacitively couple energy to one another in a direct fashion or inductively couple energy through return path loops [32]. Forward crosstalk in the direction of wave propagation results in far-end crosstalk (FEXT) at the sensitive receivers of a unidirectional bus. Reverse crosstalk in the opposite direction of signal propagation results in near-end crosstalk (NEXT) at the transmitters of a unidirectional bus. FEXT and NEXT components increase with frequency, and their energy can approach that of the signal at extreme frequencies.

Board-based interconnects have been used for relatively long-range interconnects (~ 30–100cm) between modules.

Apparently, there are two ways of increasing the bandwidth of system-level interconnects to keep pace with the speed improvement of on-chip interconnects. One way is to increase the pin-count of the IC package to scale up the data throughput of the package. The main problem is that simply adding pins to an off-chip bus does not necessarily increase the throughput in a linear manner [33]. As pins are added to a package to increase the bus size and throughput, the noise induced by the signals that are simultaneously switching leads to a significant increase in noise. This means that as signals are added, the per-pin data rate of each pin needs to be decreased to meet the system noise limits. As a result, the overall throughput of an off-chip bus approaches an asymptotic limit as signals are added [34].

Wide parallel buses undergo several signal integrity issues associated with simultaneous switching of digital signals [35, 36]. The noise induced in the package is proportional to the number of off-chip signals that are switching simultaneously. The other possible option of increasing the throughput of system-level interconnects is to drive each pin at a higher data rate, or to deploy the mechanism of signal modulation, coupled with frequency division multiplexing (FDM) so that each physical channel on the system level can support multiple virtual channels. The problem of this approach is that the electromagnetic coupling among the adjacent wires at the elevated frequencies would be too high to maintain signal integrity. One solution to this problem is to increase the number of power and ground pins in the bus configuration so that they can provide electromagnetic shield between two nearby channels. For instance, PCI-Express data bus, which is a successor of PCI parallel bus, intersperses ground pins more frequently among the I/O data pins in order to facilitate at least an order of increase in the speed of data transfer with high signal fidelity. Figure 1.18 shows the connector pin-out from 18 to 32 pins of PCI-Express ×4 data bus, where we can see that each differential pair of input and output channels is shielded by two ground pins (one on each side of the channel) to maintain signal integrity. However, the inclusion of more ground pins among the data bus increases the cost of the package because the number of I/O pins increases.

Among the existing technologies of parallel buses are [35] Rapid I/O, PCI Express, and Hyper Transport. The factors that cause an upper limit of the information transfer capacity of a parallel data bus are propagation loss in the channel due to channel resistance, channel latency (i.e., minimum rise/fall time of signal at the channel terminal) due to electrical capacitance, noise (such as thermal and flicker noise), and crosstalk due to electromagnetic coupling among adjacent channels. Among them, the resistance derives from the material properties,

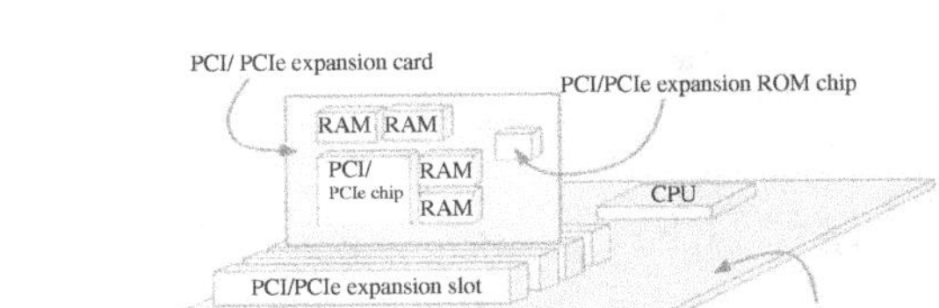

(a)

Pin number	Side B	Side A	Description
18	Ground	Ground	
19	HSOp(1)	Reserved	Transmitter Lane 1, differential pair
20	HSOn(1)	Ground	
21	Ground	HSIp(1)	Receiver Lane 1, differential pair
22	Ground	HSIn(1)	
23	HSOp(2)	Ground	Transmitter Lane 2, differential pair
24	HSOn(2)	Ground	
25	Ground	HSIp(2)	Receiver Lane 2, differential pair
26	Ground	HSIn(2)	
27	HSOp(3)	Ground	Transmitter Lane 3, differential pair
28	HSOn(3)	Ground	
29	Ground	HSIp(3)	Receiver Lane 3, differential pair
30	PWRBRK	HSIn(3)	
31	PRSNT2	Ground	
32	Ground	Reserved	

(b)

Figure 1.18 Illustration of (a) an exemplary parallel bus (PCI module expansion) on motherboard, (b) I/O pin-out configuration of PCIe-×4 data bus, showing frequent use of ground pins to prevent crosstalk among data bus.

specifically from the conductivity of metal. Whereas crosstalk mostly owes to the geometry/layout of the wires. If the metal traces are assumed to be almost ideal (i.e., metal with very large conductivity), then the propagation loss as well as channel latency would be insignificant. Under such conditions, the only limiting factor for information transfer would then be the crosstalk among neighboring wires in the bus.

1.5.3 High Power Consumption in System-level Interconnect

Despite significant research efforts, the energy cost of information transfer through interconnects did not scale down the same way as the energy consumption of transistors [37]. As a result, most of the energy dissipation inside modern electronic devices is associated with data communication through interconnects and data links rather than data processing. The energy cost of different operations in short-reach wired communication technology is shown in a pie chart in Figure 1.19 [38–40]. It shows that interconnects may cause energy dissipation three orders of magnitude larger than switching and memory operations.

With the scaling down of feature sizes of CMOS transistors, the chip itself continues to scale up both in size and complexity. While the performance of local interconnects for wiring at the first metal layer(s) of the chip between individual transistors can be improved by judicious uses of repeaters and wire engineering, the biggest challenge lies in keeping up the performance of global interconnects and system-level interconnects. The wire delay in system-level interconnects is getting aggravated, and has already exceeded gate delay by more than an order of magnitude [41], making interchip communications increasingly challenging [42].

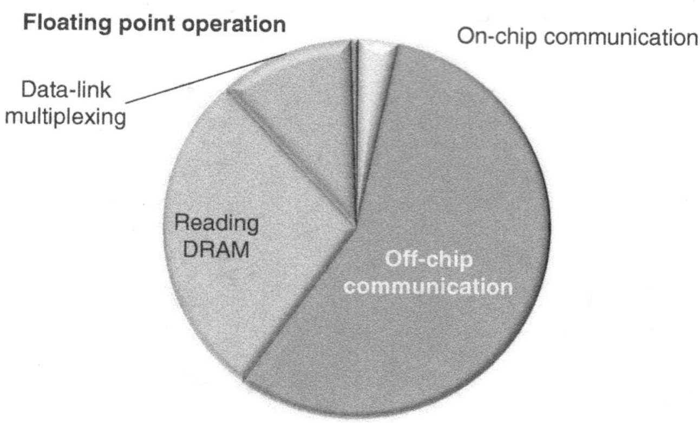

Figure 1.19 Comparison of power dissipation in different components of short-reach wired communication.

As an illustration, the bandwidth of information carried by the internet has reached 280 Tb/s by 2016 [43]. At the same time, a typical graphic processor chip has a peak data rate of 1.4 Tb/s [38], and mere 200 such chips can generate as much data as the entire global long-distance internet traffic. Thus, a major challenge in high-performance cloud computing is to meet the demand for big data exchange in-between electronic devices. While ITRS roadmap is aiming to reach 0.5 THz speed with existing silicon technology [44, 45] and there have been reports on demonstration of high-speed III-V semiconductor material-based transistors operating at THz [46–50], existing data transfer technologies cannot fully utilize the data generation capacity provided by the electronic switching devices [1].

1.6 Optical Interconnect: Evolution Toward Chip-scale Communication

Over the last three decades, optical interconnects have been increasingly adopted into communications systems, ranging from cross-country distance coverage to small-scale distances on electrical backplane. Figure 1.20 shows the commercial realization of optics over time. The approximate trend of shrinking distance and increasing bandwidth for optical interconnect indicates a staggering bandwidth-distance product exceeding 10^6 Gbps-m with single-mode solutions. To put this figure-of-merit into perspective, the electrical interconnects such as copper links have usually been limited within 100 Gbps-m bandwidth-distance product bar. If this trendline continues, many researchers, with somewhat qualified optimism,

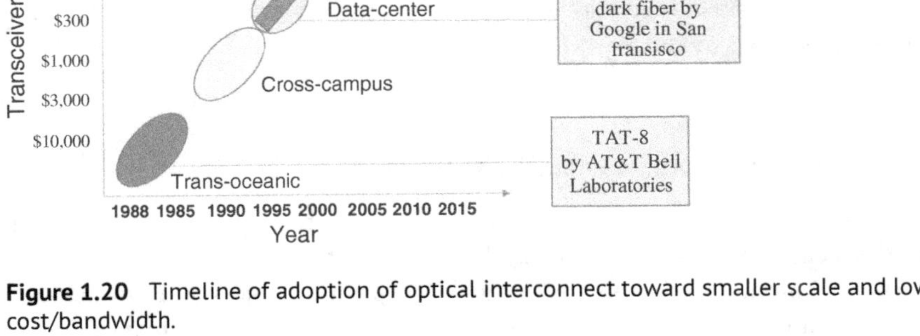

Figure 1.20 Timeline of adoption of optical interconnect toward smaller scale and lower cost/bandwidth.

believe that optical interconnect would eventually substitute interchip and intra-chip electrical interconnect, representing link distance below 1 m.

Compared to electrical links, optical channels offer two prime advantages: First, their substantially greater bandwidth and response uniformity over a wide spectrum, which allow designers to avoid the complexity and costs of channel equalization; and second, optical channels using parallel fibers and wavelength division multiplexing (WDM) can significantly increase the number of effective channels to a chip. Thus, the bandwidth of a chip package is no longer bound by the count of physical I/O pins, which reduces the complexity of the clock-data recovery problem and lowers its energy cost. Because of the inherent low material loss of optical waveguides, the main sink of power loss is the electro-optic converter, i.e., the laser driver, modulator, and the receiver circuit. Hence, the prospect of achieving low energy/bit in optical interconnects is critically dependent on designing low-power transceivers.

A low-power transmitter requires intimate integration of modulator and driver circuit components with low parasitics. Although monolithic integration is ideal for low-parasitic integration, photonic devices and electrical circuits currently have contradictory requirements on substrate, precluding a low-energy budget for short-distance coverage. In (Progress in Low-power switched optical interconnects), Krishnamoorthy et al. demonstrated a low-power driver using a 90 nm CMOS process with an effective voltage swing of up to 2 V. They achieved error-free transmission at 5 Gb/s while the total transmitter electrical power consumption was only 2 mW, yielding an energy efficiency of 400 fJ/bit.

Recently, Hsu et al. [51] in Chinese University of Hong Kong combined the use of dense wavelength division multiplexing and mode division multiplexing (MDM) with advanced modulation formats for on-chip optical interconnects. A three-mode MDM multiplexer (MUX) and demultiplexer (DEMUX) was designed and used. Each wavelength carrier had an aggregate data rate of 192 Gbit/s using 4 level pulse amplitude modulation. The mode crosstalk of the MDM MUX-DEMUX was measured to be below −20 dB in the 40 nm wavelength band from 1520 to 1560 nm. Fourteen wavelength channels are multiplexed and successfully transmitted with an aggregate transmission capacity of 2.6 Tbit/s (14 wavelengths × 3 modes × 64 Gbit/s) with all channels satisfying the hard decision forward error correction threshold.

1.6.1 Integrated Photonic Circuits on Silicon

Silicon photonics is an innovative endeavor that leverages the knowledge and expertise acquired from silicon CMOS processing techniques to advance the realms of photonics and opto-electronics. The primary objective is to achieve substantial enhancements in yield, cost-effectiveness, and time-to-market while minimizing additional investments [52]. At its core, silicon photonics revolves around manipulating light within silicon, encompassing crucial functions such as light generation, routing, modulation, processing, and detection [53], as illustrated in Figure 1.21. These integrated functionalities collectively form the optical equivalent of electronic integrated circuits, commonly referred to as photonic integrated circuits (PICs). The fundamental principles of silicon photonics can be traced back to the influential work of Bennett and Soref [54]. Their paper introduced the concept of free carrier dispersion in silicon, highlighting how the introduction or removal of carriers within the material can induce alterations in its optical properties, specifically the refractive index (n) and extinction coefficient (k). This pivotal breakthrough paved the way for integrating PN junctions into waveguide structures, enabling the development of the initial crucial component of a photonic integrated circuit: the phase modulator [55]. Subsequently, numerous academic institutions and research centers embarked on an extensive exploration, dedicated to realizing various device types essential for a comprehensive silicon PIC solution, thereby instigating a wave of innovation and progress in the field.

The development of silicon-on-insulator (SOI) wafers by Bruel enabled advancements in silicon photonics [56]. This allowed for the confinement of light in the vertical direction, enabling the definition of waveguides within the plane using lithography techniques. Initially, the SOI waveguides were relatively large, but the research community later adopted a standard size of 220 nm, resulting in low losses and tight bending radii. These standardized waveguides facilitated the implementation of various passive devices like splitters and crossings. Ongoing discussions revolve around the use of thicker silicon, which offers advantages such as lower waveguide losses and improved vertical light coupling.

In the field of silicon photonics, significant progress has been made with the emergence of various components. Directional couplers, multiplexer/demultiplexer devices, and active components like tunable micro-ring resonators

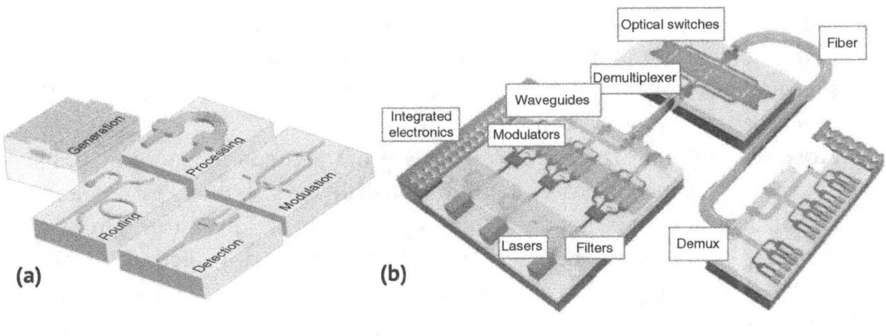

Figure 1.21 (a) Circuit components for silicon photonics; (b) a functional photonic circuit integrated with electronics on silicon.

have become prominent. High-speed Mach–Zehnder modulators capable of achieving a bandwidth of 1 GHz based on the plasma-dispersion effect have been developed. Additionally, the integration of strained SiGe on silicon into waveguide photodetectors has been successful. While the silicon photonic chip is nearing completion, the absence of a silicon laser remains a significant challenge. Efforts are being made to address this challenge through laser integration into silicon via packaging and the exploration of silicon-based lasers.

Packaging solutions for silicon photonics have gained attention as the technology progresses toward commercialization. Integrating silicon photonic chips with optical fibers poses challenges due to the refractive index mismatch, leading to reflections and low coupling efficiencies. However, innovative approaches such as high-efficiency grating couplers and edge couplers have shown promise in achieving reasonable coupling losses. It is anticipated that advancements in packaging will continue, with the expectation that 2.5D or 3D integrated photonics will be commercially deployed in the near future. 2.5D integration involves interconnecting chips through an interposer or substrate, while 3D integration entails stacking optical and electrical dies.

A significant milestone in silicon photonics was the development of commercially viable transceivers. In 2007, Luxtera introduced silicon photonics transceivers for the datacom market, debuting the world's first 40G active optical cable with a reach of up to 300 meters. In 2014, Acacia unveiled their silicon photonics solutions, including the first 100G coherent transceiver suitable for metro research applications.

However, like any emerging technology, silicon PICs are not without their shortcomings and limitations. The following account for some of the key challenges and drawbacks associated with silicon photonic integrated circuits [57].

- **Integration Complexity**: One of the primary challenges in silicon PICs lies in the complexity of integration. While silicon offers excellent compatibility with existing CMOS fabrication processes, integrating different photonic components on a single chip remains a formidable task. The integration of active and passive components, such as lasers, modulators, detectors, and waveguides, requires precise alignment and optimization, resulting in increased fabrication complexity and cost.
- **Absence of On-Chip Laser Source**: One of the fundamental limitations of silicon PICs is the lack of a monolithic on-chip laser source. Pure crystalline silicon lacks a direct bandgap, making it incapable of emitting light efficiently. As a result, the integration of a laser source with silicon PICs necessitates additional off-chip elements or hybrid integration techniques, which can compromise the overall performance, scalability, and cost-effectiveness of the integrated circuit.

- **Limited Wavelength Range**: Silicon PICs primarily operate in the near-infrared wavelength range, typically around 1.3 and 1.55 μm. This limited wavelength range restricts their applicability in certain fields, such as telecommunications, where other wavelength bands are preferred for specific applications. Expanding the operational wavelength range of silicon PICs requires the incorporation of alternative materials, such as III–V semiconductors, which introduces further integration challenges and complexities.
- **High Propagation Losses**: Silicon, despite its compatibility with CMOS processes, suffers from intrinsic optical losses due to its relatively high refractive index and material absorption characteristics. These losses become more pronounced as the dimensions of the waveguides and optical components shrink. Consequently, high propagation losses limit the overall efficiency and reach of silicon PICs, necessitating the use of additional amplification techniques or expensive low-loss fabrication platforms.
- **Thermal Management**: Silicon PICs generate significant amounts of heat due to their compact form factor and high integration density. Efficient thermal management is crucial to ensure the reliable operation and longevity of the integrated circuits. However, silicon's poor thermal conductivity poses challenges in dissipating heat effectively, leading to potential performance degradation, reliability issues, and increased power consumption.
- **Packaging and Fiber Coupling Challenges**: The successful integration of silicon PICs with external fiber-optic systems is vital for practical deployment. However, the high refractive index of silicon presents challenges in achieving efficient coupling between the chip and optical fibers. Mode and refractive index mismatches lead to significant reflections and coupling losses, reducing the overall system performance. Innovative packaging techniques, such as grating couplers and edge couplers, have shown promise but require further development and optimization.

1.6.2 Is Optical Interconnect Viable at Short Range?

By now, we have argued how crosstalk among the channels in data bus can become the ultimate limiting factor in achieving high bandwidth density, given the fact that electrical wires succumb to electromagnetic interference as we move toward higher frequencies. This may ignite the thought of replacing electrical data bus with optical data bus, since optical waveguides suppress crosstalk at high frequencies: an opposite trend of electromagnetic interference with frequencies compared to that of electrical ones. While the optical data bus can indeed outperform the electrical data bus in terms of dealing with crosstalk, it comes with a price: optical waveguides always require sources of carrier frequency, modulators, and demodulators, which add to the overhead energy budget of information transfer. Usage of a given interconnect technology over the other alternatives is justifiable when the technology in question can result in the minimum energy consumption for per bit of transferred data. Therefore, unless the rate of data transfer exceeds a threshold value, optical interconnects give in to their electrical counterparts in the ability of data transfer. It turns out that, for short-range interconnects, the rate of data traffic in a channel may vary in such a wide range over time that employing an optical interconnect may often remain underutilized of its full capacity of information transfer, and therefore cost unnecessary loss of energy attributed to its overhead power budget.

Because of the above-mentioned system complexity and overhead cost, fiber-optic communication systems couldn't make their niche in short-distance communication. They have primarily been installed in long-distance applications, where they can be used to their full transmission capacity, offsetting the set-up cost. Finally, plasmonic interconnects, which grabbed researchers' interest over the last decades due to their ability of subwavelength confinement, are plagued with metallic losses and the high attenuation factor limits the signal propagation length within a few microns.

1.6.2.1 Electrical vs Optical Interconnect: Cost Comparison

The cost per data rate in a communication link tends to increase as the distance between the transmitter and receiver becomes longer. This is primarily due to factors such as longer cables and the need for additional circuitry to compensate for signal loss. However, the cost dynamics change significantly at critical transition points dictated by packaging limitations. On-chip interconnects are cost-effective because of their integration within the chip. However, once the interconnect length exceeds the chip's typical dimensions, the cost rises as additional packaging is required. Similarly, at the board level, the cost of the communication channel gradually increases with the channel length until it surpasses the board's typical dimensions, necessitating additional connectors and receptacles. The cost pattern continues as link lengths exceed the size of a room or building. Interestingly, Figure 1.21 shows that optical interconnects, due to lower material loss compared to copper interconnects, can minimize signal compensation circuitry and thus reduce the cost per unit length. However, optical links require additional circuit components like lasers and photodetectors, which add to the overhead cost. Electrical links, on the other hand, do not require such components and remain preferable for chip-scale communication. As we move to board-level communication and longer links, the advantages of optical links become evident. The extra cost of electro-optic conversion circuitry in optical links is offset by the significant material loss experienced in electrical links. The crossover point between optical and electrical links depends on the data rate, with higher data rates leading to smaller crossover lengths, as illustrated in Figure 1.22.

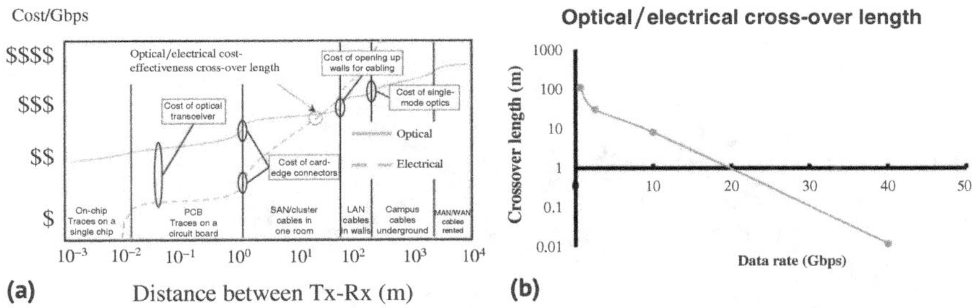

Figure 1.22 Link cost versus distance plot showing the factors determining effective optical/electrical crossover length.

In summary, electrical transmission is cost-effective for shorter chip-scale links, while optical means offer benefits for longer distances. The cost dynamics in communication links are influenced by factors such as distance, packaging limitations, material loss, and the specific requirements of optical and electrical components.

1.6.3 Wireless Network on Chip

Decades of dedicated research in the millimeter-wave (mmWave), THz, and optical frequency bands have positioned wireless technology as a viable solution for interconnecting Systems-on-Chip (SoC). By integrating numerous antennas and transceiver circuitry into intricate networks, wireless data transfer within microprocessor modules on a single chip becomes achievable, as illustrated in Figure 1.23. Particularly in the context of many-core architectures, where the speed of core-to-core communication is primarily limited by latency rather than interconnect bandwidth, the adoption of wireless NoC (WNoC) shows promise in facilitating direct and unobstructed communication among multiple processor cores, schematically shown in Figure 1.24. At the opposite end of the electromagnetic spectrum, the optical bands have emerged as an attractive option for wireless communication within chips, thanks to significant advancements in integrated silicon photonics, giving rise to microscale lasers [58], photodetectors [59], and optical antennas [60]. Furthermore, the THz band is gaining attention as a potential WNoC solution, despite being relatively nascent, as progress is being made in complementary electronic, photonic, and plasmonic approaches to bridge the THz technology gap. However, the benefits of wireless techniques in efficient and low-latency multicasting of data come at a considerable cost: the amplified occurrence of multipath interference among channels. This necessitates the deployment of complex circuitry for equalization, filtering, and the implementation of protocols for time and frequency multiplexing, raising concerns about the overall cost-effectiveness of wireless data recovery circuits.

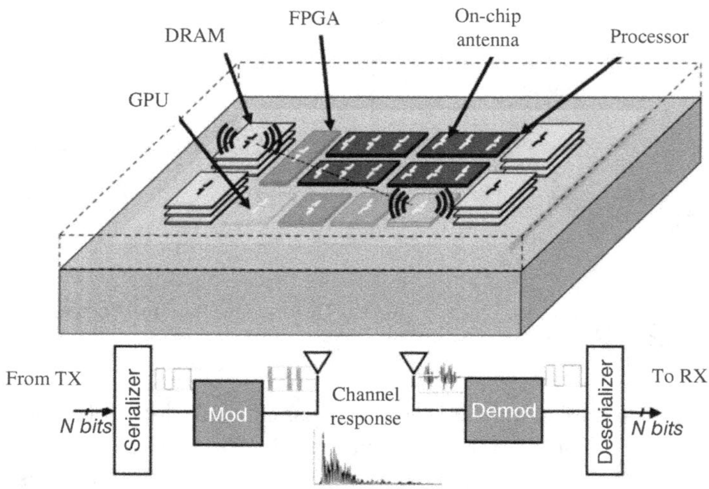

Figure 1.23 A general view on wireless communications at the chip scale within a heterogeneous computer architecture with detail on the wireless transmission process.

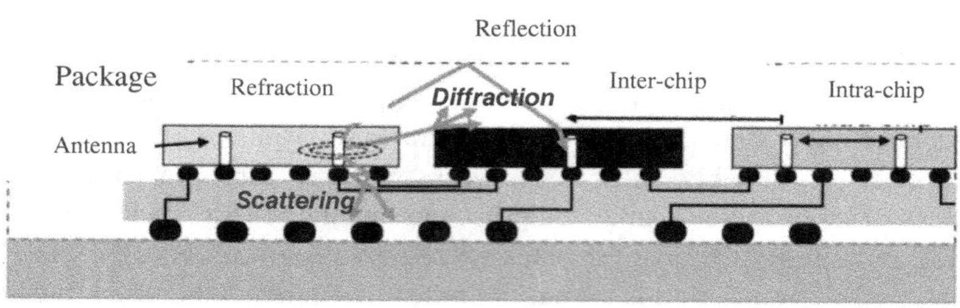

Figure 1.24 Schematic representation of wave propagation in an interposer system with flip chip package excited with vertical monopole antennas, distinguishing between intra- and interchip regions, and exemplifying different propagation phenomena.

One of the key obstacles in wireless technology within chip packages is the significant loss of signal propagation. Studies have demonstrated that even at short distances of a few millimeters, carrier frequencies in the tens of gigahertz range can experience substantial attenuation of up to 50 dB. High attenuation levels have been observed in wireless links within standard chips and packages. A visual representation in Figure 1.25 showcases FDTD simulations, illustrating that the loss of a 60 GHz carrier within a standard flip-chip package is significantly higher compared to guided wave propagation in a similar medium. In response to this challenge, researchers have explored various approaches to mitigate the excessive loss. For instance, Zhang et al. [61] have investigated the use of low-doped silicon as a means to reduce losses induced by the substrate, resulting in a 20 dB reduction in

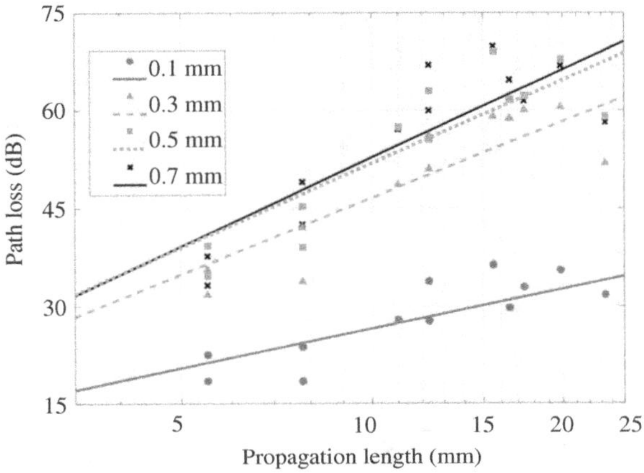

Figure 1.25 Intrachip propagation loss within a flip-chip package at 60 GHz with variable silicon thickness, heat spreader of 0.2 mm propagation loss in silicon.

loss. Similarly, incorporating an undoped silicon layer beneath the die substrate is proposed, which can achieve attenuation levels of 15–30 dB. Another alternative avenue involves leveraging metamaterials integrated on the chip, enabling stronger coupling to surface waves and offering protection against radiation losses [62]. However, it is important to note that these unconventional approaches introduce compatibility issues with standard silicon technology. While improvements in signal propagation within high-resistivity silicon can be achieved, they may come at the expense of circuit performance, such as slower transistor switching speeds. Additionally, the utilization of metamaterial-assisted waveguiding techniques poses challenges in the manufacturing process of integrated chip packaging.

1.6.3.1 Dispersive Nature of the Intrachip/Interchip Wireless Channels

The group delay dispersion is a crucial metric that affects the data transfer capability. It arises from the multiple reflections of incident waves on various interfaces within flip-chip packages. It was demonstrated that a 60 GHz carrier signal traveling through a silicon chip package can experience delay dispersion on the order of a fraction of a nanosecond over a distance of a few tens of units [63], as shown in Figure 1.26. Consequently, the bandwidth limited by dispersion within a standard package is typically a few Gbps at best. To mitigate delay dispersion, one approach is to design thinner packages or utilize carrier frequencies with longer wavelengths, which minimize the phase shift caused by multiple reflections at package interfaces. Timoneda et al. capitalized on this concept by developing a flip-chip package design that effectively controlled dispersion-induced delay

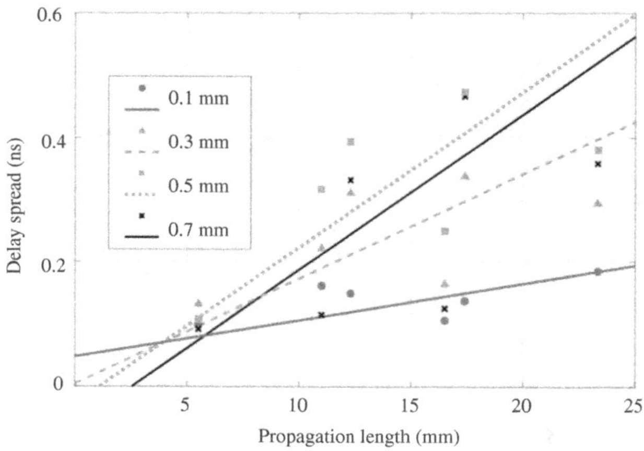

Figure 1.26 Delay dispersion in intra-chip channel at 60 GHz with variable silicon thickness. *Source*: Abadal et al. [63]/IEEE/CC BY 4.0.

spread to below 100 ps Timoneda et al. [64]. This design achieved moderate propagation loss while providing a chip-wide bandwidth exceeding 10 GHz. By reducing delay dispersion through optimized packaging, the limitations on data transfer caused by dispersion were alleviated, leading to improved performance in terms of bandwidth and data transmission capabilities.

1.6.3.2 Signal Fidelity: Gigahertz Carrier vs Optical Carrier

In order to implement low power data transfer over relatively longer distances, propagation loss has to be minimized. It has been computationally shown that the optical free space carrier (10–300 THz) is more susceptible to path loss than carrier with a few tens to a few hundred of gigahertz, as shown in Figure 1.27. Moreover, the propagation constant periodically changes over frequencies, where the periodicity derives from the frequencies between two neighboring surface wave modes. It has been shown that a thinner underfill layer and a bottom layer between the silicon substrate and the heat sink would be able to reduce the path loss. Besides, the thickness of the silicon substrate must be cautiously designed to aid signal immunity to propagation loss.

1.6.4 Carbon Nanotube Interconnect

Another material gaining attention for interconnect applications is carbon nanotubes (CNTs). As integrated circuits (ICs) continue to scale down, the size of interconnects becomes increasingly smaller, reaching nanometric dimensions. This reduction in interconnect size results in delays with copper (Cu) interconnects, as discussed earlier. CNTs have emerged as a potential replacement for

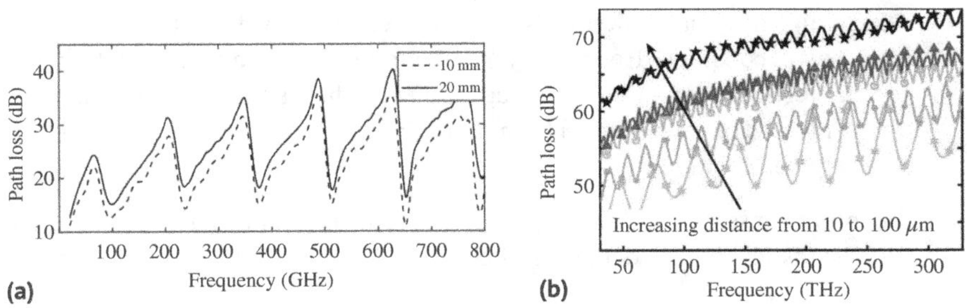

Figure 1.27 Propagation loss of the THz chip-scale wireless channel at (a) sub-THz frequency and at (b) THz frequency. *Source*: Abadal et al. [63]/IEEE/CC BY 4.0.

on-chip Cu interconnects [65]. Compared to traditional Cu interconnects, CNTs exhibit significantly higher current-carrying capabilities, with reported current densities around 10^9 A/cm^2, whereas Cu interconnects typically have current densities in the order of 10^7 A/cm^2. In 2003, ballistic transport in single-walled CNTs (SWCNTs) was reported [66]. CNTs demonstrate low resistivity, high current density tolerance, and scalability to nanoscale dimensions. Numerous research efforts have focused on CNTs as future interconnect materials. Bailey et al. provided a comprehensive summary of recent advances in CNT interconnects and highlighted the challenges that the research community needs to address, particularly in terms of modeling technology [67]. Brun et al. described CNT models for high-frequency interconnects using electromagnetic and circuit approaches [68]. CNTs have been recognized as a promising technology for sub-nanometer scales. Since their discovery in 1991 by Iijima, global research endeavors have been made to grow CNTs for specific applications. While CNTs have also been considered for mainstream microelectronics, there are various integration challenges with VLSI components that must be properly addressed to achieve desired outcomes. The use of CNTs for interconnections is still in the research phase. An equivalent transmission line model for CNTs is proposed [69], and it was found that, at the local interconnect level, CNT interconnects do not outperform Cu wires. However, at the intermediate and global interconnect levels, CNTs may exhibit equal or even better performance compared to Cu interconnects.

1.7 Complexity and Dilemma in Data Transfer

We identify two key features prevailing in interchip communications:

- Centimeter-scale distance is the most critical range of communication, where no single means of standard communication (between electronics and optics) stands out to be distinguishingly superior to the other.

- The nature of data-traffic rate statistically varies from time to time rather than being constant. Hence, any fixed mode of communication link for centimeter-scale distance of coverage may run the risk of remaining largely under-utilized or over-utilized for most of the time, resulting in a severe strain on energy dissipation.

In the following sections, we will exposit the term "last centimeter" in conjunction with "statistical nature of data-traffic" in some greater detail.

1.7.1 The Last Centimeter Barrier: The Most Critical Distance in Data Transfer

To discuss the trade-off between electrical and optical interconnect in terms of power dissipation at 10's of centimeter distance, we provide Figure 1.28a, which compares the power consumption of electrical and optical interconnects at 4 and 6 Gb/s, where the diagram is reproduced from Cho et al. [70]. It can be observed that power consumption in electrical interconnect rises with length and bit rate due to a larger attenuation and fixed noise sources. At higher bit rates, a smaller rise time would further increase noise sources.

The optical interconnect power dissipation also rises with length due to loss in optical power, although this rise is slower than that for the electrical interconnect. Beyond a critical length, optical interconnect yields lower power. This critical length reduces at higher bit rates. The "critical length" divides the space into two regions in Figure 1.28a. The region to the left of "critical length" is favorable for

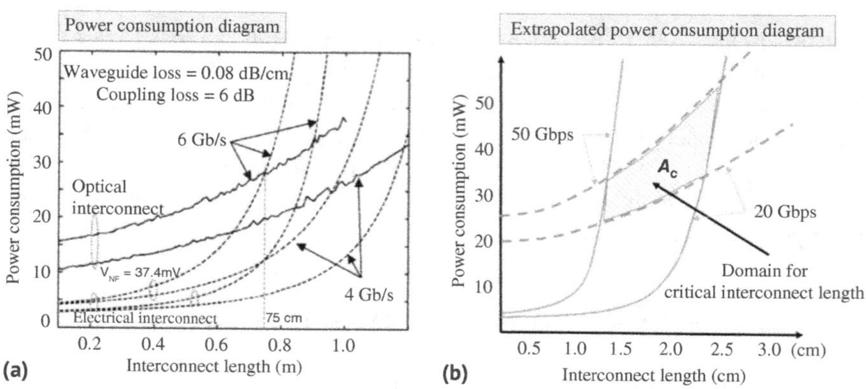

Figure 1.28 (a) Power comparison between optical and electrical interconnects, as predicted by Cho et al.; (b) power consumption diagram by extrapolating data in Cho et al. [70]/IEEE, showing that the critical interconnect length, depending on the data rate at a given time, can be anywhere in region A_c.

electronic data transmission, while the region to the right is favorable for optical data transfer. Figure 1.28b illustrates the power consumption diagram plotted by extrapolating the data taken from Figure 1.28a. It shows that, if the data rate varies in a wide range of "20–50 Gbps" over the course of time, then the "critical length" could be anywhere in the area A_c. In other words, for a given length of interconnect in the order of centimeters and given that the data rate varies with time, there is no fixed mode of interconnect technologies that can ensure minimum power consumption for all possible time.

1.7.2 Time Variation in Data Traffic: A Dilemma in Selecting Interconnect Technology

The dilemma in the selection of interconnect for energy economization becomes obvious once we consider the time-dependent statistical nature of the volume of data traffic. Traffic can be characterized by assuming that the packet arrivals obey the Poisson distribution. For example, at IT data centers, it witnesses a large temporal variation in data flow over the course of a day [71]. The time variation of traffic gets even more prominent as we move down the network hierarchy: from data center core to the individual user terminals. Benson et al. [72] showed that link utilization is one-fourth in the terminal layer than in the core layer, with the full utilization not exceeding 10%. Because of the temporal distribution of data traffic, especially at the range of centimeter-scale interconnect length at the board-level, it is most difficult to choose an optimum interconnect technology that can ensure full utilization of link capacity and minimize energy cost for a highly random volume of data flow. This is why it would be immensely beneficial to overcome the "last centimeter interconnect barrier" with a technology that, rather than supporting a fixed mode, can alternate between two different modes of communication.

1.7.3 Do We Have Any Alternative Interconnect Technology in Hand?

As discussed in the preceding sections, there remains a range of distance of 0.1–10 cm, dubbed as the "last centimeter barrier," where state-of-the-art interconnect technologies compromise on critical performance metrics such as bandwidth and energy efficiency [73]. The unresolved challenges in short-range communications between integrated circuits (ICs), therefore, warrant discoveries of new interconnect technologies having ultrafast data transfer rate, high energy-efficiency, and compatibility with CMOS technology [41, 42, 74]. As shown in Figure 1.29, between the electromagnetic spectral ranges for electronic and optical technologies, an emerging terahertz regime (0.1–10 THz) in the underexploited and underutilized electromagnetic spectral band is now garnering attention for wired and wireless communication systems [75–79].

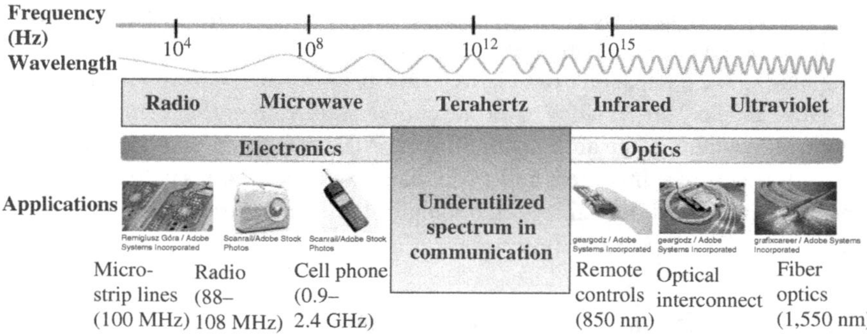

Figure 1.29 Communication device design in microwave and optical frequencies left terahertz regime underexploited.

1.7.4 Spoof Plasmon Interconnect: A New Paradigm in Communication Technology

In the framework of the current book, we have investigated the concept of implementing a communication network that can transfer information by a new type of electromagnetic mode, called spoof surface plasmon polariton (SSPP) (Figure 1.30) for the future high-speed data transfer between integrated circuit (IC) chips, overcoming the limitations of state-of-the-art interconnects. SSPP is a special, surface-bound electromagnetic wave in patterned metal [80, 81]. Artificially engineered interconnects offer a potential breakthrough by combining the benefits of both electrical and optical technologies, see Figure 1.31. On the one hand, they could provide an energy-efficient alternative for short-reach data transfer (10 GHz–10 THz) by minimizing the wire parasitics to suppress crosstalk and ground bounce, and on the other hand, they offer a scalable solution to short-reach interconnects by avoiding the extra energy overhead of transceivers.

It may be noted that the precursor of spoof plasmon technology was surface plasmon polariton (SPP) or plasmonic mode of signal propagation in the optical and near-infrared frequency range (300–800 THz).

1.7.4.1 Plasmonics: An Early Precursor for Spoof Plasmonic Technology

The emergence of plasmonics-based technology occurred in the early 2000s, aiming to harness the unique optical properties of nanoscale metallic structures for the manipulation and routing of light at the nanoscale. Researchers envisioned integrating plasmonic, electronic, and traditional photonic devices on a single chip to leverage the strengths of each technology. The idea was that metal nanostructures could possess the ideal combination of electronic and optical properties to address existing challenges and achieve significantly enhanced processing speeds.

Figure 1.30 An interchip communication network employing SSPP channels. The typical free space wavelength corresponding to spoof plasmon resonance is 300 microns.

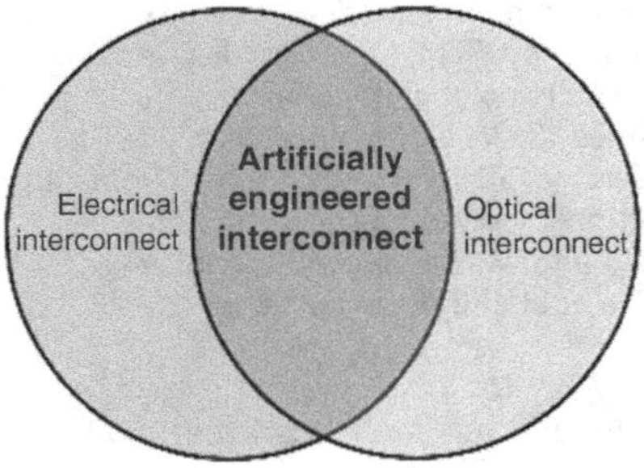

Figure 1.31 Artificially engineered interconnect can bring the advantage of both the electrical and optical interconnect technology to fill the gap for and energy-efficient, high bandwidth, low latency short-reach interconnects.

Commonly used metals for electrical interconnections, such as Cu and Al, could excite surface plasmon polaritons (SPPs), which are electromagnetic waves propagating along the interface between a metal and a dielectric, coupled to the free electrons in the metal.

From an engineering perspective, SPPs can be seen as a distinct type of light wave. The metallic interconnects supporting these waves act as miniature optical waveguides known as plasmonic waveguides. The concept that the optical mode, or "light beam," diameter perpendicular to the metal interface could be significantly smaller than the wavelength of light generated great excitement and fostered the vision of eventually interfacing nanoscale electronics with optical (plasmonic) devices of similar size.

One of the most promising applications of plasmonics is in sensing and imaging, exemplified by its use in a scanning tunneling microscope (STM). Brongersma et al. from Stanford University developed a photon scanning tunneling

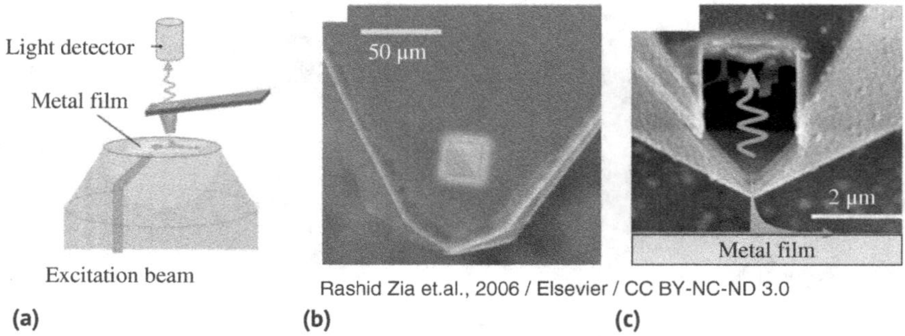

Figure 1.32 (a) Illustration of the functioning of a PSTM (photon scanning tunneling microscope), enabling the investigation of surface plasmon polariton (SPP) propagation along metal film surfaces. The arrow indicates the launch of an SPP from an excitation spot onto the surface of a metal film, achieved using a high numerical aperture microscope objective. The scanning electron microscopy (SEM) image (b) displays the near-field optical cantilever probe utilized by Brongersma et al. It comprises a microfabricated, hollow glass pyramid coated with a thick layer of aluminum (Al). The Al film on the top of the pyramid features a tiny aperture (50 nm) through which light can be collected or emitted. Additionally, a cross-sectional view (c) of the same hollow pyramidal tip is depicted after a focused ion beam (FIB) was employed to remove a significant portion of the sidewall. *Source*: (b,c) Zia et al. [82]/Elsevier/CC BY-NC-ND 3.0

microscope (PSTM) by modifying a commercially available scanning near-field optical microscope. PSTMs are widely employed for studying the propagation of SPPs along extended films and metal stripe waveguides. Figure 1.32 demonstrates how a microscope objective within the PSTM system can focus a laser beam onto a metal film at a precisely defined angle, effectively launching an SPP along the top surface of the metal. This technique for exciting SPPs utilizes the well-known Kretschmann geometry, which enables phase matching between the free space excitation beam and the SPP.

To locally tap into the guided SPP wave and scatter light toward a far-field detector, a sharp pyramidal tip coated with metal is employed (as depicted in Figure 1.32b,c). These tips possess a nanoscale aperture at the apex of the pyramid through which light can be collected. The scattered light is then detected using a photomultiplier tube. By measuring the intensity of the local light directly beneath the tip and scanning the tip across the metal surface, the propagation of SPPs can be imaged and visualized.

Because of the unique properties of surface plasmon polariton, plasmonics were advertised to be the proper conglomeration for electronics and optics. Figure 1.33 captures the envisioned role of plasmonic devices on a chip, illustrating their operating speeds and critical dimensions in comparison to other chip-scale device technologies. In the past, devices were limited by their sluggishness and bulkiness.

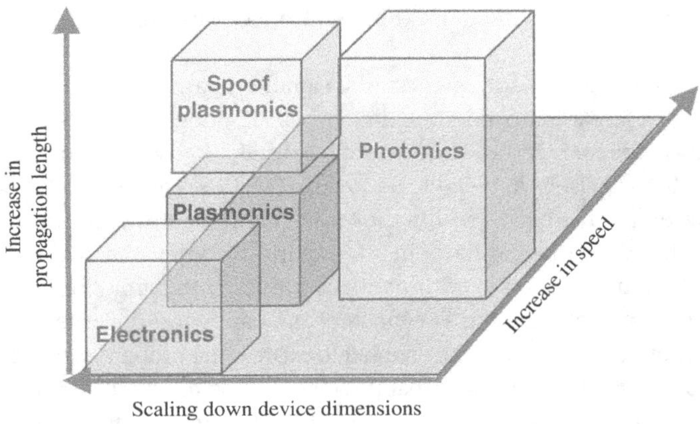

Figure 1.33 The proposed perspective in this book toward different technologies that entails a holistic view of figures of merit for evaluating the prospect of chip-scale communication. While a simple two-dimensional diagram in Figure 1.6 propounded for plasmonics technology to merge the gap between electronics and optics, a more sophisticated three-dimensional diagram incorporating signal propagation length may help understand the limit of plasmonics and how we can push the boundary further by adopting spoof plasmonics, a version originated from but different from its precursor, plasmonics.

However, the semiconductor industry has made remarkable progress in scaling electronic devices to nanoscale dimensions. Despite this achievement, challenges arise in terms of interconnect delay times, hindering the realization of purely electronic circuits operating above approximately 10 GHz.

In contrast, photonic devices possess immense bandwidth, enabling high-capacity data transmission. Nevertheless, dielectric photonic components face limitations in size due to the laws of diffraction, impeding their scalability comparable to electronics. This is where plasmonics were hoped to offer a distinct advantage, combining the size characteristics of electronics with the speed capabilities of photonics. Plasmonic devices have the potential to naturally interface with photonic devices operating at similar speeds and electronic components of comparable sizes. By bridging the gap between these two technologies, which currently struggle to communicate effectively, plasmonics were envisioned to serve as the interface between two disparate technologies.

Although plasmonic technology can potentially enhance the capability of nanoimaging, the potential of plasmonics for light propagation and routing appears rather limited. What makes it so is the inherent material loss associated with plasmonic systems. The metals commonly used for plasmonic devices, such as gold and silver, exhibit considerable absorption and scattering losses, limiting the efficiency and performance of these devices. Material loss directly impacts the signal intensity and propagation distance, reducing the overall effectiveness of plasmonic systems.

To comprehensively investigate and quantify the signal propagation properties of plasmonic systems, a research team in Mazumder Lab (K. Song, P. Mazumder) embarked on a study in the early 2000s. The focus of Mazumder Lab was on analyzing the power flow characteristics in metallic nanowires (MNWs) composed of silver (Ag). The choice of nanowire radii within the tens of nanometers range aimed to ensure compatibility with interconnects in VLSI chips. In order to assess the performance of MNWs in conjunction with nanoelectronics, a transmission line model was employed, transforming the TM mode of the nanowires governed by Maxwell's electrodynamics. This approach enabled to express the properties of the MNWs using standard signal transmission parameters such as resistance, inductance, capacitance, and conductance (referred to as RLCG parameters) within the SPICE circuit model. As a result, Mazumder Lab successfully characterized the propagation wave vector and attenuation constant of Ag MNWs. Through a comprehensive investigation, two significant findings emerged:

First, Mazumder Lab found that the frequencies at which the electromagnetic mode couples with plasmonic nanowires to create nonradiative surface-bound modes were situated in the ultraviolet range, with angular frequencies on the order of 10^{15} rad/s, see Figure 1.34. This frequency range proved incompatible with both baseband transmission in gigahertz electronics and the 1.55 μm technology commonly utilized in optical fiber transmission.

Second, Mazumder Lab observed that the length of signal propagation, limited to −3 dB power transmission, barely extended to a few hundred nanometers.

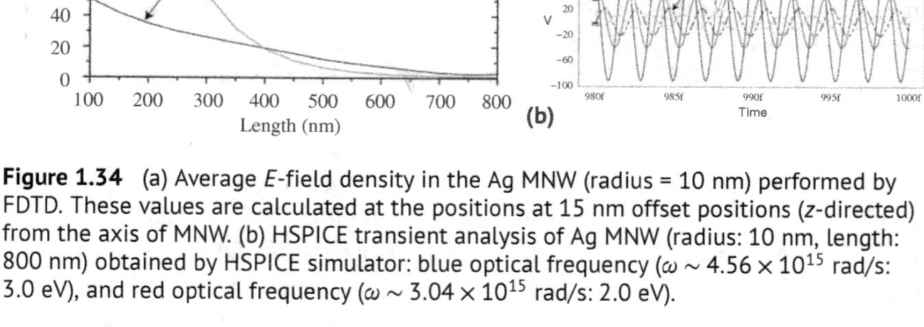

Figure 1.34 (a) Average E-field density in the Ag MNW (radius = 10 nm) performed by FDTD. These values are calculated at the positions at 15 nm offset positions (z-directed) from the axis of MNW. (b) HSPICE transient analysis of Ag MNW (radius: 10 nm, length: 800 nm) obtained by HSPICE simulator: blue optical frequency ($\omega \sim 4.56 \times 10^{15}$ rad/s: 3.0 eV), and red optical frequency ($\omega \sim 3.04 \times 10^{15}$ rad/s: 2.0 eV).

This distance fell at least 5000 times shorter than the size (1 mm) of a contemporary chip.

These discoveries revealed the shortcomings of plasmonic technology to replace or enhance the bandwidth capacity of nanoscale electronics in VLSI. Consequently, it became evident that a true candidate for high-performance data transmission in chip-scale applications needed to be sought in a regime where metallic loss is more manageable and less detrimental. This realization led Mazumder and his colleagues to develop a keen interest in spoof surface plasmon polaritons, an electromagnetic mode inspired by plasmonics but superior in many ways, particularly in terms of manageable material loss.

In this book, the authors aim to introduce the potential of the alternative technology based on spoof surface plasmon polariton, particularly in the domain of chip-scale ultrafast data transfer, as illustrated in Figure 1.30. SSPP waves are strongly confined subwavelength modes [80]. Therefore, the parallel data bus of SSPP can be densely packed, and data can be safeguarded from crosstalk interference.

In addition, SSPP modes can propagate a considerable distance (a few centimeters) by circumventing ohmic loss resulting from penetration of the mode into the metal. On one hand, SSPP interconnect possesses the advantage of CMOS compatibility [83] and its energy-economic quasi-electrical mode can transfer data without requiring excessive energy. On the other hand, its quasi-optical mode with deep subwavelength confinement can provide high bandwidth overcoming the conventional 1/RC-limit. In our research, we demonstrated that spoof-plasmon-based data bus can provide nearly 300 Gbps data transfer rate per channel with 1 Tbps/mm bandwidth density, while the signal can propagate up to a wire length of 1 cm [84]. These results indicate a remarkable potential of SSPP interconnects to become the most viable technology for the next-generation high-speed interchip short-range communications, as illustrated in Figure 1.35. Therefore, in this book, we propose a visual illustration in Figure 1.33, which is an extended version of the vision of the early developers of plasmonic technologies represented in Figure 1.36, yet simple enough to understand and appreciate how different technologies posit in terms of their speed, size, and signal propagation properties, so that we can find any gap in between and work toward mitigating the technology gap. In particular, Figure 1.33 expands the two-dimensional (2D) view of Figure 1.36 into a 3D view by adding one more critical parameter for signal transmission in the other axis, namely signal propagation length. The elaborated view of Figure 1.33 points to the limitation of plasmonics within 1 μm of signal propagation, and shows how we can push the frontier even further by adopting spoof plasmonic technology.

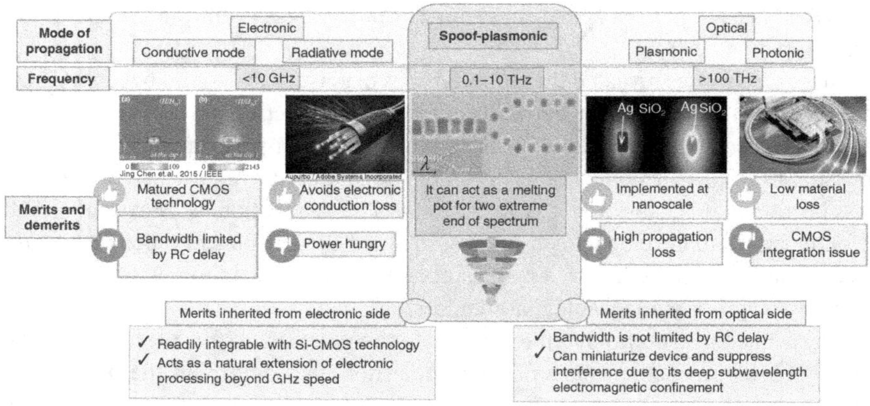

Figure 1.35 Illustration of how the spoof plasmonic technology posits in a spot between electronics and optics, bringing in the advantages of both the domain, and merge the technology gap in between.

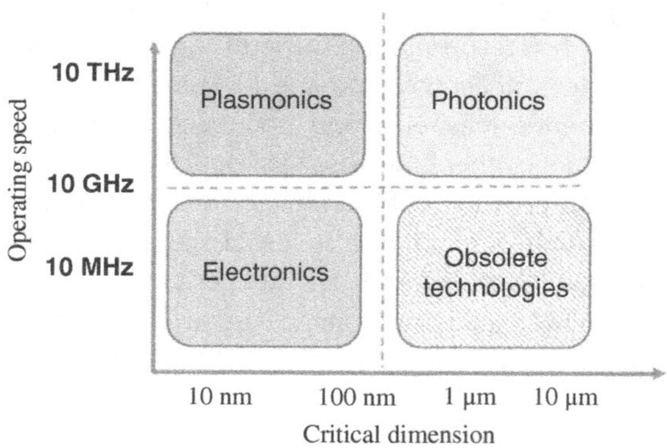

Figure 1.36 Early adopters of plasmonic technologies championed the idea that plasmonics might bring together the advantages of both nanoscale electronics and tera-speed photonics. Operating speeds and critical dimensions of various chip-scale device technologies, highlighting the strengths of the different technologies. *Source*: Zia et al. [82]/with permission of Elsevier.

1.8 Research on Spoof Plasmon Wave: Toward CMOS Compatibility

Spoof plasmon wave has become a new exciting field of research [85–90], as illustrated in Figure 1.37. Researchers in Mazumder Laboratory have developed the theory and demonstrated multifaceted applications of SSPP-based systems [81, 91–99]. Pioneered by Sir John Pendry's group at the Imperial College,

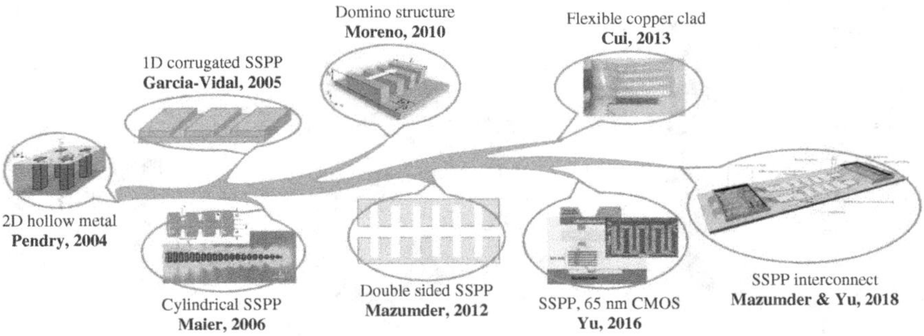

Xiaopeng Shen et al., 2012/Proceedings of the National Academy of Sciences PNAS

Figure 1.37 Brief illustration of the evolution in SSPP research. The concept and realization of THz surface wave evolved from complex structures toward on-chip, CMOS technology-compatible thin-film structures.

UK [80, 100], several researchers such as Stefan Maier at Imperial College, UK [101–104], Federico Capasso at Harvard University, US [86, 89, 105], Harry Atwater at California Institute of Technology, US [106], Tei Cui at Southeast University, China [83, 90, 107, 108], Pinaki Mazumder at University of Michigan, US [81, 92, 98, 99], and Hao Yu at Nanyang Technological University, Singapore [83, 109, 110] have advanced the field of THz surface-wave engineering.

Starting from a large 2D periodic hollow metal structure proposed by Pendry et al. [80], THz SSPP morphed into 1D corrugated structures by the work of Garcia-Vidal et al. [87] (on rectangular SSPP structures) and Maier in 2006 (on cylindrical SSPP structures) Maier et al. [101]. Subsequently, a single-sided SSPP structure was designed by Martin-Cano et al. [111] and a double-sided SSPP structure was introduced by Xu et al. [112] to increase the length of signal propagation with low attenuation. In order to adapt the spoof THz surface wave technology (SSPP) to thin-film interconnect technology, surface-wave-based structures are fabricated on a silicon substrate using commercial 65 nm CMOS technology [83, 84]. Their work has now opened a greater opportunity to improve the bandwidth of the interconnect technology by an order of magnitude, while making THz surface-wave propagation on an SSPP structure compatible with commercial CMOS technology.

1.8.1 Leading Researchers on Spoof Surface Plasmon Technology

Important milestones in the research on spoof plasmonics: spoof plasmonics, a groundbreaking subfield of photonics, has emerged as a powerful alternative to traditional plasmonic systems. By mimicking the plasmonic behavior usually found in the ultraviolet regime at low frequencies through 2D hollow metal

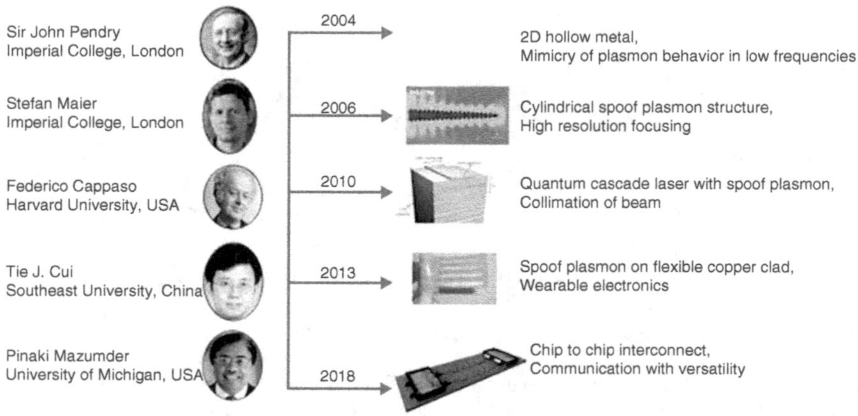

Figure 1.38 Some of the key researchers who pioneered the field of spoof plasmonics.

structures, researchers have unlocked a myriad of applications in high-resolution imaging, laser technology, wearable electronics, and chip-to-chip communication. In the following, we will mention some of the notable contributions of leading researchers in spoof plasmonics to date and explore how their work has led to the progress in this field, refer to Figure 1.38.

- **John Pendry—Imperial College London**: In 2004, John Pendry, based at Imperial College London, proposed the concept that plasmonic behavior, typically confined to the ultraviolet regime, could be emulated at low frequencies using a 2D hollow metal structure. This marked a pivotal moment in the field of spoof plasmonics, as it laid the foundation for subsequent research and development of practical applications.
- **Stefan Maier—Imperial College London**: Building upon Pendry's work, Stefan Maier, also from Imperial College London, introduced a cylindrical spoof surface plasmon structure in 2006. This innovative design showcased promising potential for high-resolution imaging and focusing. Maier's contributions solidified the viability of spoof plasmonics in real-world applications.
- **Federico Capasso—Harvard University**: In 2010, Federico Capasso, a professor at Harvard University, demonstrated an augmentation of the photoemission efficiency of quantum cascade lasers. His laboratory achieved this feat by incorporating spoof plasmons on the emitting surface of the laser. This research is expected to bridge the gap between traditional plasmonics and practical applications, opening new avenues for laser technology.
- **Tie J. Cui—Southeast University China**: Tie J. Cui, a professor at Southeast University China, and his lab demonstrated in 2013 that spoof plasmons could be supported on a flexible 2D copper clad. This demonstration

opened the prospect of wearable electronics aided by spoof plasmon technology. Cui's work showed how this technology could be integrated to potentially transform the landscape of electronic wearables.

- **Mark Stockman—Georgia State University**: Mark Stockman has played a critical role in developing the fundamental understanding of the way in which metallic nanostructures can be used to concentrate and manipulate light at the nanoscale through the excitation of surface plasmons. Research work on plasmonic waveguides, sources, and plasmonic resonator antennas was inspired by his ability to formulate basic plasmonic design principles that can capture the essence of their operation. His research on plasmon hybridization in metallic nanoparticle dimers continues to stimulate new nanophotonics developments. In these systems, hybridization can lead to a desirable redistribution of optical fields and an intense light concentration. The extreme light confinement renders the dimer's optical response also very sensitive to minute changes in gap size.
- **Nader Engheta—University of Pennsylvania**: In 2016, Nader Engheta, a professor at the University of Pennsylvania, showed theoretically and numerically, that the structural dispersion of waveguides, when augmented with the presence of thin metallic wires judiciously located at interfaces, may provide a useful platform, particularly in the microwave, millimeter-wave, and THz regimes, to demonstrate a variety of plasmonic phenomena, which in the infrared and optical wavelengths can be done with natural materials with negative or near-zero permittivity that may be lossy. However, since the proposed structures involve only positive dielectric materials and metallic walls and wires, which exhibit very little loss at these lower frequencies, it could enable achieving various plasmonic features with a low level of dissipation. This may open doors to possibilities in the design of low-loss metamaterials that offer exciting phenomena that have otherwise been limited by material loss. His laboratory also introduced the concept of "waveguide metatronics," an advanced form of optical metatronics that uses structural dispersion in waveguides to obtain the materials and structures required to construct this class of circuitry. Using numerical simulations, his lab has demonstrated that the design of a metatronic circuit can be carried out by using a waveguide filled with materials with positive permittivity. This includes the implementation of all "lumped" circuit elements and their assembly in a single circuit board. In doing so, they extended the concepts of optical metatronics to frequency ranges where there are no natural plasmonic materials available.
- **Andrea Alu—University of Texas at Austin**: Andrea Alu and his group have contributed to the theoretical development of THz metamaterial-based applications. Starting from the theoretical and methodological motivations

of their research and covering macro-scale performance-driven design of volumetric and planar metamaterials, they introduced advanced task-oriented modeling approaches. Their approaches include specific reference to their multi-scale/multi-physics customization in recent metamaterial science and engineering. In the introduction of these concepts, they illustrated the physical mechanisms and phenomena at the basis of the field manipulation capabilities enabled by metamaterials. These could benefit a wide set of recent applications in microwave, terahertz, photonics, and optics scenarios, highlighting the logical and methodological interconnections among the different domains.

- **Vladimir Shalaev—Purdue University**: Vladimir Shalaev and his group have demonstrated dynamically controllable optical properties of metasurfaces in both space and time. Gigahertz-to-terahertz-speed optical transistors for telecom (spanning the visible to mid-infrared wavelengths), beam-steering devices for smart cars and lidar, and active cloaking for defense applications are but a few examples of devices that need dynamically configurable optical properties. For metasurfaces, such dynamic control is achieved by tuning the surface's dielectric permittivity or changing its topology (geometry). Amplitude modulation—where the intensity of light is changed by controlling the metasurface's reflectance, transmittance, and absorbance—finds use in optical computation and data transfer. They demonstrated wavefront manipulation through phase control, which could allow beam-steering for lidar and autonomous vehicles, flat-tunable lenses for 3D imaging and endoscopy, and holograms for augmented and virtual reality (AR/VR).
- **Pinaki Mazumder—University of Michigan, US**: In 2019, Soumitra Joy and Pinaki Mazumder at the University of Michigan demonstrated a spoof plasmon-based interconnect system for chip-to-chip communication. This demonstration gave a proof-of-concept validation of the feasibility of employing spoof plasmons in chip-to-chip communication systems, spanning over several centimeters. This demonstration can open the possibility of on-package integration of silicon chips interfaced by high-speed short-reach interconnects to enable spoof-plasmon-based VLSI technology for high-performance computing.

The advancements made by these leading researchers in spoof plasmonics have reshaped the landscape of photonics and paved the way for a myriad of applications in various domains. From high-resolution imaging and laser technology to wearable electronics and chip-to-chip communication, spoof plasmonics has become a versatile and promising field of study. As the technology continues to evolve, future researchers will be able to build upon these contributions, further pushing the boundaries of what is possible with spoof plasmons. This book would encourage the research communities and pave the path toward solving the longstanding problem of "last centimeter barrier" of interconnect by revealing the full potential of spoof plasmonics.

1.9 Summary of the Chapter

This chapter presents a comprehensive overview of interconnect technologies and their significance in advancing next-generation computing. It delved into various hierarchical levels of interconnect research, identifying key areas of focus. The chapter evaluated the performance and advantages of conventional interconnect technologies while exploring lesser-known research frontiers that hold promise for future developments. Additionally, the chapter introduces readers to a novel interconnect technology centered around spoof surface plasmon polaritons. This innovative technology has the potential to address longstanding interconnect challenges and complement existing solutions toward addressing the "Last Centimeter Barrier," a term indicative of the critical length of interconnect where neither electronics nor optics can provide an energy-efficient optimal solution. By combining theoretical insights with practical applications, the chapter shed light on the transformative impact of this emerging interconnect approach in shaping the future of computing.

References

1 Ammar Karkar, Terrence Mak, Kin-Fai Tong, and Alex Yakovlev. A survey of emerging interconnects for on-chip efficient multicast and broadcast in many-cores. *IEEE Circuits and Systems Magazine*, 16(1):58–72, 2016.

2 J Michel and LC Kimerling. Electronics and photonics: Convergence on a silicon platform. In *33rd European Conference and Exhibition of Optical Communication*, pages 1–2, VDE - VERLAG, 2007. https://doi.org/10.1049/ic:20070261.

3 Yutaka Urino, Takanori Shimizu, Makoto Okano, Nobuaki Hatori, Masashige Ishizaka, Tsuyoshi Yamamoto, Takeshi Baba, Takeshi Akagawa, Suguru Akiyama, Tatsuya Usuki, et al. First demonstration of high density optical interconnects integrated with lasers, optical modulators, and photodetectors on single silicon substrate. *Optics Express*, 19(26):B159–B165, 2011. https://doi.org/10.1364/OE.19.00B159.

4 M Haurylau, G Chen, H Chen, J Zhang, NA Nelson, DH Albonesi, EG Friedman, and PM Fauchet. On-chip optical interconnect roadmap: Challenges and critical directions. *IEEE Journal of Selected Topics in Quantum Electronics*, 12(6):1699–1705, 2006. https://doi.org/10.1109/JSTQE.2006.880615.

5 Gerd Keiser. Optical fiber communications. In *Wiley encyclopedia of telecommunications*. John Wiley & Sons, Inc., 2003. https://doi.org/10.1002/0471219282.eot158.

6 René-Jean Essiambre, Gerhard Kramer, Peter J Winzer, Gerard J Foschini, and Bernhard Goebel. Capacity limits of optical fiber networks. *Journal of Lightwave Technology*, 28(4):662–701, 2010. https://www.osapublishing.org/abstract.cfm?uri=jlt-28-4-662.

7 AEJ Lim, J Song, Q Fang, C Li, X Tu, N Duan, KK Chen, RPC Tern, and TY Liow. Review of silicon photonics foundry efforts. *IEEE Journal of Selected Topics in Quantum Electronics*, 20(4):405–416, 2014. https://doi.org/10.1109/JSTQE.2013.2293274.

8 Sudeep Pasricha and Nikil Dutt. *On-chip communication architectures: System on chip interconnect*. Morgan Kaufmann, 2010.

9 Wayne Wolf. Multiprocessor system-on-chip technology. *IEEE Signal Processing Magazine*, 26(6):50–54, 2009.

10 Krishna Sekar, Kanishka Lahiri, Anand Raghunathan, and Sujit Dey. Dynamically configurable bus topologies for high-performance on-chip communication. *IEEE Transactions on Very Large Scale Integration (VLSI) Systems*, 16(10):1413–1426, 2008.

11 Isiaka A Alimi, Romil K Patel, Oluyomi Aboderin, Abdelgader M Abdalla, Ramoni A Gbadamosi, Nelson J Muga, Armando N Pinto, and António L Teixeira. *Network-on-chip topologies: Potentials, technical challenges, recent advances and research direction*. IntechOpen, 2021.

12 Tao Li, Jie Hou, Jinli Yan, Rulin Liu, Hui Yang, and Zhigang Sun. Chiplet heterogeneous integration technology—status and challenges. *Electronics*, 9(4):670, 2020.

13 Paolo Mantovani. *Scalable system-on-chip design*. Columbia University, 2017.

14 Hristomir Yordanov and Peter Russer. Wireless inter-chip and intra-chip communication. In *2009 European Microwave Conference (EuMC)*, pages 145–148. IEEE, 2009.

15 A Vijaya Bhaskar and TG Venkatesh. Performance analysis of network-on-chip in many-core processors. *Journal of Parallel and Distributed Computing*, 147:196–208, 2021.

16 Yongrong Shi, Dengyun Shao, Wenjie Feng, Junzhi Zhang, and Ming Zhou. Silicon interposer package for MMIC heterogeneous integration based on gold/solder ball flip-chip technique. *IEEE Transactions on Components, Packaging and Manufacturing Technology*, 9(8):1659–1662, 2019.

17 Jacob Murray, Paul Wettin, Partha Pratim Pande, and Behrooz Shirazi. *Sustainable wireless network-on-chip architectures*. Morgan Kaufmann, 2016.

18 Md Shahriar Shamim, Jagan Muralidharan, and Amlan Ganguly. An interconnection architecture for seamless inter and intra-chip communication

using wireless links. In *Proceedings of the 9th International Symposium on Networks-on-Chip*, pages 1–8. Association for Computing Machinery, 2015.

19 Sanjukta Mandal, Sujit Kumar Mandal, and Ashis Kumar Mal. On-chip antennas using standard CMOS technology: A brief overview. In *2017 International Conference on Innovations in Electronics, Signal Processing and Communication (IESC)*, pages 74–78. IEEE, 2017.

20 Mohammad Alibakhshikenari, Esraa Mousa Ali, Mohammad Soruri, Mariana Dalarsson, Mohammad Naser-Moghadasi, Bal S Virdee, Caslav Stefanovic, Anna Pietrenko-Dabrowska, Slawomir Koziel, Stanislaw Szczepanski, et al. A comprehensive survey on antennas on-chip based on metamaterial, metasurface, and substrate integrated waveguide principles for millimeter-waves and terahertz integrated circuits and systems. *IEEE Access*, 10:3668–3692, 2022.

21 Jiaxin Chen, Jinzhao Wang, Jiewen Li, Yong Yao, Yunxu Sun, Jiajun Tian, Yi Zou, Xiangjie Zhao, and Xiaochuan Xu. Subwavelength structure enabled ultra-long waveguide grating antenna. *Optics Express*, 29(10):15133–15144, 2021.

22 Francesco Monticone, Christos Argyropoulos, and Andrea Alu. Optical antennas: Controlling electromagnetic scattering, radiation, and emission at the nanoscale. *IEEE Antennas and Propagation Magazine*, 59(6):43–61, 2017.

23 Arindam Dasgupta, Marie-Maxime Mennemanteuil, Mickaël Buret, Nicolas Cazier, Gérard Colas-des Francs, and Alexandre Bouhelier. Optical wireless link between a nanoscale antenna and a transducing rectenna. *Nature Communications*, 9(1):1992, 2018.

24 M Fukuda, Y Tonooka, T Inoue, and M Ota. Feasibility of plasmonic circuits for on-chip interconnects. *Solid-State Electronics*, 156:33–40, 2019.

25 Shaloo Rakheja and Vachan Kumar. Comparison of electrical, optical and plasmonic on-chip interconnects based on delay and energy considerations. In *Thirteenth International Symposium on Quality Electronic Design (ISQED)*, pages 732–739. IEEE, 2012.

26 Isaac Suárez, Albert Ferrando, Jose Marques-Hueso, Antonio Díez, Rafael Abargues, Pedro J Rodríguez-Cantó, and Juan P Martínez-Pastor. Propagation length enhancement of surface plasmon polaritons in gold nano-/micro-waveguides by the interference with photonic modes in the surrounding active dielectrics. *Nanophotonics*, 6(5):1109–1120, 2017.

27 Yihao Yang, Yuichiro Yamagami, Xiongbin Yu, Prakash Pitchappa, Julian Webber, Baile Zhang, Masayuki Fujita, Tadao Nagatsuma, and Ranjan Singh. Terahertz topological photonics for on-chip communication. *Nature Photonics*, 14(7):446–451, 2020.

28 Yiwen Sun, Zhijie Mei, Xuejiao Xu, Qingxuan Xie, Shuting Fan, Zhengfang Qian, and Xudong Liu. The perspective of topological photonics for on-chip terahertz modulation and sensing. *APL Photonics*, 8(11): 110901, 2023.

29 Xiang Yi, Cheng Wang, Zhi Hu, Jack W Holloway, Muhammad Ibrahim Wasiq Khan, Mohamed I Ibrahim, Mina Kim, Georgios C Dogiamis, Bradford Perkins, Mehmet Kaynak, et al. Emerging terahertz integrated systems in silicon. *IEEE Transactions on Circuits and Systems I: Regular Papers*, 68(9):3537–3550, 2021.

30 Mark Horowitz, Chih-Kong Ken Yang, and Stefanos Sidiropoulos. High-speed electrical signaling: Overview and limitations. *IEEE Micro*, 18(1):12–24, 1998.

31 Koji Yamada, Yutaka Urino, Takahiro Nakamura, and Yasuhiko Arakawa. Integrated silicon-based optical interconnect for fast, compact, energy-efficient electronic circuit systems. *NTT Technical Review*, 11(2):41–47, 2013. https://www.ntt-review.jp/archive/ntttechnical.php?contents=ntr201302fa7_s.html.

32 Yi-Ting Tsou, I-Huai Wang, and Sung-Mao Wu. Discussion of the signal transmission crosstalk. In *2018 IEEE 20th Electronics Packaging Technology Conference (EPTC)*, pages 785–788. IEEE, 2018.

33 Actel Application Note. Simultaneous switching noise and signal integrity. *Actel Corporation*, 2003.

34 Chunjie Duan, Brock J LaMeres, and Sunil P Khatri. *On and off-chip crosstalk avoidance in VLSI design*. Springer, 2010.

35 Bryan K Casper, Matthew Haycock, and Randy Mooney. An accurate and efficient analysis method for multi-gb/s chip-to-chip signaling schemes. In *2002 Symposium on VLSI Circuits. Digest of Technical Papers (Cat. No. 02CH37302)*, pages 54–57. IEEE, 2002.

36 Hyungsoo Kim, Youchul Jeong, Jongbae Park, Seok Kyu-Lee, Jong Kuk-Hong, Youngsoo Hong, and Joungho Kim. Significant reduction of power/ground inductive impedance and simultaneous switching noise by using embedded film capacitor. In *Electrical performance of electrical packaging (IEEE Cat. No. 03TH8710)*, pages 129–132. IEEE, 2003.

37 David AB Miller. Attojoule optoelectronics for low-energy information processing and communications. *Journal of Lightwave Technology*, 35(3):346–396, 2017.

38 Stephen W Keckler, William J Dally, Brucek Khailany, Michael Garland, and David Glasco. Gpus and the future of parallel computing. *IEEE Micro*, 31(5):7–17, 2011.

39 Keren Bergman, John Shalf, and Tom Hausken. Optical interconnects and extreme computing. *Optics and Photonics News*, 27(4):32–39, 2016.

40 Sébastien Rumley, Robert P Polster, Simon D Hammond, Arun F Rodrigues, and Keren Bergman. End-to-end modeling and optimization of power consumption in HPC interconnects. In *2016 45th International Conference on Parallel Processing Workshops (ICPPW)*, pages 133–140. IEEE, 2016.

41 Krishna C Saraswat and Farrokh Mohammadi. Effect of scaling of interconnections on the time delay of VLSI circuits. *IEEE Transactions on Electron Devices*, 29(4):645–650, 1982.

42 John D Owens, William J Dally, Ron Ho, DN Jayasimha, Stephen W Keckler, and Li-Shiuan Peh. Research challenges for on-chip interconnection networks. *IEEE Micro*, 27(5):96–108, 2007.

43 V Cisco. *The zettabyte era: trends and analysis*. CISCO, 2017.

44 Wolfgang Arden. Future semiconductor material requirements and innovations as projected in the ITRS 2005 roadmap. *Materials Science and Engineering: B*, 134(2–3):104–108, 2006.

45 Michel Brillouet. Emerging technologies on silicon. In *IEDM Technical Digest. IEEE International Electron Devices Meeting, 2004*, pages 17–24. IEEE, 2004.

46 Robert Chau, Suman Datta, and Amlan Majumdar. Opportunities and challenges of III-V nanoelectronics for future high-speed, low-power logic applications. In *IEEE Compound Semiconductor Integrated Circuit Symposium, 2005. CSIC'05*, pages 4. IEEE, 2005.

47 Gilbert Dewey, Mantu K Hudait, Kangho Lee, Ravi Pillarisetty, Willy Rachmady, Marko Radosavljevic, Titash Rakshit, and Robert Chau. Carrier transport in high-mobility III–V quantum-well transistors and performance impact for high-speed low-power logic applications. *IEEE Electron Device Letters*, 29(10):1094–1097, 2008.

48 Suman Datta, G Dewey, JM Fastenau, MK Hudait, D Loubychev, WK Liu, M Radosavljevic, W Rachmady, and R Chau. Ultrahigh-speed 0.5 v supply voltage $in_{0.7}ga_{0.3}as$ quantum-well transistors on silicon substrate. *IEEE Electron Device Letters*, 28(8):685–687, 2007.

49 Ravindiran Munusami and Shankar Prabhakar. Group III–V semiconductor high electron mobility transistor on Si substrate. In *Different Types of Field-Effect Transistors: Theory and Applications*, page 89, Intechopen, 2017.

50 Jun Wu. *Vertical III-V/High-k nanowire MOS capacitors and transistors*. Lund University, 2016.

51 Yung Hsu, Chun-Yen Chuang, Xinru Wu, Guan-Hong Chen, Chin-Wei Hsu, Yuan-Chia Chang, Chi-Wai Chow, Jyehong Chen, Yin-Chieh Lai, Chien-Hung Yeh, et al. 2.6 tbit/s on-chip optical interconnect supporting mode-division-multiplexing and pam-4 signal. *IEEE Photonics Technology Letters*, 30(11):1052–1055, 2018.

52 David Thomson, Aaron Zilkie, John E Bowers, Tin Komljenovic, Graham T Reed, Laurent Vivien, Delphine Marris-Morini, Eric Cassan, Léopold Virot, Jean-Marc Fédéli, et al. Roadmap on silicon photonics. *Journal of Optics*, 18(7):073003, 2016.

53 Siqi Yan, Jeremy Adcock, and Yunhong Ding. Graphene on silicon photonics: Light modulation and detection for cutting-edge communication technologies. *Applied Sciences*, 12(1):313, 2021.

54 Brian R Bennett and Richard A Soref. Electrorefraction and electroabsorption in InP, GaAs, GaSb, InAs, and InSb. *IEEE Journal of Quantum Electronics*, 23(12):2159–2166, 1987.

55 Chia-Chi Wang, Marc Currie, Sotiris Alexandrou, and Thomas Y Hsiang. Ultrafast, all-silicon light modulator. *Optics Letters*, 19(18):1453–1455, 1994.

56 Michel Bruel. Separation of silicon wafers by the smart-cut method. *Materials Research Innovations*, 3(1):9–13, 1999.

57 Shawn Yohanes Siew, Bo Li, Feng Gao, Hai Yang Zheng, Wenle Zhang, Pengfei Guo, Shawn Wu Xie, Apu Song, Bin Dong, Lian Wee Luo, et al. Review of silicon photonics technology and platform development. *Journal of Lightwave Technology*, 39(13):4374–4389, 2021.

58 Stephanie Law, Viktor Podolskiy, and Daniel Wasserman. Towards nano-scale photonics with micro-scale photons: The opportunities and challenges of mid-infrared plasmonics. *Nanophotonics*, 2(2):103–130, 2013.

59 Guanyu Chen, Yu Yu, Yang Shi, Nanxi Li, Wei Luo, Lin Cao, Aaron J Danner, Ai-Qun Liu, and Xinliang Zhang. High-speed photodetectors on silicon photonics platform for optical interconnect. *Laser & Photonics Reviews*, 16(12):2200117, 2022.

60 Pablo Ginel-Moreno, Alejandro Sánchez-Postigo, José de Oliva-Rubio, Abdelfettah Hadij-ElHouati, Winnie N Ye, J Gonzalo Wangüemert-Pérez, Íñigo Molina-Fernández, Jens H Schmid, Pavel Cheben, and Alejandro Ortega-Moñux. Millimeter-long metamaterial surface-emitting antenna in the silicon photonics platform. *Optics Letters*, 46(15):3733–3736, 2021.

61 Yachao Zhang, Shiteng Liu, Jinbang Ma, Yixin Yao, Baiqi Wang, Shengrui Xu, Bin Hou, Zhihong Liu, Jincheng Zhang, and Yue Hao. More than 60% RF loss reduction and improved crystal quality of GaN-on-SI achieved by in-situ doping tert-butylphosphorus. *Journal of Crystal Growth*, 625:127443, 2024.

62 Lei Zhu, Fan-Yi Meng, Liang Dong, Jia-Hui Fu, and Qun Wu. Low-loss magnetic metamaterial at THz frequencies by suppressing radiation losses. *IEEE Transactions on Terahertz Science and Technology*, 3(6):805–811, 2013.

63 Sergi Abadal, Chong Han, and Josep Miquel Jornet. Wave propagation and channel modeling in chip-scale wireless communications: A survey from millimeter-wave to terahertz and optics. *IEEE Access*, 8:278–293, 2019.

64 Xavier Timoneda, Sergi Abadal, Albert Cabellos-Aparicio, Dionysios Manessis, Jin Zhou, Antonio Franques, Josep Torrellas, and Eduard Alarcón. Millimeter-wave propagation within a computer chip package. In *2018 IEEE International Symposium on Circuits and Systems (ISCAS)*, pages 1–5. IEEE, 2018.

65 Kyung-Hoae Koo, Hoyeol Cho, Pawan Kapur, and Krishna C Saraswat. Performance comparisons between carbon nanotubes, optical, and Cu for future high-performance on-chip interconnect applications. *IEEE Transactions on Electron Devices*, 54(12):3206–3215, 2007.

66 Phaedon Avouris, Joerg Appenzeller, Richard Martel, and Shalom J Wind. Carbon nanotube electronics. *Proceedings of the IEEE*, 91(11):1772–1784, 2003.

67 Chris Bailey, Stoyan Stoyanov, Hua Lu, Tim Tilford, Chunyan Yin, and Nadia Strusevich. Modelling technologies and applications. *Nanopackaging: Nanotechnologies and electronics packaging*, pages 45–82. Cham, Springer, 2018.

68 Christophe Brun, Chin Chong Yap, Dunlin Tan, Stéphane Bila, Sebastien Pacchini, Dominique Baillargeat, and Beng Kang Tay. Flip chip based on carbon nanotube–carbon nanotube interconnected bumps for high-frequency applications. *IEEE Transactions on Nanotechnology*, 12(4):609–615, 2013.

69 Carlo Forestiere, Antonio Maffucci, Sergey A Maksimenko, Giovanni Miano, and Gregory Ya Slepyan. Transmission-line model for multiwall carbon nanotubes with intershell tunneling. *IEEE Transactions on Nanotechnology*, 11(3):554–564, 2012.

70 Hoyeol Cho, Pawan Kapur, and Krishna C Saraswat. Power comparison between high-speed electrical and optical interconnects for interchip communication. *Journal of Lightwave Technology*, 22(9):2021, 2004. https://www.osapublishing.org/jlt/abstract.cfm?uri=jlt-22-9-2021.

71 Brandon Heller, Srinivasan Seetharaman, Priya Mahadevan, Yiannis Yiakoumis, Puneet Sharma, Sujata Banerjee, and Nick McKeown. Elastictree: Saving energy in data center networks. In *NSDI*, volume 10, pages 249–264, USENIX, 2010.

72 Theophilus Benson, Ashok Anand, Aditya Akella, and Ming Zhang. Understanding data center traffic characteristics. *ACM SIGCOMM Computer Communication Review*, 40(1):92–99, 2010.

73 QJ Gu. THz interconnect: The last centimeter communication. *IEEE Communications Magazine*, 53(4):206–215, 2015. https://doi.org/10.1109/MCOM.2015.7081096.

74 Man Lung Mui, Kaustav Banerjee, and Amit Mehrotra. A global interconnect optimization scheme for nanometer scale VLSI with implications for latency, bandwidth, and power dissipation. *IEEE Transactions on Electron Devices*, 51(2):195–203, 2004.

75 JM Chamberlain. Where optics meets electronics: Recent progress in decreasing the terahertz gap. *Philosophical Transactions of the Royal Society of London A: Mathematical, Physical and Engineering Sciences*, 362(1815):199–213, 2004. https://doi.org/10.1098/rsta.2003.1312.

76 Carlo Sirtori. Applied physics: Bridge for the terahertz gap. *Nature*, 417(6885):132–133, 2002. https://doi.org/10.1038/417132b.

77 R Han and E Afshari. Filling the terahertz gap with sand: High-power terahertz radiators in silicon. In *2015 IEEE Bipolar/BiCMOS Circuits and Technology Meeting – BCTM*, pages 172–177, IEEE, 2015. https://doi.org/10.1109/BCTM.2015.7340574.

78 Daryoosh Saeedkia. *Handbook of Terahertz Technology for Imaging, Sensing and Communications*. Elsevier, 2013.

79 Thomas Kleine-Ostmann and Tadao Nagatsuma. A review on terahertz communications research. *Journal of Infrared, Millimeter, and Terahertz Waves*, 32(2):143–171, 2011.

80 JB Pendry, L Martín-Moreno, and FJ Garcia-Vidal. Mimicking surface plasmons with structured surfaces. *Science*, 305(5685):847–848, 2004. https://doi.org/10.1126/science.1098999.

81 Mikhail Erementchouk, Soumitra Roy Joy, and Pinaki Mazumder. Electrodynamics of spoof plasmons in periodically corrugated waveguides. *Proceedings of the Royal Society A: Mathematical, Physical and Engineering Sciences*, 472(2195):20160616, 2016.

82 Rashid Zia, Jon A Schuller, Anu Chandran, and Mark L Brongersma. Plasmonics: The next chip-scale technology. *Materials Today*, 9(7-8):20–27, 2006.

83 Yuan Liang, Hao Yu, Jincai Wen, Anak Agung Alit Apriyana, Nan Li, Yu Luo, and Lingling Sun. On-chip sub-terahertz surface plasmon polariton transmission lines with mode converter in CMOS. *Scientific Reports*, 6:30063, 2016. https://doi.org/10.1038/srep30063.

84 SR Joy, M Erementchouk, H Yu, and P Mazumder. Spoof plasmon interconnects—communications beyond RC limit. *IEEE Transactions on Communications*, 67(1):599–610, 2019. https://doi.org/10.1109/TCOMM.2018.2874242.

85 Anastasia Rusina, Maxim Durach, and Mark I Stockman. Theory of spoof plasmons in real metals. In *SPIE NanoScience+ Engineering*, pages 77572R–77572R. International Society for Optics and Photonics, 2010.

86 Mikhail A Kats, David Woolf, Romain Blanchard, Nanfang Yu, and Federico Capasso. Spoof plasmon analogue of metal-insulator-metal waveguides. *Optics Express*, 19(16):14860–14870, 2011. https://doi.org/10.1364/OE.19.014860.

87 FJ Garcia-Vidal, L Martin-Moreno, and JB Pendry. Surfaces with holes in them: New plasmonic metamaterials. *Journal of Optics A: Pure and Applied Optics*, 7(2):S97, 2005.

88 Linfang Shen, Xudong Chen, and Tzong-Jer Yang. Terahertz surface plasmon polaritons on periodically corrugated metal surfaces. *Optics Express*, 16(5):3326–3333, 2008. https://doi.org/10.1364/OE.16.003326.

89 Nanfang Yu, Qi Jie Wang, Mikhail A Kats, Jonathan A Fan, Suraj P Khanna, Lianhe Li, A Giles Davies, Edmund H Linfield, and Federico Capasso. Designer spoof surface plasmon structures collimate terahertz laser beams. *Nature Materials*, 9(9):730–735, 2010. https://doi.org/10.1038/nmat2822.

90 Hui Feng Ma, Xiaopeng Shen, Qiang Cheng, Wei Xiang Jiang, and Tie Jun Cui. Broadband and high-efficiency conversion from guided waves to spoof surface plasmon polaritons. *Laser & Photonics Reviews*, 8(1):146–151, 2014. https://doi.org/10.1002/lpor.201300118.

91 Zhao Xu, Kyungjun Song, and Pinaki Mazumder. Analysis of doubly corrugated spoof surface plasmon polariton (DC-SSPP) structure with sub-wavelength transmission at THz frequencies. *IEEE Transactions on Terahertz Science and Technology*, 2(3):345–354, 2012b.

92 K Song and P Mazumder. Active terahertz spoof surface plasmon polariton switch comprising the perfect conductor metamaterial. *IEEE Transactions on Electron Devices*, 56(11):2792–2799, 2009. https://doi.org/10.1109/TED.2009.2030838.

93 Z Xu and P Mazumder. Bio-sensing by Mach Zehnder interferometer comprising doubly-corrugated spoofed surface plasmon polariton (DC-SSPP) waveguide. *IEEE Transactions on Terahertz Science and Technology*, 2(4):460–466, 2012. https://doi.org/10.1109/TTHZ.2012.2202811.

94 K Song and P Mazumder. Dynamic terahertz spoof surface plasmon–polariton switch based on resonance and absorption. *IEEE Transactions on Electron Devices*, 58(7):2172–2176, 2011. https://doi.org/10.1109/TED.2011.2135370.

95 K Song and P Mazumder. Design of highly selective metamaterials for sensing platforms. *IEEE Sensors Journal*, 13(9):3377–3385, 2013. https://doi.org/10.1109/JSEN.2013.2260143.

96 N Zheng, M Aghadjani, K Song, and P Mazumder. Metamaterial sensor platforms for Terahertz DNA sensing. In *2013 13th IEEE International Conference on Nanotechnology (IEEE-NANO 2013)*, pages 315–320, 2013. https://doi.org/10.1109/NANO.2013.6720831.

97 Kyungjun Song and Pinaki Mazumder. Nonlinear spoof surface plasmon polariton phenomena based on conductor metamaterials. *Photonics and Nanostructures - Fundamentals and Applications*, 10(4):674–679, 2012. https://doi.org/10.1016/j.photonics.2012.06.008.

98 M Aghadjani, M Erementchouk, and P Mazumder. Spoof surface plasmon polariton beam splitter. *IEEE Transactions on Terahertz Science and Technology*, 6(6):832–839, 2016. https://doi.org/10.1109/TTHZ.2016.2599289.

99 Soumitra Roy Joy, Mikhail Erementchouk, and Pinaki Mazumder. Spoof surface plasmon resonant tunneling mode with high quality and purcell factors. *Physical Review B*, 95(7):075435, 2017. https://doi.org/10.1103/PhysRevB.95.075435.

100 Anders Pors, Esteban Moreno, Luis Martin-Moreno, John B Pendry, and Francisco J Garcia-Vidal. Localized spoof plasmons arise while texturing closed surfaces. *Physical Review Letters*, 108(22):223905, 2012.

101 Stefan A Maier, Steve R Andrews, L Martín-Moreno, and FJ García-Vidal. Terahertz surface plasmon-polariton propagation and focusing on periodically corrugated metal wires. *Physical Review Letters*, 97(17):176805, 2006. https://doi.org/10.1103/PhysRevLett.97.176805.

102 Zhen Liao, Yu Luo, Antonio I Fernández-Domínguez, Xiaopeng Shen, Stefan A Maier, and Tie Jun Cui. High-order localized spoof surface plasmon resonances and experimental verifications. *Scientific Reports*, 5:9590, 2015.

103 AI Fernandez-Dominguez, L Martin-Moreno, FJ Garcia-Vidal, Steven Andrews, and SA Maier. Spoof surface plasmon polariton modes propagating along periodically corrugated wires. *IEEE Journal of Selected Topics in Quantum Electronics*, 14(6):1515–1521, 2008.

104 Jianqiang Gu, Ranjan Singh, Xiaojun Liu, Xueqian Zhang, Yingfang Ma, Shuang Zhang, Stefan A Maier, Zhen Tian, Abul K Azad, Hou-Tong Chen, et al. Active control of electromagnetically induced transparency analogue in terahertz metamaterials. *Nature Communications*, 3(1):1–6, 2012.

105 David Woolf, Mikhail A Kats, and Federico Capasso. Spoof surface plasmon waveguide forces. *Optics Letters*, 39(3):517–520, 2014. https://doi.org/10.1364/OL.39.000517.

106 Victor W Brar, Min Seok Jang, Michelle Sherrott, Josue J Lopez, and Harry A Atwater. Highly confined tunable mid-infrared plasmonics in graphene nanoresonators. *Nano Letters*, 13(6):2541–2547, 2013.

107 Xiaopeng Shen, Tie Jun Cui, Diego Martin-Cano, and Francisco J Garcia-Vidal. Conformal surface plasmons propagating on ultrathin and flexible films. *Proceedings of the National Academy of Sciences of the United States of America*, 110(1):40–45, 2013.

108 Xi Gao and Tie Jun Cui. Spoof surface plasmon polaritons supported by ultrathin corrugated metal strip and their applications. *Nanotechnology Reviews*, 4(3):239–258, 2015.

109 Yuan Liang, H Yu, Junfeng Zhao, Wei Yang, and Yuangang Wang. An energy efficient and low cross-talk CMOS sub-THz I/O with surface-wave modulator and interconnect. In *2015 IEEE/ACM International Symposium on Low Power Electronics and Design (ISLPED)*, pages 110–115, 2015a. https://doi.org/10.1109/ISLPED.2015.7273499.

110 Yuan Liang, Hao Yu, Hao Chi Zhang, Chang Yang, and Tie Jun Cui. On-chip sub-terahertz surface plasmon polariton transmission lines in CMOS. *Scientific Reports*, 5:14853, 2015b. https://doi.org/10.1038/srep14853.

111 D Martin-Cano, ML Nesterov, AI Fernandez-Dominguez, FJ Garcia-Vidal, L Martin-Moreno, and Esteban Moreno. plasmons for subwavelength terahertz circuitry. *Optics Express*, 18(2):754–764, 2010. https://doi.org/10.1364/OE.18.000754.

112 Z Xu, K Song, and P Mazumder. Analysis of doubly corrugated spoof surface plasmon polariton (DC-SSPP) structure with sub-wavelength transmission at THz frequencies. *IEEE Transactions on Terahertz Science and Technology*, 2(3):345–354, 2012a. https://doi.org/10.1109/TTHZ.2012.2188558.

2

Spoof Plasmonics: Origin and State-of-the-Art Development

2.1 Slow Wave Structure: A Historical Perspective

The emergence of spoof surface plasmon polariton (SSPP) owes its origin to two different, apparently unrelated technologies: one is in the field of microwave engineering in hundreds of megahertz to a few gigahertz, the other is in the field of nanophotonics, applied in the field of hundreds of terahertz (THz). The first predecessor of spoof plasmonics can be traced back to slow wave structures. Slow wave structures are vital components in microwave devices, enabling efficient interaction between electron beams and electromagnetic waves. Their origins can be traced back to the early 20th century when researchers sought to enhance microwave devices' efficiency. These structures are commonly used in traveling wave tubes (TWTs) and backward wave oscillators (BWOs) to amplify or generate microwave signals (see Figure 2.1).

During World War II, microwave technology played a vital role in radar systems, leading to advancements in magnetron tubes for generating microwaves. After the war, researchers focused on improving microwave devices for broader applications, such as satellite communications and scientific research. In the post-World War II era, microwave amplification technologies like the klystron and TWT emerged, both utilizing slow wave structures to facilitate efficient interaction between electron beams and microwave signals. Klystrons found applications in radar systems, while TWTs were used in communication and broadcasting. Figure 2.2 shows the inventors of the TWT amplifier in microwave technology.

The need for slow wave structures arose from the constraints of waveguides, where the electromagnetic wave's phase velocity is greater than the speed of light in free space. This mismatch caused weak interaction between electron beams and microwave signals in conventional tubes, leading to low energy transfer. Slow

Artificial Plasmonics for VLSI Interconnects: Bridging the Gap between Electronics and Optics, First Edition. Soumitra R. Joy and Pinaki Mazumder.

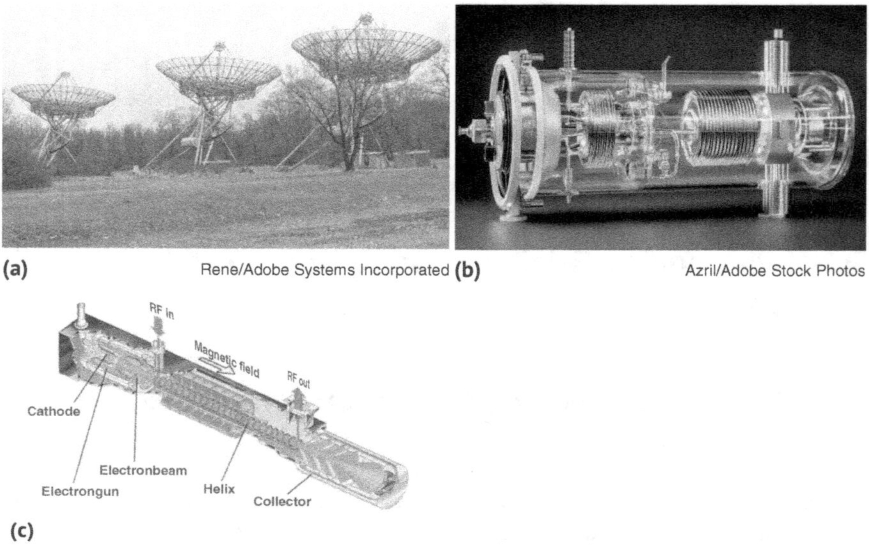

Figure 2.1 (a) Radar technology in the United Kingdom during World War II; (b) the interior of a magnetron, designed for high-power RF signal generation; and (c) the interior design of a TWT, designed for RF wave amplification through slow-wave structure.

Figure 2.2 Inventors and Pioneers in traveling-wave tube amplifier in microwave technology. John Pierce (left) and Rudolf Kompfner (center), along with Harry Nyquist (right), taken in 1960 (David and Julia Bart, 2011/IEEE).

wave structures were designed to slow down the phase velocity of electromagnetic waves, enhancing the interaction time between waves and electron beams. This interaction allows for efficient energy transfer from the electron beam to the microwave signal, leading to amplification or oscillation.

2.1.1 Variants on Slow Wave Structures

Various types of slow wave structures are used in microwave technology, each with unique characteristics and applications (see Figure 2.3). Helix structures consist of cylindrical helices winding around a central axis, ideal for high-power amplification in TWTs. Coupled cavity structures, formed by a series of resonant cavities, enable amplification or oscillation in BWOs. Folded waveguide structures provide an extended interaction path in limited spaces, and interdigital structures use interleaved metallic fingers to slow down electromagnetic waves, common in distributed amplifiers and filters. These diverse designs cater to specific application requirements and contribute to the versatility of slow wave structures in modern microwave technology.

In addition to being utilized as a slow wave structure in microwave amplifiers, patterned metallic structures have been employed in diverse fields of context and applications, including signal filtering and optical grating, refer to Figure 2.4. For several decades, metallic structures patterned with periodic teeth-shaped features have been utilized for signal processing, particularly in microwave and optical frequencies tailored at 1.55-micron wavelength. Researchers [1] have leveraged these structures to design compact microstrip bandpass filters in the 1–10 GHz range with wide stopbands, high selectivity, and tunable passbands. Such filters have proven to be essential components in wireless communication systems due to their high out-of-band rejection, low insertion loss, and excellent return loss

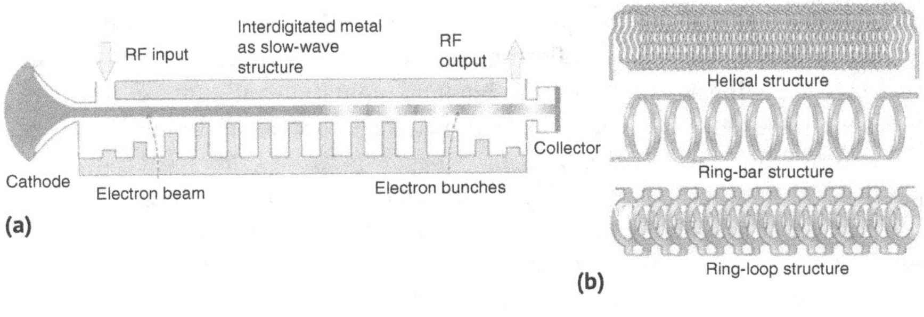

Figure 2.3 Various shapes of slow wave structure utilized inside a TWT: (a) interdigital structures; and (b) helix and ring structures.

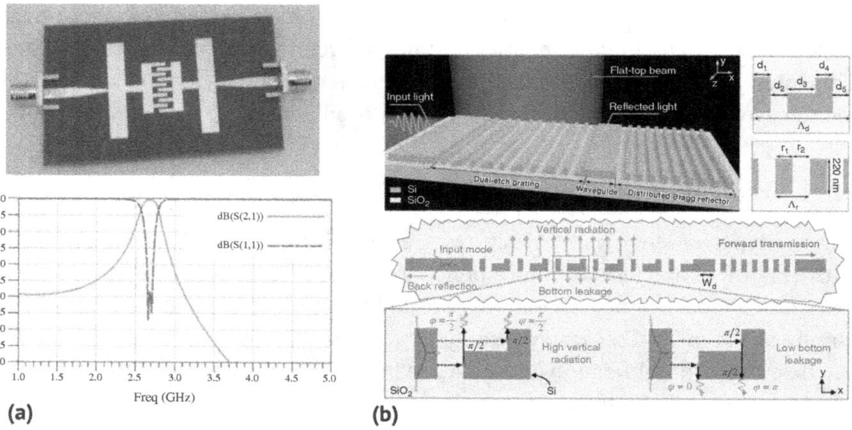

Figure 2.4 Diverse usage of patterned metallic structure in disparate domains of engineering: (a) design of high-quality filter at the microwave regime by staggered teeth-shaped patterned metal (Mohamad Farhat et.al., 2019/Science Publication PTY LTD); and (b) optical grating design and 1.55-micron wavelength by periodic corrugation to generate out-of-plane beam.

characteristics. On the other hand, Tao et al. [2] demonstrated double-sided teeth-shaped nanoplasmonic waveguide filters to function as a two-wavelength filter with linearly correlated wavelengths to the depths of the teeth, and also exhibit a wide and sharp bandgap. These ultracompact filters, with lengths of a few hundred nanometers, offer the potential for constructing nanoscale waveguide filters in high-density nanoplasmonic integration circuits. In addition to signal filtering, corrugated metal structures have played a pivotal role in optical grating applications. For instance, Tian et al. [3] introduced a grating coupler capable of generating high-quality flat-top beams with narrow beamwidths from photonic integrated circuits into free space. By manipulating the structural parameters of the dual-etch grating and distributed Bragg reflector, the researchers achieved remarkable performance, enabling advanced optical signal manipulation and routing.

Despite the extensive use of patterned metallic structures in signal processing and optical grating, it took John Pendry's paper in 2004 [4] to reveal a new perspective. He showed that by varying the geometric structure of patterned metals, it is possible to imitate the plasmonic behavior of strong surface confinement of light at a designer's chosen low frequency. This idea opened a realm of possibilities, giving birth to the field of spoof plasmonics. Following Pendry's discovery, slow wave structures have found newer niches of application in cutting-edge technologies. They play a crucial role in particle accelerators, where they assist in producing high-frequency electromagnetic fields for accelerating charged particles to high energies [5]. THz technology benefits from slow wave structures to generate, manipulate, and detect THz radiation [6], enabling applications in spectroscopy, high-resolution subwavelength imaging, and communication. Additionally, slow wave structures are utilized in advanced high-frequency communication systems, enabling higher data rates and improved signal stability in modern wireless networks. These evolving applications demonstrate the continued significance and adaptability of slow wave structures in addressing contemporary challenges from microwave to THz technology.

After discussing the early slow wave structure akin to the one supporting the electromagnetic mode of SSPP, our focus will shift to another crucial physical phenomenon that can be considered the closest predecessor of the SSPP mode. This phenomenon is known as surface-plasmon polariton, to which the SSPP owes its existence.

What kind of "surface wave" are we referring to?

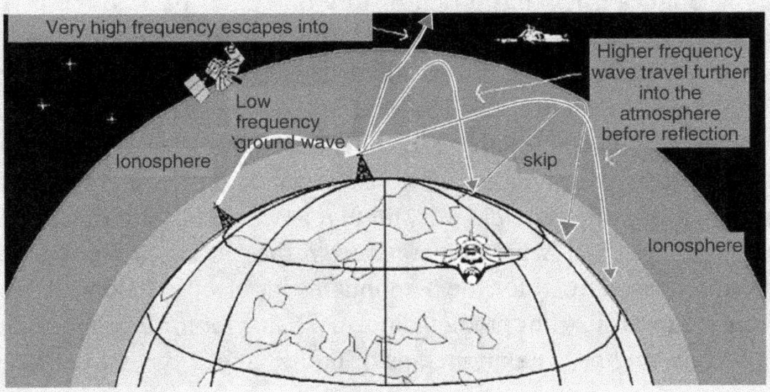

Figure 2.5 Wave propagation through the atmosphere of Earth. The relatively low frequency in the microwave regime can graze through the curvature of Earth, known as the Sommerfeld wave.

The term "surface wave" can be often ambiguous, particularly in the field of microwave engineering, where it may have multiple meanings. For instance, the Sommerfeld wave, often referred to as a low-frequency grazing field, stands out as a phenomenon where emitted waves from an antenna station can travel along the Earth's surface, following its curvature to reach distant receiver antennas. This wave's origin lies in the refraction of the electromagnetic wave within the Earth's atmosphere, which comprises layers with varying densities as we move away from the surface, refer to Figure 2.5.

When the electromagnetic wave is incident at an angle greater than the Brewster's angle, it leads to a curved propagation trace, giving rise to the Sommerfeld wave. Conversely, in the absence of an atmosphere, such as on the Moon, the electromagnetic wave would propagate in a line-of-sight manner.

While the term "surface wave" is commonly used throughout this book, it is essential to differentiate between the Sommerfeld wave, extensively used in microwave engineering for long-distance communication, and the spoof plasmonic surface waves envisioned for chip-scale communication, as they originate from distinct physics, leading to applications very different from each other.

2.2 Surface Plasmon Polariton in Metal

Surface plasmon resonance (SPR) has a rich history that dates back to the early 20th century. In 1902–1912, R.W. Wood observed unusual dark and light bands in reflected light when polarized light was shone onto a metal-backed diffraction grating [7, 8], refer to Figure 2.6. However, a clear explanation for this phenomenon was not provided at the time. In 1907, Lord Rayleigh made the first theoretical treatment of these anomalies with his "dynamical theory of the grating" [9]. He found that the scattered field was singular at wavelengths corresponding to the Wood anomalies [10], which emerged from the grating at the grazing angle. Rayleigh's theory accounted for the S anomalies for electric field polarization perpendicular to the rulings but predicted normal behavior for P polarization near the Rayleigh wavelengths. The anomalies of both S and P types were later

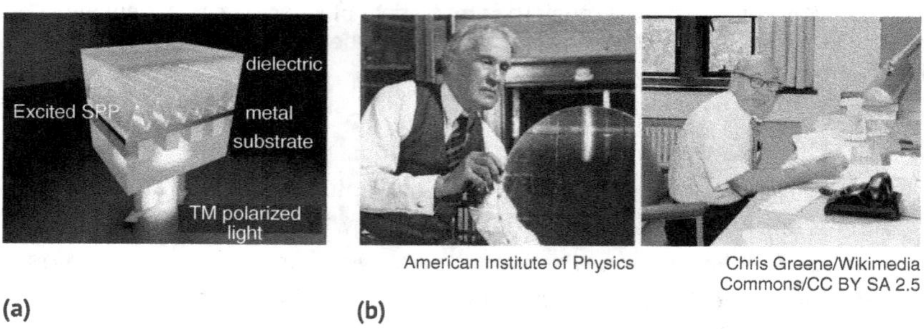

Figure 2.6 (a) Artistic renditions for surface plasmon polariton (SPP) at metal–dielectric interface, courtesy of [1]; and (b) the pioneers in the discovery of SPR in metal: R.W. Wood, experimental physicist at Johns Hopkins University (left), Ugo Fano, theoretical physicist at the University of Chicago, Carnegie Institute (right).

demonstrated, with P anomalies mainly observed on deeply ruled gratings. Theoretical analysis by Fano [11] linked these anomalies to surface waves, now known as surface plasmons, supported by the grating network (see Figure 2.6).

In the 1950s and 1960s, researchers, including Raether [12], conducted experimental studies and provided further theoretical explanations, advancing the understanding of the interaction between light and metallic thin films. The term "surface plasmon resonance" was coined by Powell and Swan in 1962 when they conducted experimental studies on the interaction between light and thin metal films, demonstrating the excitation of surface plasmons and the enhanced absorption of light [13].

The development of SPR continued with research by Klaus B. Wolf and Wilhelm Lukosz in the late 1960s and early 1970s, who explored the use of SPR for sensing applications [14]. This led to SPR becoming a powerful tool for studying molecular interactions, biomolecular binding, and chemical sensing.

Between 1980–1990, Pharmacia's interest in SPR led to the establishment of Pharmacia Biosensor AB, focused on developing a functional SPR machine [15], refer to Figure 2.7. By 1990, the first BIAcore, utilizing a wedge-shaped laser beam and diode array, was sold [16]. Over the years, various models were introduced, including BIAlite [17] and BIAcore X, 2000, 3000, and Q, catering to different applications and sensitivity levels. In 2005, the BIAcore A100 with dynamic addressing was launched, and the Biacore 8K/8K+ system with 16 flow cells for high-throughput screening and characterization of small

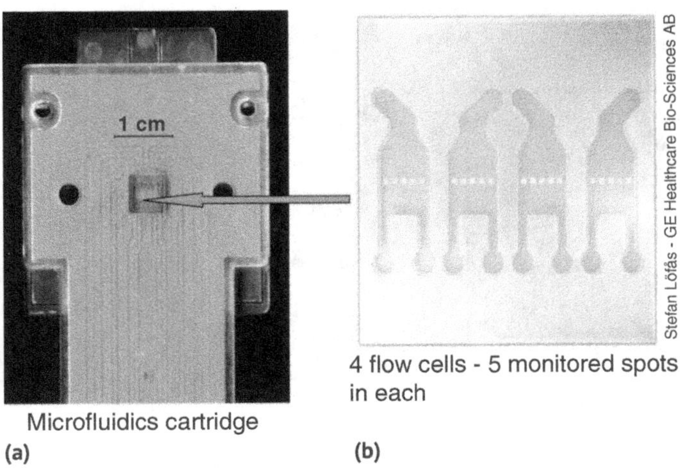

Figure 2.7 Commercial biosensor design using SPP phenomenon: Biacore A100 integrated fluidic cartridge (a) microfluidic flow path; and (b) design of flow cell with detector spots. Courtesy of [5].

molecules was reported in drug discovery programs [18]. Other manufacturers have also developed SPR systems, such as Sensata Technologies' Spreeta Evaluation module [19], Bio-Rad's ProteOn XPR36, and Bruker's Sierra SPR-32, each offering unique capabilities for simultaneous measurements and addressing multiple interactions. Some systems, like IAsys Biosensor and Horiba SPRi, utilize a resonant mirror for resonance angle determination and image reflectivity of the sensor surface, enabling monitoring of numerous interactions simultaneously.

The field of SPR has seen advancements in modern times, with the emergence of nanofabrication and characterization tools driving a vast interest in its study and exploitation. SPR finds applications in diverse fields, including biosensing, medical diagnostics, environmental monitoring, and materials science.

2.2.1 Extraordinary Transmission by Surface Plasmonics

In a paper published in 2001 by Martin-Moreno, Garcia-Vidal (University of Zaragoza, Spain), JB Pendry (Imperial College, London), and TB Ebbesen (Louis Pasteur University, France), the authors explored a structured surface consisting of a metal film with a two-dimensional (2D) periodic array of subwavelength holes. This configuration was found to exhibit extraordinary transmission of p-polarized light. Contrary to the standard aperture theory, which predicted less than 1% transmission, the authors' theory and experiments demonstrated that transmission rates of $\geq 15\%$ could be achieved, despite the holes constituting just $\geq 11\%$ of the total metal area. This phenomenon was aptly termed "extraordinary transmission, refer to Figure 2.8."

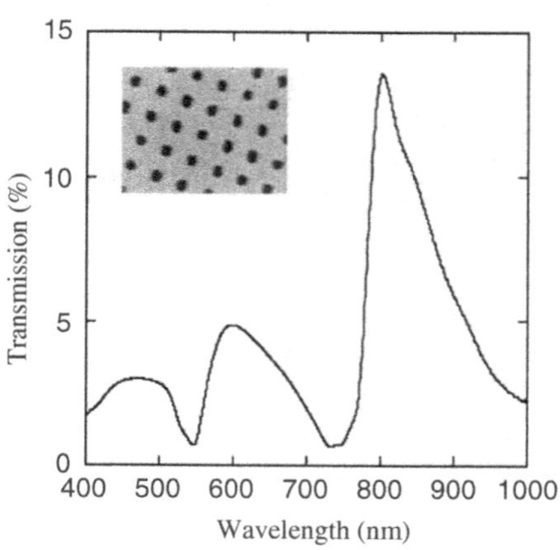

Figure 2.8 Experimental zero-order power transmittance at normal incidence for a square array of holes of lattice constant 750 nm in silver film.

The mechanism behind this extraordinary transmission lies in the conversion between free space modes and SPR modes at the metal–dielectric interfaces. The periodic hole array acts as a grating, converting the propagating mode into an in-plane surface-bound mode, which subsequently decays into another propagating mode on the other side of the metal film. This discovery sparked considerable theoretical and experimental interest in understanding the underlying mechanisms responsible for the remarkable transmissivity and exploring ways to enhance it even further. The structure of a metal film with a periodic array of subwavelength holes has become one of the most extensively studied and well-known examples of optical metamaterials, paving the way for numerous advancements in the field.

SPPs, electromagnetic surface waves sustained at the interface between a conductor and a dielectric, offer the unique ability to confine electromagnetic energy below the diffraction limit of classical optics. While the fundamental physics of SPPs has been described in several papers [20, 21] across the twentieth century, recent advancements in nanofabrication and characterization tools have significantly increased interest in their exploration and application. Plasmonics, a field aiming to capitalize on subwavelength light confinement achieved by exciting SPPs, particularly focuses on the optical regime where strong confinement occurs at frequencies comparable to the metal's intrinsic plasma frequency. The motion of conduction electrons at the interface is dephased with respect to the driving electromagnetic fields, leading to reduced phase and group velocities and, consequently, strong localization. Most SPP research has focused on noble metals like Ag, Au, and Cu, showing plasma frequencies in the UV. However, achieving light localization at subwavelength dimensions in the microwave or THz frequencies has been a desirable goal. The addition of a subwavelength corrugation, such as arrays of holes, to the metal surface has been known to produce enhanced surface impedance and bind a surface mode to the interface, even in the perfect conductivity limit. In 2004, Pendry and co-workers discovered an additional mechanism for periodicity-induced binding of electromagnetic fields to the surface, describing the effective surface layer with a dielectric permittivity of the Drude form, with the plasma frequency given by the cutoff frequency of the hole waveguide. This led to the concept of SSPPs, where the surface electromagnetic modes supported by corrugated metal surfaces can be entirely controlled by geometry. Since then, the concept of spoof SPPs has been applied to various geometries, including planar, cylindrical, and more complex structures like helically grooved wires, V-grooves, wedges, and domino structures [22–25]. In the subsequent sections of the chapter, we will be explicating the details of these discoveries.

2.3 Surface Plasmon Polariton: Explanation Through Drude's Model

Surface plasmons are a result of the interaction between electromagnetic fields and the conduction of electrons within a metal. These electrons, when subjected to an electromagnetic field, collectively oscillate, giving rise to charge density oscillations. These oscillations can be sustained within the bulk of a metal or confined at the interface between a metal and a dielectric medium. Surface plasmons offer the valuable ability of strong subwavelength localization and enhancement of electromagnetic fields near metal/dielectric interfaces.

Two distinct types of surface plasmons exist, depending on the interface's geometry: propagating SPPs and localized surface plasmons (LSPs). SPPs represent surface-coupled electromagnetic waves propagating on a flat metal surface. They find applications in the efficient guiding of waves in one-dimensional (1D) plasmonic waveguides and bridging the realms of photonics and electronics, promising more compact optoelectronic devices. On the other hand, LSPs are supported by metal nanoparticles with dimensions smaller than the wavelength at optical frequencies. These LSPs facilitate the efficient transfer of energy between near and far fields, playing the role of nanoantennas.

The presence of both SPPs and LSPs characterizes the electromagnetic response of metals. The optical response of a material is typically described by its dielectric permittivity, $\epsilon(\omega)$, which is a function of the incident light frequency. The Drude model treats a metal as a background of fixed core ions combined with a free electron gas to model the behavior of conduction electrons. In this model, the electric permittivity can be written as

$$\epsilon(\omega) = \epsilon_\infty \left(1 - \frac{\omega_p^2}{\omega^2 + i\gamma\omega}\right) \tag{2.1}$$

where ω_p is the plasma frequency, γ is the damping constant of electrons, and ϵ_∞ is a high-frequency offset. The bulk plasma frequency is characteristic of each metal and reads as:

$$\omega_p = \sqrt{\frac{ne^2}{m_e \epsilon_0}} \tag{2.2}$$

With n being the electron density, e and m_e the electron charge and effective mass, and ϵ_0 the vacuum permittivity. For most metals, ω_p lies in the ultraviolet (UV) regime, $\omega_p \sim 10^{16}$ Hz. The real and imaginary parts of the electric permittivity for silver are plotted in Figure 2.9.

Below the plasma frequency (i.e., $\omega < \omega_p$), the real part of $\epsilon(\omega)$ is negative, and its imaginary part is nonzero, indicating the relaxation time of electrons in the

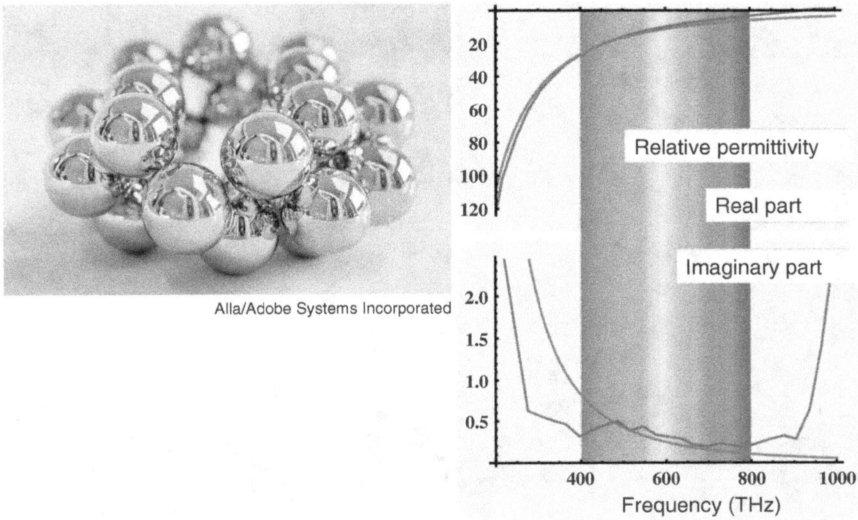

Figure 2.9 The complex relative permittivity of silver (Ag), contrasting between the theoretical value from Drude's dispersion model and experimental data.

lattice. The Drude model, however, fails to account for the interband transition-induced increase of the imaginary part of permittivity at the $\omega \geq \omega_p$. A more accurate modeling of $\epsilon(\omega)$ for higher frequencies is given by the Drude–Lorentz model.

SPPs are p-polarized (magnetic field component lying in the plane of the interface), and their field intensity decays exponentially in the direction perpendicular to the interface. A sketch of an SPP propagating through the interface separating a metal and a dielectric is shown in Figure 2.2. SPPs are solutions of Maxwell's equations bound to an interface,

$$\nabla \times \nabla \times \mathbf{E}(\mathbf{r}, \omega) - \frac{\omega^2}{c^2}\epsilon(\mathbf{r}, \omega)\,\mathbf{E}(\mathbf{r}, \omega) = 0 \tag{2.3}$$

Their dispersion relation can be written as

$$k_x(\omega) = k_0\sqrt{\frac{\epsilon_m(\omega)\epsilon_d(\omega)}{\epsilon_m(\omega) + \epsilon_d(\omega)}} \tag{2.4}$$

where $k_0 = \frac{\omega}{c}$ is the free space wavenumber (c is the velocity of light), k_x is the wave vector component along the propagation direction, and ϵ_m and ϵ_d is the electric permittivity of metal and dielectric, respectively. SPPs are evanescent modes propagating along the interface: they have $Re\,(k_x) > Im(k_x)$ since generally $Re\,(\epsilon_m) \gg Im(\epsilon_m)$, with $Re\,(k_x) > k_0$, and complex wave vector component normal to the interface, $k_{j,z} = \sqrt{\epsilon_j(\omega)k_0^2 - k_x^2}$. Hence, for surface waveguiding condition,

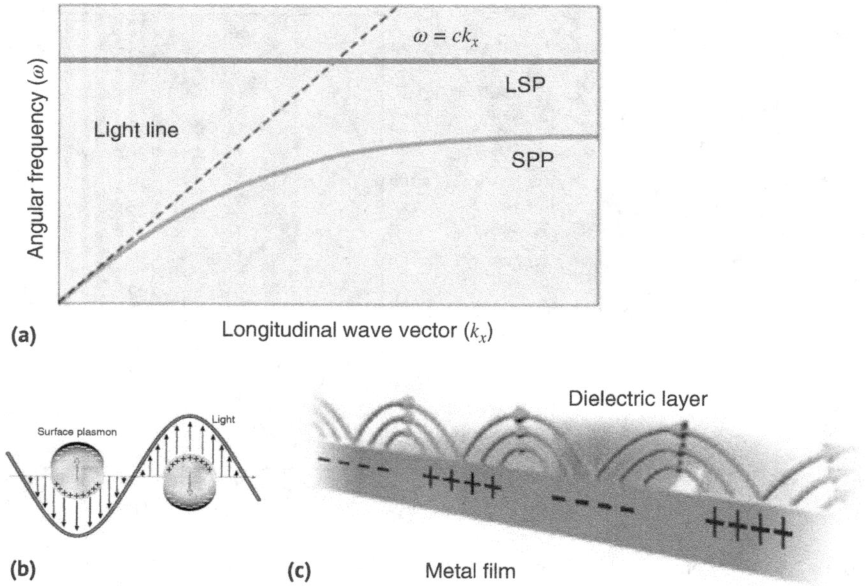

(a) (b) (c)

Figure 2.10 Presence of two resonance phenomena in metal near its plasma frequency: (a) the dispersion relation of LSP mode and propagating surface plasmon mode; (b) illustration of LSP; and (c) illustration of SPP at metal–dielectric interface.

it requires $\epsilon_m \epsilon_d < 0$ and $\epsilon_m(\omega) + \epsilon_d(\omega) < 0$. The dispersion relation given by Eq. (2.1) is plotted in Figure 2.10 for a silver–air interface.

According to Eq. (2.1), for a lossless metal ($\gamma = 0$), at frequencies $\omega < \omega_{sp}$ where $\omega_{sp} = \omega_p/\sqrt{1+\epsilon_\infty^{-1}}$, we have $k_x > k_0$, indicating that the mode is below the light line ($\omega = ck$), and hence, is surface bound. At very low frequencies, $|\epsilon_m(\omega)| \gg 1$, hence $k_x \approx k_0$, which means very weak modal confinement to the interface. As the frequency approaches ω_{sp}, $|\epsilon_m(\omega)|$ decreases, and the value of $(k_x - k_0)$ increases asymptotically. The imaginary part of the parallel wave vector of the SPP mode, $Im[k_x]$ accounts for the finite propagation length of the SPP,

$$L_p = \frac{1}{Im\,[k_x]}$$

There is a trade-off between modal confinement and propagation. The closer the frequency ω to the characteristic frequency ω_{sp}, the stronger the confinement and the smaller the propagation length. Therefore, surface plasmon polariton waves are subjected to high metallic loss, and are limited within few microns of propagation. Metallic losses could be reduced if we could achieve plasmonic behavior at a lower frequency. Unfortunately, at lower frequencies, such as at THz or sub-THz range, the modal confinement of surface-plasmon polariton significantly reduces. This dilemma is addressed by the introduction of the concept of SSPP, where the plasmonic behavior of metals is mimicked at much lower frequency by patterning the metal surface. In the next section, we shall elaborate on the concept.

2.4 SSPP in Planar Geometry

2.4.1 1D Groove Pattern

2.4.1.1 Textured Surfaces

In this section we analyze the electromagnetic modes bound to textured perfect electric conductor (PEC) surfaces by considering the two simple geometries for which the spoof SPP concept was first developed: 1D arrays of grooves and 2D arrays of square dimples [26]. Guided modes in these two structures are given by the nonvanishing solutions of the matching equations describing the continuity of the fields at the metal–vacuum interface. First, we consider the case of 1D arrays of grooves. The inset of Figure 2.11 shows the geometrical parameters of the structure: the array period, d, and the grooves' width and depth, a and h, respectively. As mentioned before, the PEC approximation makes all lengths scalable.

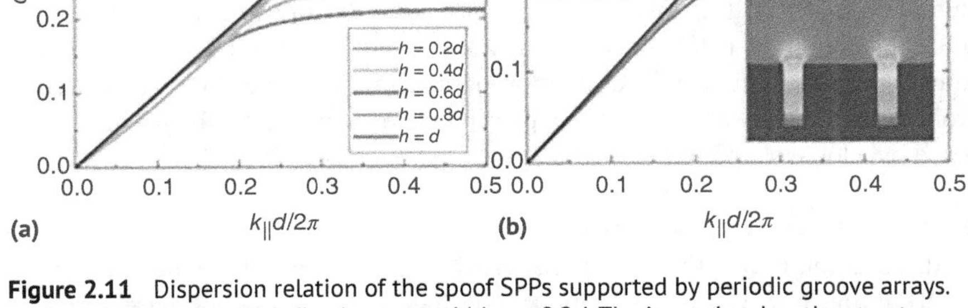

Figure 2.11 Dispersion relation of the spoof SPPs supported by periodic groove arrays. (a) Dependence on h for a fixed groove width $a = 0.2d$. The inset sketches the structure considered;and (b) bands for h = $0.6d$ at different groove widths. The inset depicts the electric field amplitude for $h = 0.6d$ and $a = 0.2d$ evaluated at the band edge ($k_{||} = \pi/d$).

Thus, from now on, we take the system periodicity, d, as the unit of length. The approximate spoof SPP band is given by

$$k_{\|} = k_0\sqrt{\left\{1 + \left(\frac{a}{d}\right)^2 \tan^2 (k_0 h)\right\}} \tag{2.5}$$

According to Eq. (2.5), enlarging the groove depth translates into the deviation of the dispersion relation from the light line toward larger wave vectors, and that when $a = 0$ or $h = 0$ (flat surface), $k_{\|} = k_0$, and no confined modes are supported by the structure.

Figure 2.11 displays the normalized frequency (d/λ) versus wave vector ($k_{\|}d/2\pi$) for the spoof SPP modes, in groove arrays of width $a = 0.2d$ and depths ranging from $h = 0.2d$ to $h = d$. As predicted by Eq. (2.5), the bands shift to lower frequencies when the depth of the grooves is increased. This result can be understood in terms of cavity resonances occurring inside the indentations. Note that the lowest waveguide mode supported by 1D apertures is always propagating. Thus, electromagnetic fields completely explore the groove depth, making them strongly dependent on h. This is reflected through the tangent function in Eq. (2.5), which diverges when the Fabry–Perot condition, $\sin(k_0 h) = 0$, is satisfied. We can interpret this behavior as resulting from the fact that spoof SPPs in 1D blind indentations have a hybrid nature with characteristics of both surface and cavity electromagnetic modes.

The previous calculations have been done within the PEC approximation, and, therefore, the propagation length of the corresponding spoof SPP modes is infinite. This length is reduced to finite values in a real metal due to absorption. Shen and co-workers [27] have studied the propagation length of geometrically modified SPP modes supported by 1D arrays of grooves in the GHz regime. As expected, the loss is large at frequencies close to the asymptotic frequency.

Recently, two different types of applications based on the propagation characteristics of spoof SPP modes in 1D arrays of grooves have been proposed. In the first one by Lu et al. [28], the slowing down of THz waves based on 1D arrays of grooves with graded depths can be achieved by a proper design of the dispersion curves and asymptotic frequencies. Another proposal is that of active THz switches, controlled by an electro-optical material placed inside the grooves, which can be activated by a low-voltage control signal [29].

2.4.1.2 SSPP in 2D Dimple Arrays

In this section, we analyze the characteristics of the spoof SPP modes supported by 2D dimple arrays. For simplicity, we consider only the case of square arrays of square apertures; see the inset of Figure 2.12. The geometry of the system is now given by the array period, d, taken as a reference length, and the side and depth

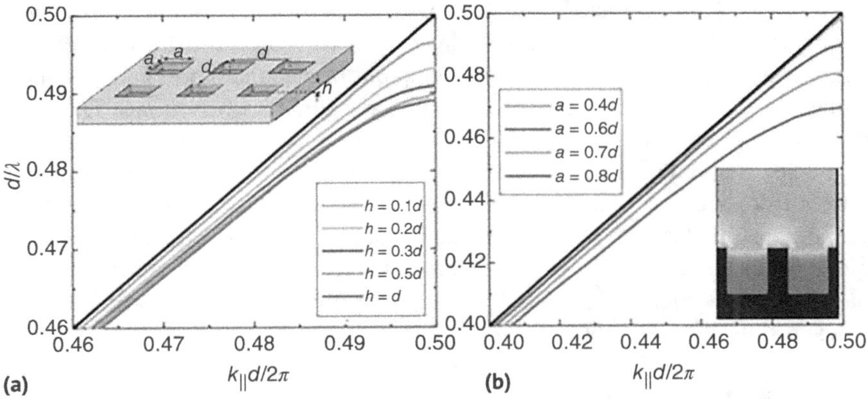

Figure 2.12 Dispersion relation of the spoof SPPs sustained by periodic dimple arrays. (a) Bands for $a = 0.2d$ and several dimple depths h. The inset shows a schematic of the system; and (b) bands for $h = 0.6d$ and different a. The inset displays the electric field amplitude at the band edge for the case $h = 0.6d$ and $a = 0.6d$.

of the dimples, a and h, respectively. The analytical expression for the dispersion relation of the modes has the form

$$k_{||} = k_0 \sqrt{\left(1 + \left(\frac{2\sqrt{2}a}{\pi d}\right)^4 \frac{k_0^2}{\left(\frac{\pi}{a}\right)^2 - k_0^2} \tanh^2 \sqrt{\left[\left(\frac{\pi}{a}\right)^2 - k_0^2\right]} h\right)} \tag{2.6}$$

where the coupling factor between the resonance mode inside the dimple and the free space propagation is $S = \frac{2\sqrt{2}}{\pi}\frac{a}{d}$ in the limit of deep subwavelength indentations, and the propagation constant for the lowest waveguide mode (TE_{11}) supported by 2D square apertures is $q = i\sqrt{\left(\frac{\pi}{a}\right)^2 - k_0^2}$. In the range of validity of the approximation used in this case, $\lambda \gg a$, and the fields decay evanescently within perforations. This evanescent character of the fields is expressed in the hyperbolic tangent dependence on h in Eq. (2.6).

Figure 2.12 plots the exact spoof SPP bands for dimples of side $a = 0.6d$. The depth of the indentations is varied from $h = 0.1d$ to $h = d$. As in this figure, the mode frequency is lowered when the depth of the perforations is increased. However, the displacement of the bands is much smaller than in the 1D case. This difference is related to the evanescent nature of the fields within the dimples and can be understood through Eq. (2.6), where the tanh function leads to a low sensitivity of the modes to variations in h.

In Figure 2.12b, the dependence of the spoof SPP bands on the dimple area is analyzed. Dimples of depth $h = 0.6d$ and sides between $a = 0.4d$ and $a = 0.8d$

are considered. As predicted by Eq. (2.6), the dispersion relation bends at lower frequencies when a is enlarged. Similar to groove arrays, this effect is due to an increase in the electromagnetic coupling of diffracted and dimple waveguide modes, which in our analytical approach is proportional to the ratio a/d. The inset depicts the electric field amplitude evaluated at the band edge for $a = 0.6d$ and $h = 0.5d$. As predicted by Eq. (2.6), the electric field is mostly located at the system interface ($z = 0$) and decays into both the indentations and the vacuum superstrate. In Williams et al. [30], the propagation and confinement of THz radiation in planar copper surfaces pierced by square arrays of square dimples have been experimentally analyzed, obtaining results in accordance with the analytical results described above.

2.4.1.3 SSPP in Cylindrical Geometries

Figure 2.13 shows the simplest structure supporting cylindrical spoof SPPs, a PEC wire milled with a periodic array of rings. The geometrical parameters of the system are: the wire radius, R; the array period, d; and the rings' width and depth, a and h, respectively.

We restrict our analysis to azimuthally θ independent modes, which present the lowest frequency and for which light polarizations are decoupled.

This allows us to consider only p-polarized modes in our expansion basis. The dispersion relation of SSPP mode in such structures is:

$$\sum \frac{k_0}{\kappa_n} \frac{K_1(\kappa_n R)}{K_0(\kappa_n R)} |\langle k_n|\mathbf{0}\rangle|^2 = -\frac{J_1(k_0 R) - \gamma_0 N_1(k_0 R)}{J_0(k_0 R) - \gamma_0 N_0(k_0 R)} \tag{2.7}$$

where we have used the fact that $q_\alpha = k_0$ for the TM0 waveguide mode. The expressions for the overlap integrals $\langle k_0|0\rangle$ can be found in [46]. In the left panel of Figure 2.14, the azimuthally independent spoof SPP bands obtained from Eq. (2.7) for three different ring arrays are plotted. We take d as the unit of length.

Thus, the wire radius is $R = 2d$, and the ring width $a = 0.2d$. The three ring depths are: $h = 1.6d$, $h = 0.8d$, and $h = 0.4d$. The dispersion relations deviate farther from

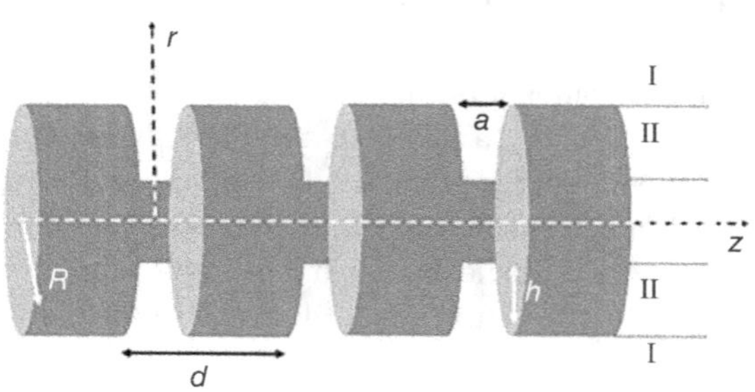

Figure 2.13 Schematic of the simplest structure supporting cylindrical spoof SPPs: a PEC wire of radius R perforated with an array of rings of period d, width a, and depth h.

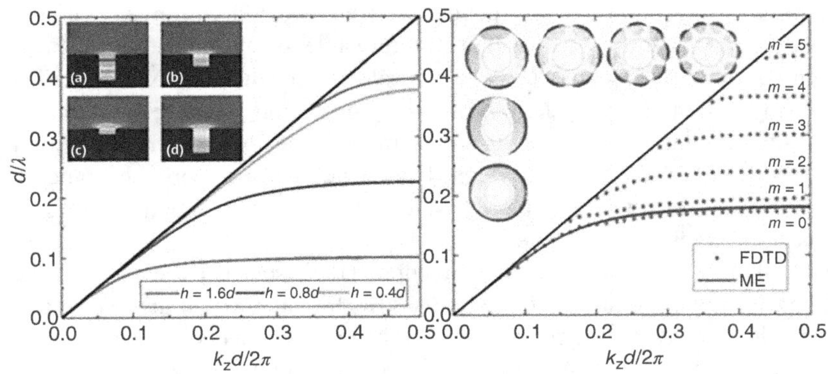

Figure 2.14 Spoof SPP dispersion relation for wires of radius $R = 2d$ perforated with rings of width $a = 0.2d$. Left panel: θ-independent bands for different ring depths. The inset shows the electric field amplitude ($r > R - h$) at the edge of the bands. Right panel: FDTD bands for higher azimuthal orders (m) for $h = 0.5d$. The insets plot the electric field patterns at the band edge ordered with increasing m from the left bottom corner to the right top corner of the figure.

the light line when h is increased, in a similar way to what was observed in 1D groove arrays.

At low frequencies $\lambda \gg d, a$, and for wires much thicker and rings much shallower than the array period ($R, R - h \gg d$), we can neglect diffraction orders in the G term and obtain an analytical expression for the dispersion relation of the guided modes. By introducing the asymptotic expansions of the different Bessel functions involved in Eq. (2.7), we have

$$k_z = k_0\sqrt{1 + \left(\frac{a}{d}\right)^2 \tan^2(k_0 h)} \tag{2.8}$$

Note that Eq. (2.8) coincides with Eq. (2.5), which gives the spoof SPP bands for 1D arrays of grooves of width a and depth h. This agrees with the fact that, as in the planar case, the key parameter governing the binding of the mode is the depth of the rings, h.

2.4.1.4 SSPP in Milled Wires

The lack of lateral confinement of spoof SPPs in 1D and 2D indentations prevents their use as waveguides, aiming to transport electromagnetic energy within small transverse cross-sections. Taking this into account, probably the most straightforward spoof SPP waveguide consists of a metallic wire periodically corrugated with an array of rings.

Taking advantage of the strong dependence of the spoof SPP confinement on the wire geometry, it is feasible to design a structure able to concentrate electromagnetic energy at one of its ends, such as a conical wire in which the external radius is gradually decreased along the direction of propagation, keeping the depth of the rings fixed. Figure 2.15a plots the guided mode bands for ring arrays of period 100 μm with $h = 30$ μm and $a = 50$ μm. Four structures with wire radii ranging between 140 and 40 μm are considered. Note that as R is decreased, the spoof surface plasmon bands deviate further from the light line. The inset renders the radial component of the electric field, E_r, as a function of the distance to the wire surface, $r - R$, for the four structures evaluated at 0.6 THz. We can see that the lowering of the bands leads to a stronger confinement of the modes with decreasing R.

Figure 2.15b shows the electric field pattern corresponding to a 2 mm-long wire whose external radius is gradually reduced from 140 to 40 μm. The structures are milled with ring arrays with the same dimensions. At 0.6 THz, the guided modes are tightly bound to cylindrical wires, even for the smallest radius considered. However, at larger frequencies, 1.2 THz, modes are not supported by wires with $R \leq 140$ μm. This has the consequence that electromagnetic radiation is guided along the cone and focused at its tip at 0.6 THz. On the other hand, the absence of guided modes at 1.2 THz leads to electromagnetic waves being scattered out of the structure without reaching the end of the wire. Remarkably, the high confinement of electromagnetic fields featured by cylindrical spoof SPPs allows the concentration of THz waves into deep subwavelength volumes in conical geometries.

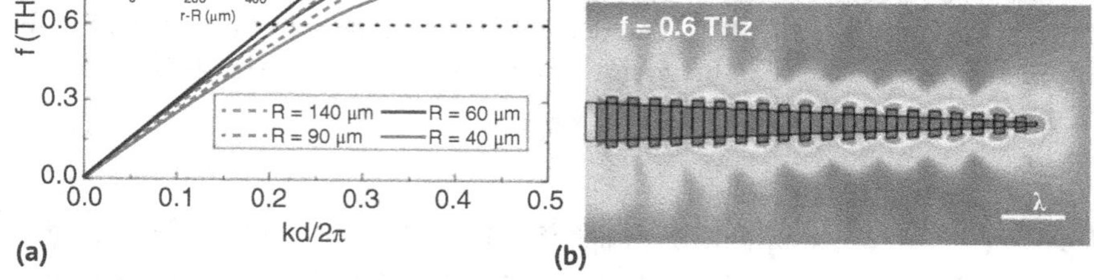

Figure 2.15 (a) Frequency versus propagation wave vector, kz, for the guided modes supported by four corrugated wires of different radii. The inset plots the radial component of the electric field versus r − R (f = 0.6 THz) for the four structures. (b) Electric field amplitude at 0.6 and 1.2 THz for a 2 mm corrugated cone whose radius is reduced from 140 to 40 μm.

2.4.1.5 SSPP in Domino Structure

THz waveguiding based on the concept of spoof SPPs in the domino structure (see inset of Figure 2.16) was introduced in Nesterov et al. [31]. This structure consists of a periodic array of metallic parallelepipeds standing on top of a metallic surface, resembling a chain of domino pieces. The properties of its guided modes,

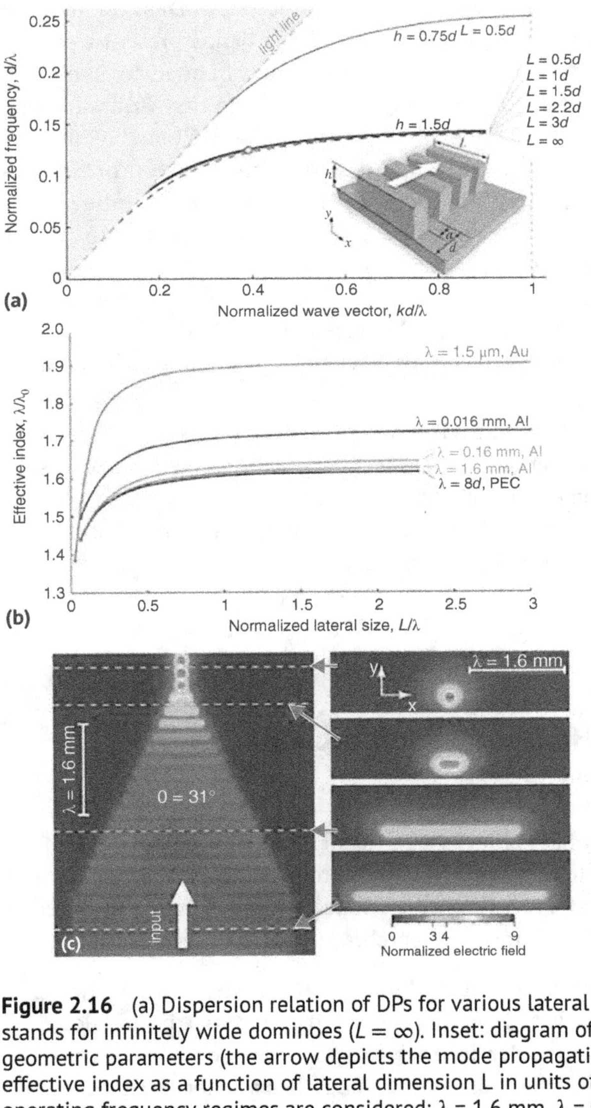

Figure 2.16 (a) Dispersion relation of DPs for various lateral widths L. The dashed line stands for infinitely wide dominoes ($L = \infty$). Inset: diagram of the domino structure and geometric parameters (the arrow depicts the mode propagation direction). (b) DP modal effective index as a function of lateral dimension L in units of wavelength. Various operating frequency regimes are considered: $\lambda = 1.6$ mm, $\lambda = 0.16$ mm, $\lambda = 0.016$ mm, and $\lambda = 1.5$ μm. A realistic description of the metals is used. (c) Electric field distribution at various planes of the structure.

the so-called *domino plasmons* (DPs), are governed by the geometric parameters defining the dominoes: periodicity (d), height of the boxes (h), lateral width (L), and interdomino spacing (a).

Here we analyze how the dispersion relation of DPs changes with the lateral width, L. To gain physical insight, as in previous cases, it is better to model the metal first as a PEC. Within the PEC approach we choose the periodicity d as the unit of length. The value of a is not critical for the properties of DPs and, in these simulations, is set equal to $a = 0.5d$. Domino plasmon bands present a typical plasmonic character, i.e. they approach the light line for low frequencies and reach a horizontal frequency limit at the edge ($k_{\text{edge}} = \pi/d$) of the first Brillouin zone (Figure 2.16a; note that only fundamental modes are plotted). While the limit frequency of SPPs for large k is related to the plasma frequency, the corresponding value for DPs is controlled by the geometry. In particular, the influence of the height h is clear: the band frequency rises for short dominoes ($h = 0.75d$) as compared with that of taller ones ($h = 1.5d$). The most striking characteristic of DPs is their behavior when the lateral width L is changed. All bands in the range $L = 0.5d, \ldots, 3d$ lie almost on top of each other (see Figure 2.16a). In other words, the modal effective index, $n_{\text{eff}} = \frac{k}{k_0}$, is rather insensitive to lateral width. Remarkably, the bands remain almost unchanged even for $L = 0.5d$, whose modal size is well inside the subwavelength regime. The described behavior is to be contrasted with that of conventional plasmonic modes in the optical regime for which subwavelength lateral confinement is not a trivial issue.

The important message of Figure 2.16b is that, although a variation of n_{eff} begins to be noticeable when the lateral dimension L goes below $\lambda/2$, DP bands are fairly insensitive to L in the range $L = 0.5d, \ldots, 24d$ when operating at low frequencies. Such a key property does not appear in spoof SPP modes in corrugated wedges, channels, or wires.

2.5 SSPP-based THz Circuits: Research in Mazumder Laboratory

The Mazumder Laboratory has achieved significant research accomplishments in the field of THz spoof plasmonics, leading to numerous publications, patents, and doctoral theses, refer to Figure 2.17. Over the last decade, they have contributed to four broad aspects of spoof plasmonics research:

1. **Fundamental Physics and Electrodynamic Analysis:** The lab developed analytical frameworks based on transversal transfer matrix and longitudinal transfer matrix in order to analyze spoof plasmonic modes in patterned metastructures. They also formulated theories for electrodynamic transport through homo- and heterojunction spoof plasmon waveguide

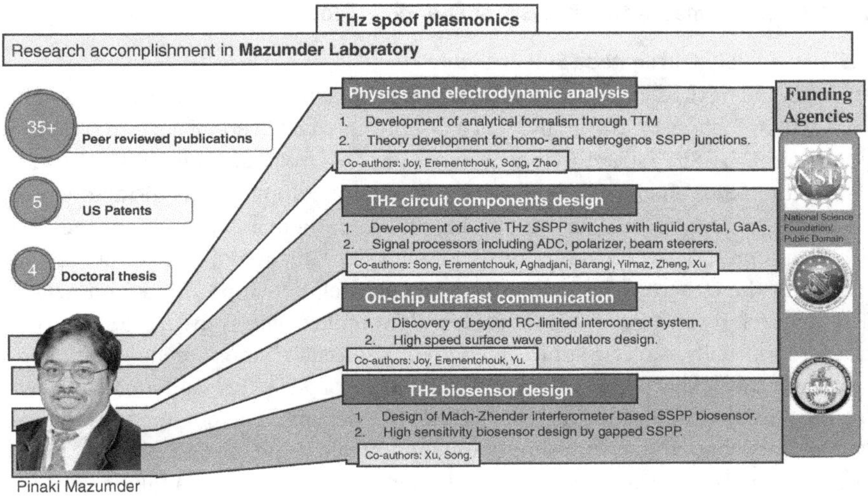

Figure 2.17 Highlights on research on THz SSPP in the Mazumder laboratory.

junctions. These theories provide a generalized understanding of electromagnetic coupling between optical resonance and free-space light on metamaterial surfaces.

2. **THz Circuit Components Design:** The lab designed active THz SPP switches using materials like liquid crystal, GaAs, and optical Kerr materials. These components enabled the integration between electronics and THz spoof plasmonics. Various signal processors like analog-to-digital converters (ADCs), polarizers, and beam steerers were also developed.
3. **On-chip Ultrafast Communication Systems**: Through their research, the lab demonstrated how THz spoof plasmonics can surpass the speed limits imposed by traditional RC-delay, improving the performance of interconnect systems. They designed ultrafast surface wave interconnects that enhanced signal integrity within densely packed multi-chip modules while minimizing cross-talk.
4. **THz Spoof Plasmonic Biosensor Design:** The lab created an SSPP biosensor by implementing Mach–Zehnder interferometry on a metamaterial platform. They also introduced novel metamaterial structures to enhance the device's sensitivity, enabling label-free sensing.

Overall, the Mazumder Laboratory has led to advancements in spoof plasmonics research by combining theoretical development, engineering design, prototype creation, fabrication, and proof-of-concept demonstrations. This multidisciplinary approach has expanded the understanding and application of THz spoof plasmonics over the past decade.

2.5.1 Electromagnetic Analysis of the SSPP Mode

2.5.1.1 Doubly Corrugated SSPP Structure

In 2012, Zhao Xu, Kyungjun Song, and Pinaki Mazumder published their work [32], which focuses on the analysis of a doubly corrugated spoof surface plasmon polariton (DC-SSPP) structure operating in the THz frequency range. The utilization of the SSPP effect in THz devices has potential applications in rapid signal processing and storage. The authors present a mathematical model of the DC-SSPP structure, derived from a full-field analysis. This model enables characterization of the structure, including its dispersion relationships and S-parameters. The research reveals the presence of slow light modes within the DC-SSPP structure. Additionally, discrete subwavelength transmission bands are identified, and, in certain cases, sharp transmission peaks with strong electromagnetic field localization can be achieved through specific device designs. The obtained results are validated by comparison with simulations conducted using the CST software package, demonstrating a generally accurate agreement between the analytical model and the simulation software. The chapter concludes by highlighting the utility of the proposed simulation in calculating dispersion relationships for various SSPP modes in the DC-SSPP structure. The observed subwavelength transmission bands and sharp transmission peaks have potential applications in passive frequency filters and active THz switches, making the DC-SSPP structure a promising candidate for such devices. The simplicity and efficiency of the presented model make it an effective tool for the initial design stages of any device based on the DC-SSPP structure. The authors showed that the electromagnetic modes inside the double-corrugated SSPP structure can be written as follows:

$$\begin{bmatrix} H_x^I \\ H_y^I \\ H_z^I \end{bmatrix} = \sum_{n=-\infty}^{\infty} \rho n^{e^{jk_{x,I}^{(n)}x}} \begin{bmatrix} j\dfrac{k_y k_{x,I}^{(n)}}{v_y^2} \cos(k_y y) \sin\left(k_{z,I}^{(n)} z\right) \\ \sin(k_y y) \sin\left(k_{z,I}^{(n)} z\right) \\ \dfrac{k_y k_{z,I}^{(n)}}{v_y^2} \cos(k_y y) \cos\left(k_{z,I}^{(n)} z\right) \end{bmatrix}$$

$$\begin{bmatrix} E_x^I \\ E_y^I \\ E_z^I \end{bmatrix} = \frac{1}{\omega\epsilon} \frac{k_0^2}{v_y^2} \sum_{n=-\infty}^{\infty} \rho n^{e^{jk_{x,I}^{(n)}x}} \times \sin(k_y y) \begin{bmatrix} -jk_{z,I}^{(n)} \cos\left(k_{z,I}^{(n)} z\right) \\ 0 \\ -k_{x,I}^{(n)} \sin\left(k_{z,I}^{(n)} z\right) \end{bmatrix}$$

$$\begin{bmatrix} H_x^{II} \\ H_y^{II} \\ H_z^{II} \end{bmatrix} = \begin{bmatrix} -\frac{k_y k_{x,II}}{v_y^2}(A^+ e^{jk_{z,II}z} + A^- e^{-jk_{z,II}z})\cos(k_y y)\sin(k_{x,II}x) \\ (A^+ e^{jk_{z,II}z} + A^- e^{-jk_{z,II}z})\sin(k_y y)\cos(k_{x,II}x) \\ j\frac{k_y k_{z,II}}{v_y^2}(A^+ e^{jk_{z,II}z} - A^- e^{-jk_{z,II}z})\cos(k_y y)\cos(k_{x,II}x) \end{bmatrix}$$

$$\begin{bmatrix} E_x^{II} \\ E_y^{II} \\ E_z^{II} \end{bmatrix} = \frac{1}{\omega\epsilon}\frac{k_0^2}{v_y^2} \times \sin(k_y y) \begin{bmatrix} k_{z,II}(A^+ e^{jk_{z,II}z} - A^- e^{-jk_{z,II}z})\cos(k_{x,II}x) \\ 0 \\ -jk_{z,II}(A^+ e^{jk_{z,II}z} + A^- e^{-jk_{z,II}z})\sin(k_{x,II}x) \end{bmatrix}$$

By setting appropriate boundary conditions, the above set of equations results in the following dispersion relation:

$$1 - \sum_n \frac{k_{z,II}}{k_{z,I}} |T_n|^2 \tan(k_{z,I}t)\tan(k_{z,II}h) = 0 \quad (2.9)$$

Figure 2.18 shows the structure of the proposed double corrugated SSPP waveguide and the analyzed electromagnetic modes and their corresponding transmission properties. The results from the theoretical analysis are well consistent with full-wave numerical simulation from commercial software.

2.5.2 THz SSPP Switch Design

2.5.2.1 Switching with Liquid Crystal

In 2009, Song and Mazumder [29] presented an active THz switch using artificially corrugated perfect conductor metamaterials. The study focuses on manipulating THz SSPP modes within a grooved metallic structure by changing the refractive index of dielectrics. Incorporating nematic liquid crystal (N-LC)

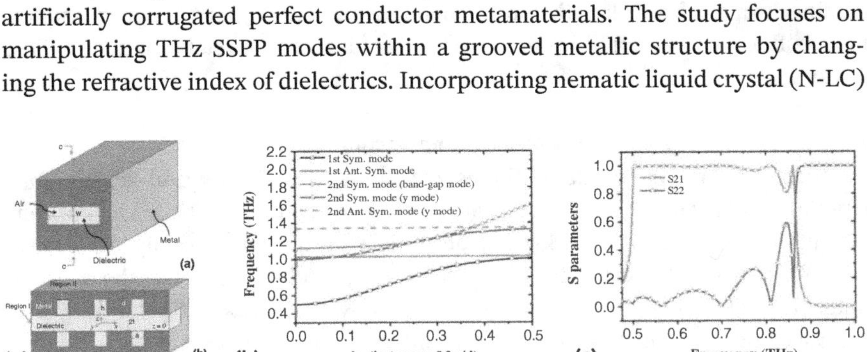

Figure 2.18 (a) The structure of double corrugated SSPP waveguide; (b) the dispersion diagram from the theoretical analysis; and (c) the scattering matrix analyzed from theoretical formalism.

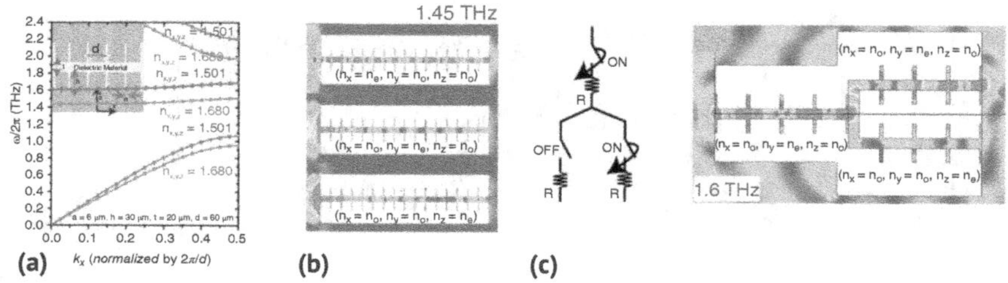

Figure 2.19 The concept of active THz SSPP switch. (a) Dispersion graphs of the spoof surface plasmon supported by the sandwiched conductor metamaterial; (b) spatial distributions of the *E*-field along the active THz switch devices for three different fundamental LC orientations at 1.45 THz; and (c) Equivalent circuit of the THz Y-junction switch with three control signals, and spatial distributions of the *E*-field along the active Y-junction THz switch.

enhances the compact and efficient THz switch, which operates with a low-voltage control signal. The optimized SSPP switch design achieves strong subwavelength localization, high extinction ratio, and minimal damping attenuation.

The proposed concept of the SSPP-based active switch is shown in Figure 2.19. The work controls THz SSPP signals through refractive index modulation using N-LC material. It highlights challenges such as the slower switching speed of N-LC compared to conventional electro-optical materials and complexities in device implementation. The study suggests potential solutions like nonlinear electro-optical methods or optically controlled materials. Despite the challenges, the proposed approach holds promise for compact active THz devices. Notably, the linear Y-junction switching devices serve as fundamental building blocks for digital systems. The chapter concludes that SSPP modes' unique attributes, encompassing strong subwavelength confinement, low attenuation, and high extinction ratio, offer new possibilities for next-generation subwavelength active THz devices.

2.5.2.2 Switching Based on Resonance and Absorption

In 2011, Song and Mazumder [33] presented a theoretical and numerical demonstration of a THz switch utilizing perfect conductor metamaterials. The core concept involves a THz logic block constructed using a waveguide–cavity–waveguide (WCW) configuration, leading to a significant accumulation of the electromagnetic field within a small cavity. This study underscores the capability of subwavelength metallic cavities to efficiently confine light within an extremely limited space, making them well-suited for the creation of a versatile THz switch. The authors also propose the potential use of dielectric materials like n-GaAs or Si-GaAs for THz switching, and they explore dynamic THz switching utilizing

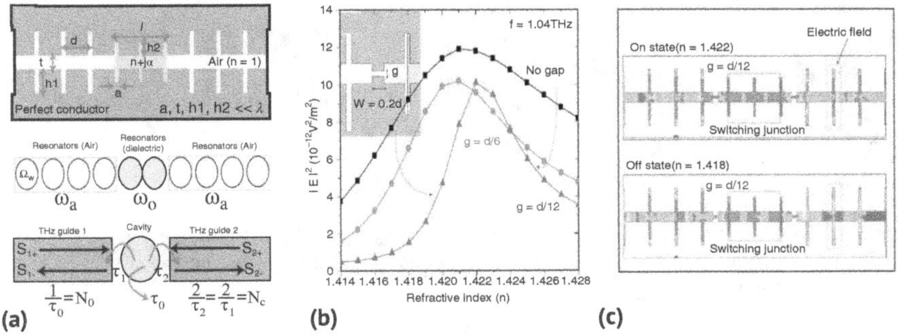

Figure 2.20 (a) THz WCW, loaded with dielectric, and the underlying physical mechanism; (b) the magnitude of electric field inside the groove corresponding to refractive index modulation inside the switching junction; and (c) spatial distribution of field during off- and on-states of the switch.

GaAs. Figure 2.20 shows the core concept and the corresponding switching mechanism of the proposed design.

The robustness of the switch design against variations in geometrical or material parameters is emphasized, attributed to a subwavelength tunneling mechanism with a t value of approximately $\lambda/9$. The importance of considering resonant matching issues ($\omega = \omega_a = \omega_0$) due to a small bandwidth is noted, especially for achieving a high Q factor, crucial for the ON–OFF switching function. The authors envision the broader application of this concept in future THz devices, such as Mach–Zehnder interferometers (MZIs), amplifiers, nonlinear switches, and biosensors. They also suggest the potential of SSPP (surface plasmon–polariton) architectures for slow-light devices operating at radio-to-microwave frequencies, opening up new possibilities in this domain.

2.5.2.3 WCW-based All-optical Switch

The work of Aghadjani and Mazumder [34] introduces a dynamic THz switch utilizing a WCW structure. The switch is composed of a GaAs dielectric slab waveguide with periodic perfect conductor patches on its upper and lower surfaces. The operational principle involves modulating the refractive index of the GaAs dielectric by changing the carrier density of the semiconductor beneath the patches through optical signal pumping. This results in the switch transitioning between on and off states.

The optical signal is strategically pumped into the region between the conductor patches, enabling the manipulation of the dielectric's carrier density. This, in turn, adjusts the effective refractive index of the dielectric, facilitating the switch's functionality. Among various candidate semiconductors for the dielectric slab,

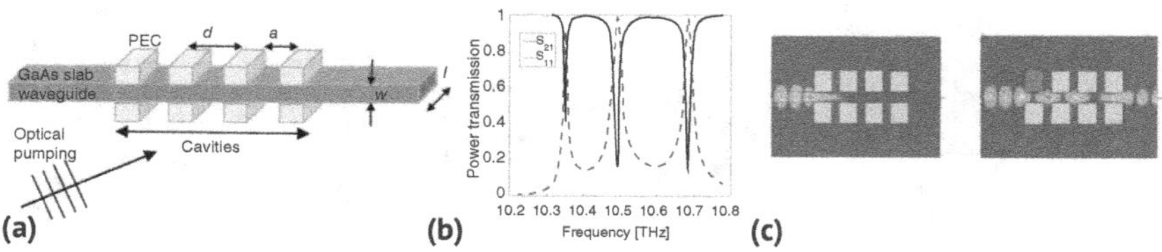

Figure 2.21 (a) Conceptual GaAs SSPP THz switch by optical pumping; (b) the transmission property of the structure; and (c) wave propagation during the off-state and on-state of optical pumping.

GaAs stands out due to its advantageous characteristics, including low intrinsic free carrier density and higher carrier mobility. The carrier concentration can be described as a function of optical pump intensity, as follows:

$$\frac{dn}{dt} = \frac{1}{h}|E|^2\chi''(\omega_0) \tag{2.10}$$

$$\alpha(N, \omega) = \frac{4\pi\omega}{n_b c}\chi''(n, \omega_0) \tag{2.11}$$

in which E is the magnitude of the electric field, n_b is refractive index of surrounding media, c is speed of light and $\chi''(\omega_0)$ is an imaginary part of permittivity.

Figure 2.21 shows the concept and transmission characteristics of the all-optical THz SSPP switch on GaAs slab. The study presents a novel THz switch designed to manipulate THz signals effectively. Notably, the switch incorporates high-Q resonators and employs optical stimuli for transitioning between on and off states. The proposed concept holds the potential for integration into THz circuitry, Boolean gates, and even MZIs.

2.5.3 THz SSPP Circuit Component Design

2.5.3.1 Nonlinear Bistable SSPP Device Design

The study by Song and Mazumder [35] explores nonlinear phenomena concerning SSPP modes within the context of conductor metamaterials. The investigation centers on the emergence of these modes below the THz spectrum due to structured conducting metastructures. This leads to the establishment of SSPP modes, analogous to visible spectrum surface plasmon polariton (spectrum SPP) modes. The research identifies the potential for inducing nonlinear behavior through strategically designed subwavelength metallic gap structures.

The viability of this approach lies in the distinctive characteristics of SSPP architectures. Subwavelength metallic gap structures are conducive to nonlinear behavior within the THz domain. As a result of elements such as slow-light propagation, electromagnetic confinement, and a high quality factor (Q), the conditions

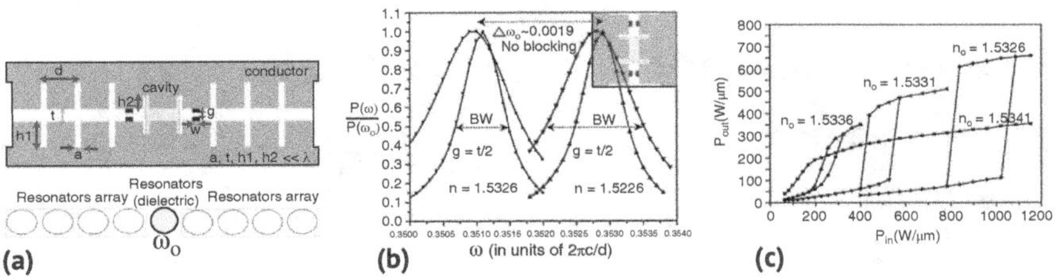

Figure 2.22 (a) Design of SSPP WCW structure with a blocking object; (b) calculated transmitted power flow for Kerr material with variable refractive index; (c) the ratio of output and input power with a Kerr material of $n_2 = 1.510(-18)\,\text{m}^2/\text{W}$. The input signal is a Gaussian wave with carrier frequency $\omega_p = 0.352\pi c/d$.

are conducive for nonlinear effects to manifest. This paves the way for the exploration of nonlinear-bistable curves as indicators of nonlinearity. Figure 2.22 illustrates the conceptual SSPP structure with cavity and blocking object. The cavity is loaded with Kerr material, and the bistable states of the device are observed.

Further investigation involves the integration of Kerr nonlinear materials within SSPP architectures. By manipulating the refractive index through the control of input signal power, the research showcases the potential of this approach. Hysteresis curves elucidate the nonlinear behavior arising from this modulation. The dependence of the refractive index on the Kerr coefficient (n_2) and the linear refractive index (n_o) establishes a framework for controlled nonlinear manipulation.

Nonetheless, it is important to acknowledge the limitations ahead. The study acknowledges the inherent weakness of nonlinear effects within conventional THz materials, leading to minimal observable outcomes. However, the research points to the significance of liquid materials like CS_2, CCl_4, and $CHCl_3$ as potential sources of nonlinear phenomena in the THz domain. These materials, despite their low Kerr coefficients, provide an avenue for exploring nonlinear behavior.

In conclusion, Song and Mazumder's research provides insights into nonlinear phenomena within the THz spectrum, specifically within SSPP architectures. The exploration of nonlinear behavior through these structures offers prospects for advancement in a challenging domain. By considering the interplay of SSPP attributes, nonlinear materials, and modulation strategies, the study contributes to the understanding of nonlinear effects within conductor metamaterials.

2.5.3.2 THz Beam Steering by SSPP Structure

The paper authored by Xu and Mazumder [36] presents two main contributions: the design and analysis of a THz beam steering structure utilizing the DC-SSPP architecture and the study of a high-speed THz ADC design.

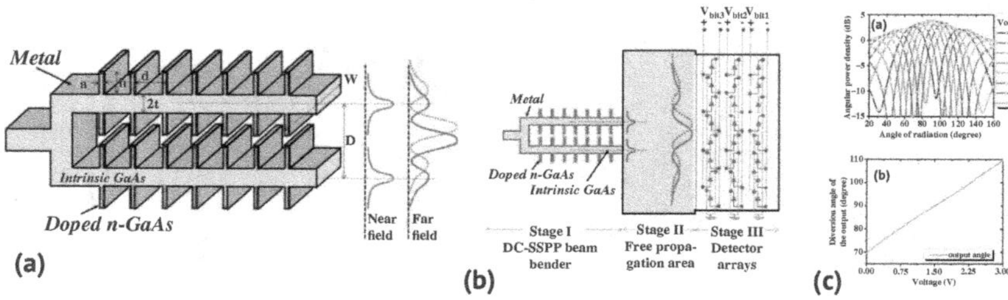

Figure 2.23 (a) Design of SSPP structure for THz beam steering; (b) the design of diode array for THz beam detection; and (c) angular profile of beam detection as a function of applied bias.

The THz beam steering structure employs DC-SSPP waveguides with metal-covered GaAs, incorporating localized doping in the groove region, refer to Figure 2.23. The modulation of free carrier density in the doped *n*-GaAs within the interferometric structure causes relative phase differences between THz beams, resulting in deflected radiation in the far field. The research employs mathematical modeling and finite-element simulations for verification and optimization. Different operating modes, enhancement and depletion, are explored in detail, each involving distinct free carrier modulation mechanisms. The study demonstrates that the modulation-induced deflection angle changes can be corrected using differential controls in the double arms of the MZI structure.

Regarding the high-speed THz ADC design, the research leverages the THz beam bender within a spatially resolved binary-encoded sensor array with THz response time, refer to Figure 2.24. The symmetric deflection angle change is achieved with high linearity by applying a differential voltage input to the MZI arms. Detectors with carefully designed gains and saturation powers enable the reading of multiple bits. By utilizing the beam deflector operating in the depletion mode, a 3-bit ADC is realized, demonstrating a remarkable operating speed exceeding 60 GS/s. The ADC's speed is constrained by carrier transit time across the space-charge region and the parasitic RC constant of the Schottky contact. Enhancing the initial doping concentration of *n*-GaAs is suggested to minimize these constraints and enhance the operating bandwidth of the proposed ADC design.

In summary, this chapter proposes a design of the THz beam steering structure through the DC-SSPP architecture for beam manipulation. Simultaneously, the high-speed THz ADC design pushes the boundaries of THz signal conversion. These contributions expand the capabilities of THz applications and pave the way for further advancements in the field.

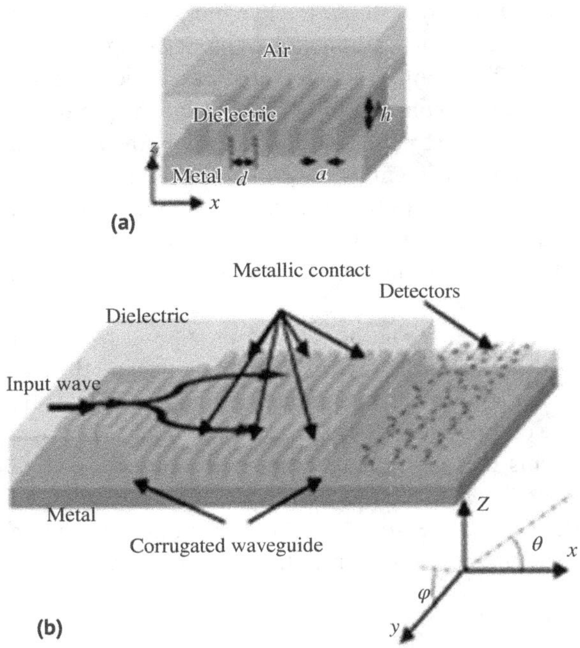

Figure 2.24 (a) The schematic view of the SS-SSPP waveguide; and (b) overall view of the ADC.

2.5.3.3 SSPP-based ADC

The ADCs based on spatial detection have been widely used by numerous optical designers throughout the literature [37]. The key mechanism behind spatial detection is the deflection of output signal at different angles with changing analog input. In order to realize this functionality, SSPP architecture is employed. Such structure features SSPP mode with a slow-wave nature and strong field confinement. Aghadjani and Mazumder [38] designed an ADC utilizing the single-sided spoof surface plasmon polariton (SS-SSPP) waveguide is presented, refer to Figure 2.24. The proposed ADC device consists of three stages: the first stage is the MZI with two arms that acts as a beam-steering device, the second is free-space propagation stage, and the third is the detector arrays to convert analog signal into digital input. The SSPP waveguide includes periodic grooves, which are filled with a thin layer of doped semiconductor. In order to alter the refractive index of the semiconductor, the depletion mode has been used. It is demonstrated that the ADC can operate with a sampling speed of around 47 GS/s. The SSPP waveguides include periodic metallic grooves, which are filled with a thin layer of doped semiconductor such as GaAs, InSb, or Si. The authors have shown that, because of its strong field localization, it is enough to tune the refractive index of a thin layer of dielectric inside the groove in order to alter the phase of propagating wave. In their proposed ADC, GaAs is used as a semiconductor, as it has higher electron mobility and lower intrinsic free carrier concentration. To describe the permittivity of GaAs versus the frequency and the free carrier concentration, the modified Drude model has been used. The enhancement mode and the depletion mode are two working modes for doped semiconductors that work based on injection and depletion of free carriers. By inserting electric voltage into the electrodes on top of the waveguide, the depletion layer shapes and the effective refractive index of the dielectric inside the grooves changes.

By implementing the Schottky effect, the authors tuned the output pattern at the end stage of the two-arm MZI. Each MZI arm acts as an antenna, and the two arms can be assumed as an array antenna. At the steady state, when there is no voltage applied to the arms, the output pattern lies in parallel with the MZI arms. By applying the voltage to the arms, the phase change is inserted to the propagating wave, and this phased difference between the two arms results in the shift of the maximum of the output pattern, refer to Figure 2.25. The output pattern can be calculated using a well-known equation for the array antenna pattern [8]:

$$F \approx F_0 N_a \text{sinc}\left(\frac{N_a(kD\cos\theta + \Delta\Phi_{1,2})}{2}\right) \tag{2.12}$$

where F_0 is the output pattern of each arm, N_a is the number of arms, D and $\Delta\Phi_{1,2}$ are distance and the phase difference between the two arms, respectively. The relation between the phase difference of the two arms and the differential applied voltage can be found as:

$$\Delta\Phi_{V1,V2} = \frac{2\pi}{\lambda_0}\frac{n_i - n_d}{D_{\text{tot}}}\sqrt{\frac{2\epsilon_s\epsilon_0}{eN}}l\sqrt{V_1} \tag{2.13}$$

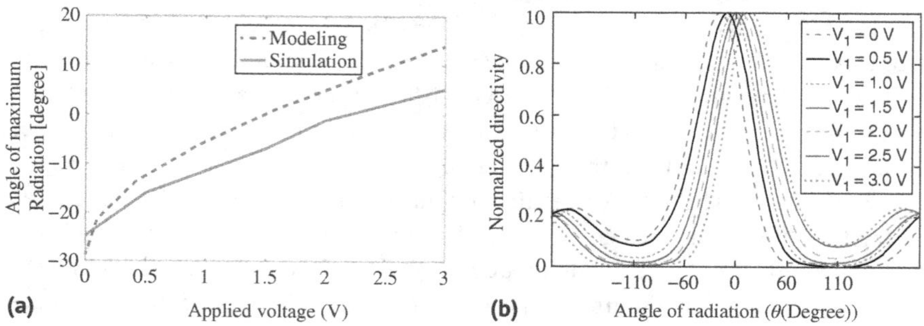

Figure 2.25 (a) Angle of maximum radiation versus input voltage in case of one-arm excitation; and (b) normalized directivity versus θ for different applied voltages.

where V_1, V_2 are applied voltage to each arm, N is carrier concentration of the GaAs semiconductor, and e is the charge of electron. D_{dep} and D_{tot} represent depletion depth and total depth of GaAs inside the groove. Under an applied bias of V, the depletion thickness is calculated by the following relation,

$$D_{\text{dep}} = \sqrt{\frac{2\epsilon_s(V_{\text{bi}} + V)}{eN}} \tag{2.14}$$

in which, N is carrier concentration, $V_{\text{bi}} = 0.75V$ is the built-in voltage of semiconductor-metal junction, ϵ_s is permittivity of GaAs, and e is electron charge. The authors showed that the designed MZI using single-sided corrugated waveguide can be used as a beam-steering device. The linearity of the ADC is improved using differential excitation.

2.5.3.4 Phase-Accumulation in SSPP for DAC

The study authored by Aghadjani et al. [39] introduces an approach to active beam steering via an SSPP waveguide. The study involves a structured design comprising a MZI and a free space propagation stage.

The SSPP waveguide employs periodic grooves embedded with a doped semiconductor layer, such as GaAs. By altering electric voltage applied to the waveguide's electrodes, the thickness of the depletion layer and the effective refractive index within the grooves can be modified.

This chapter presents the analytical derivation of the effective refractive index within the grooves' dielectric and computes the resulting phase change in the propagation wave based on input voltage. It explores the relationship between input signal and output beam radiation angle, focusing on linearity improvements. It showed that at far field, the electric field at distance r from the aperture,

$$E = \frac{jk\frac{e^{-ikr}}{2\pi r}A(1+\cos\theta)}{2}(f_y(\theta,\phi)\,\hat{\theta} - f_z(\theta,\phi)\,\hat{\phi}) \tag{2.15}$$

The dependence of the phase difference on two different arms of the SSPP structure on external voltage can be expressed as

$$\Delta\Phi_{V1,V2} \approx \frac{2\pi}{\lambda_0}\frac{\left(1-\frac{n_i}{n_d}\right)^2}{D_{\text{tot}}}\sqrt{\frac{2\epsilon_s\epsilon_0}{eN}}(a_1V_1 + b_1) \tag{2.16}$$

where

$$a_1 = \left(-1 + \frac{V_m}{4} + \frac{V_{bi}}{4}\right),\ b_1 = \left(\frac{V_m}{2} - \frac{V_m^2}{8} - \frac{V_{bi}V_m}{4}\right),\ \text{and } V_m = V_1 + V_2 \tag{2.17}$$

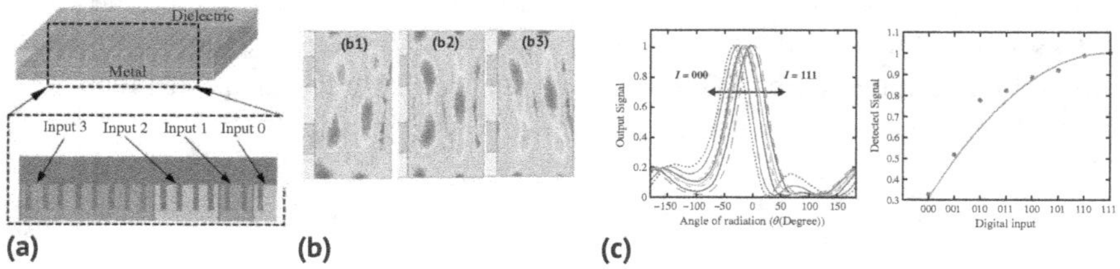

Figure 2.26 Design for SSPP phase-accumulator for DAC.

The dispersion equation of the structure comprising composite layer of intrinsic and doped GaAs of refractive index n and n' can be written as:

$$\langle -x\,|D_{\text{tot}}|\,x\rangle = 0, \tag{2.18}$$

where,

$$D_{\text{tot}}(\omega) = \sin\left(2\chi\left(n'\right)\right) + \chi(n) + 2\eta\sin\left(\chi\left(n'\right)\right)\cos\left(\chi(n) + \chi\left(n'\right)\right) \\ -\eta^2\sin 2(\chi(n'))\sin(\chi(n))) \tag{2.19}$$

The study achieves an electromagnetic beam deflection angle of up to 42° using the MZI beam steering section. The concept can be applied in spatially accumulated phase difference structures, relevant for DACs, refer to Figure 2.26.

An illustrative scenario involves applying fixed voltages to specific groove groups, with the binary representation dictating the overall structure's state and influencing the phase difference between arms. Fundamentally, the study manipulates THz signals by controlling the phase constant through the semiconductor depletion mode within the grooves. The substantial deflection angle achievement underscores the design's adaptability for diverse applications. The concept of the beam bender extends to both depletion and enhancement modes, broadening practical utility.

In essence, this chapter presents a method utilizing spoof plasmon-based MZIs for THz beam steering. The engineered SSPP waveguide and voltage modulation have the potential to impact various technological realms, particularly in enhancing digital devices, including ADCs and DACs.

2.5.3.5 THz SSPP Polarizer

Aghadjani and Mazumder [40] focus on the development of a THz polarizer controller based on a cylindrical spoof surface plasmon polariton (C-SSPP). The study capitalizes SSPP devices with their ability to transmit THz signals with minimal dispersion and low attenuation. The paper proposes a mathematical model achieved through full-field analysis of a periodically corrugated cylindrical waveguide. This structure is shown to support the propagation of localized SSPP modes for electromagnetic waves. The research investigates the impact of different groove dimensions on the band diagrams, demonstrating how the SSPP signals' propagation can be controlled by modulating the refractive index of the dielectric material within the structure.

Based on Maxwell's equations, mathematical expressions of the TM-polarized waves propagating in direction along grooves are explicitly written out for different sections of the cylindrical SSPP structure as follows:

$$\begin{bmatrix} D_\rho \\ D_\varphi \\ D_z \end{bmatrix} = \begin{bmatrix} \frac{1}{i\omega\varepsilon}\frac{n\pi}{a}k_{\rho 2}\left(A^+H_N^{2'}(k_{\rho 2}\rho)+A^-H_N^{1'}(k_{\rho 2}\rho)\right)\cos(N_\varphi)\cos\left(\frac{n\pi}{a}z\right) \\ \frac{1}{i\omega\varepsilon\rho}\frac{n\pi}{a}\left(A^+H_N^2(k_{\rho 2}\rho)+A^-H_N^1(k_{\rho 2}\rho)\right)\cos(N_\varphi)\sin\left(\frac{n\pi}{a}z\right) \\ \frac{1}{i\omega\varepsilon}\left(k^2-\left(\frac{n\pi}{a}\right)^2\right)\left(A^+H_N^2(k_{\rho 2}\rho)+A^-H_N^1(k_{\rho 2}\rho)\right)\cos(N_\varphi)\cos\left(\frac{n\pi}{a}z\right) \end{bmatrix}$$

$$\begin{bmatrix} H_\rho \\ H_\varphi \\ H_z \end{bmatrix} = \begin{bmatrix} \frac{-N}{\rho}\left(A^+H_N^2(k_{\rho 2}\rho)+A^-H_N^1(k_{\rho 2}\rho)\right)\sin(N_\varphi)\cos\left(\frac{n\pi}{a}z\right) \\ -ik_\rho\left(A^+H_N^{2'}(k_{\rho 2}\rho)+A^-H_N^{1'}(k_{\rho 2}\rho)\right)\cos(N_\varphi)\cos\left(\frac{n\pi}{a}z\right) \\ 0 \end{bmatrix}$$

By solving the boundary condition and eliminating and the dispersion relation for the SSPP modes, it is shown to be as follows:

$$\sum_{m=-M}^{m=M}\left(\frac{k_\rho}{k_{\rho 2}}\right)\frac{I_N(k_\rho R_1)}{I_{N'}(k_\rho R_1)}\frac{\left(H_N^1(k_{\rho 2}R_2)H_N^{2'}(k_{\rho 2}R_1)-H_N^2(k_{\rho 2}R_2)H_N^{1'}(k_{\rho 2}R_1)\right)}{\left(H_N^1(k_{\rho 2}R_2)H_N^2(k_{\rho 2}R_1)-H_N^2(k_{\rho 2}R_2)H_N^1(k_{\rho 2}R_1)\right)}\frac{C_m^2}{\frac{\pi}{2a}}=1$$

Figure 2.27 shows the proposed THz SSPP cylindrical structure for beam polarization. The conclusion of the paper highlights the insights gained from this investigation. The researchers derived mathematical equations for the dispersion of the SSPP signals along the periodically grooved cylindrical waveguide with anisotropic dielectric properties. They demonstrate that the proposed structure can act as a polarization rotator by engineering the refractive index of the interior dielectric material. This capability has significant implications for designing THz Boolean logic gates, such as AND or XOR gates, utilizing the polarization rotator. The paper's findings underscore the potential for applications in THz signal control and logic gate design by harnessing the capabilities of the cylindrical spoof surface plasmon polariton (C-SSPP) structure.

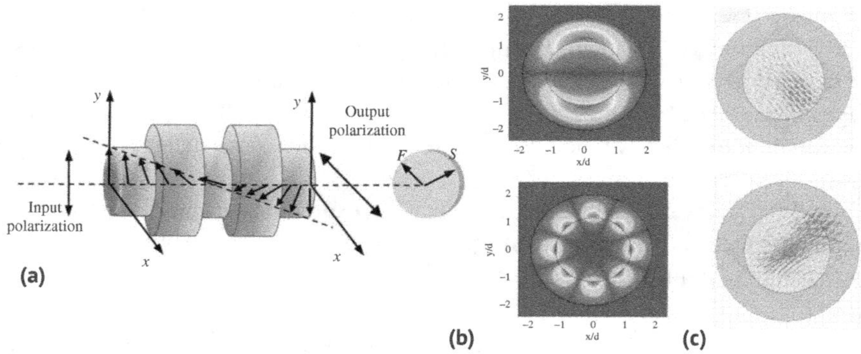

Figure 2.27 (a) THz SSPP polarizer structure with cylindrical shape; (b) the field distribution; and (c) the rotating vector of electric field inside the SSPP.

2.5.3.6 THz SSPP Beam Splitter

In 2016, Aghadjani et al. [41] investigate the SSPP beam splitter as a component in optical setups. The study focuses on the propagation of THz waves using SSPP modes, known for their ability to guide THz waves with minimal dispersion. The SSPP beam splitter presents unique challenges, unlike conventional free space implementations, as it must consider both the splitting ratio and backscattering of incoming waves into the input channels. The paper explores three different designs representing diverse approaches to addressing the relationship between these critical parameters. A mathematical formalism for studying dispersion diagrams is proposed, and the frequency-dependent behavior of the splitting ratio, scattering parameters, and backscattering for each beam splitter design is thoroughly investigated.

Figure 2.28 shows the proposed structure for SSPP-based beam splitting and the corresponding ratio of beam division for different operating frequencies. The conclusion of the paper highlights the analysis of fundamental SSPP beam splitter designs, with a focus on key characteristics: the splitting ratio and the intensity of backscattering. The research identifies that a simple design featuring two connected SSPP waveguides in a contra-propagating setup offers the best solution for balanced splitting, with near-equal splitting for most frequencies except within SSPP band gaps. However, this design exhibits high-intensity backscattering, which can be problematic for large-scale optical setups. The authors explore more complex designs to mitigate this issue, including the use of destructive interference in the design with two connectors or reducing scattering by minimizing the mismatch of the SSPP propagation constant in the design with shared grooves.

The study also delves into the applications of these beam splitters in quantum optical setups, using the homodyne detection scheme as an example. The paper

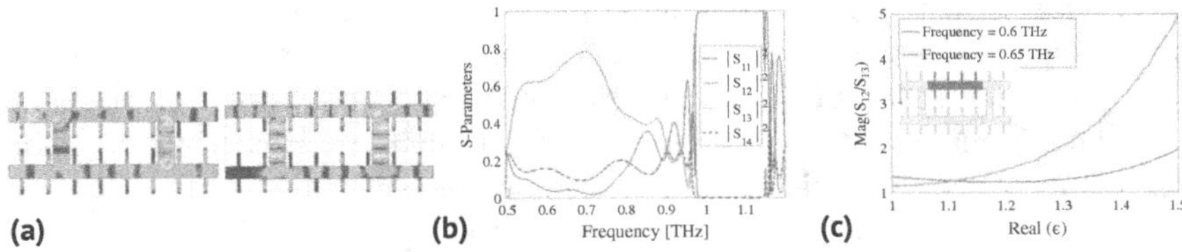

Figure 2.28 (a) The proposed structure of SSPP beam splitter and the field distribution; (b) scattering matrix analysis; and (c) the beam splitting ratio in a tunable beam splitter with the tuning of the filler dielectric.

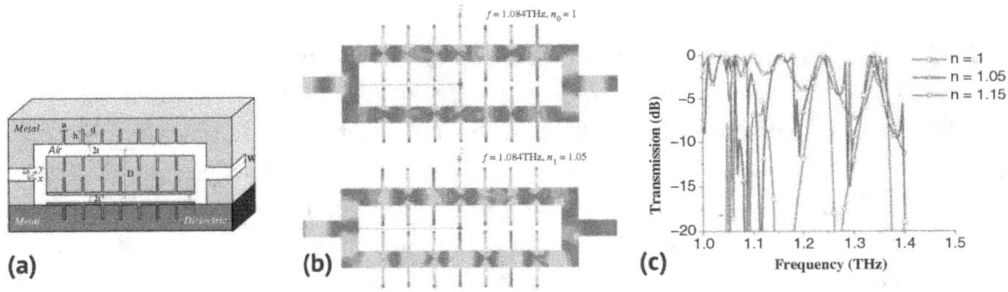

Figure 2.29 Biosensor structure design based on MZI principle in SSPP waveguide. (a) Double corrugated SSPP waveguide, loaded with DNA sample on one arm of the structure; (b) the electric field distribution at the output port due to interference between two waves undergoing different effective optical lengths of the two arms of the structure; and (c) resultant transmission spectrum for different refractive index of the DNA sample.

extends the standard description of the beam splitter to incorporate the effects of backscattering, emphasizing that the critical parameter is the splitting ratio. Furthermore, the filtering properties of the SSPP beam splitter are demonstrated by considering the variation of the splitting ratio with changes in the dielectric constant of the material filling one arm of the beam splitter.

2.5.4 THz Biosensor Design

2.5.4.1 MZI-SSPP-based Biosensor

Xu and Mazumder [42] introduced an innovative MZI with a DC-SSPP waveguide. They employ a mathematical model and HFSS simulations to predict and verify THz signal behavior in the MZI arms. The proposed SSPP-based MZI structure is illustrated in Figure 2.29. Key findings include a significant shift in transmission maxima and minima, along with heightened quality factors for transmission peaks in the MZI compared to single-armed SSPP waveguides. This suggests its potential in tag-free genetic molecular sensing applications. The study also highlights the localized electromagnetic field near SSPP resonance frequencies, which could reduce required sample amounts without compromising selectivity.

The researchers conclude that the fusion of DC-SSPP's spectral response with MZI's interference effects enhances frequency selectivity, supported by consistent mathematical and simulation results. They propose deploying the THz MZI DC-SSPP structure for biomolecular sensing, differentiating denatured and hybridized DNA through refractive index changes. Furthermore, the study suggests strategic sample loading near the groove region to optimize utilization without affecting spectral selectivity. Overall, this research has potential implications for genetic molecular sensing techniques.

2.5.4.2 SSPP Gap Structure from High Sensitivity Biosensing

Xu et al. [43] investigates the utilization of an SSPP architecture modified with gap blocks for high- sensitivity biosensing in the THz domain. The research demonstrates that the introduction of gap blocks enhances the sensitivity to refractive index (n) changes in THz sensing. This enhancement is achieved through the sharpening of the transmission peak as a function of n, facilitated by the additional cavity resonance. The performance of this gapped SSPP structure is reliant on the geometric dimensions of the block structure and the choice of probe frequencies. The study highlights the potential of noninvasive THz biosensing as a promising alternative to conventional tagging-based sensing approaches. To meet the demand for lower detection limits, the SSPP gap structure is effective in reducing sample usage by enabling localized sample deposition within the gap cavity. The authors demonstrate the differentiation of DNA molecules with distinct binding states using this structure, showing that a conformational change in a thin layer (1 micron) of immobilized DNA significantly alters the waveguide transmittance.

The conclusion of the paper emphasizes the significance of the gapped SSPP architecture in detecting refractive index changes. The additional cavity resonance significantly enhances the selectivity of the structure, leading to much sharper transmission peaks with respect to refractive index variations. The effectiveness of this enhancement is governed by the coupling rate between the cavity and the waveguide, which can be controlled by adjusting the gap structure dimensions and operating frequencies.

Figure 2.30 shows the sample deposition pattern on the proposed gapped SSPP structure and their corresponding sensitivity to the dielectric properties of the sample. In the context of biosensing applications, the heightened sensitivity of the structure allows the use of minute sample quantities, which can be precisely deposited in the proximity of areas with highly concentrated electromagnetic fields. This localized sample deposition approach results in a substantial reduction (two orders of magnitude) in sample usage while remaining compatible with

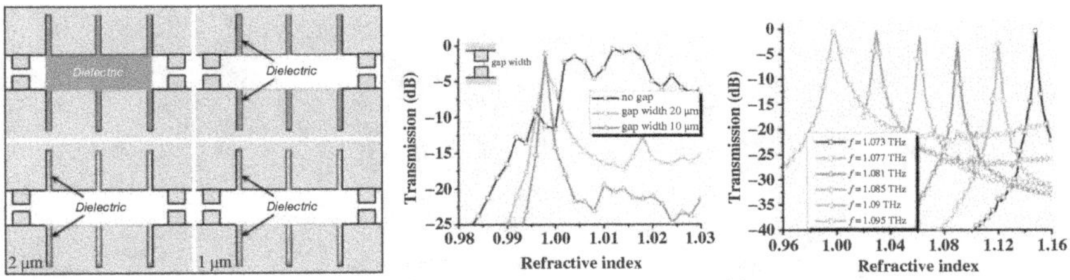

(a) (b) (c)

Figure 2.30 (a) Different patterns of sample deposition in SSPP structure; (b) transmission of gapped SSPP structure; and (c) transmission of the gapped SSPP at different probe frequencies.

conventional sample immobilization processes. The proposed sensor design, involving the deposition of a thin layer of samples on the groove wall, enables the detection of multiple samples, such as DNA molecules with different binding states.

2.6 Particle-Motion Control by SSPP Waveguide

Besides the applications of the optical and THz waves in plasmonic and SSPP waveguides, there has been significant interest in electromagnetic forces in dielectric and waveguide structures in order to utilize the electromagnetic force in optomechanical or micro-electromechanical MEMs devices. Pernice et al. [44] studied the optical forces in parallel waveguides. Optical forces between metallic nanoparticles have been investigated in Coronado et al. [45]. Povinelli et al. [46] suggested that the optical force can be increased by optical resonances. The optical forces have also been investigated in a wide variety of structures, such as photonic crystals and plasmonic waveguides [47, 48], for different applications and situations, such as force on nanoparticles, between two single-sided spoof plasmon waveguides, etc. Aghadjani et al. [49] emulated the concept of resonantly enhanced electromagnetic force to investigate the mechanical action of SSPP inside corrugated waveguides, refer to Figure 2.31.

The total force acting on a small particle can be calculated by the following formula:

$$\mathbf{F}_0 = \frac{\epsilon_r \epsilon_0}{2} Re(\alpha) \nabla(\mathbf{E}_0.\mathbf{E}_0^*) \tag{2.20}$$

When the operating frequency approaches the resonant frequency of the grooves, the field concentration inside the grooves increases, which provides the means to control the field concentration at regions near the openings of the

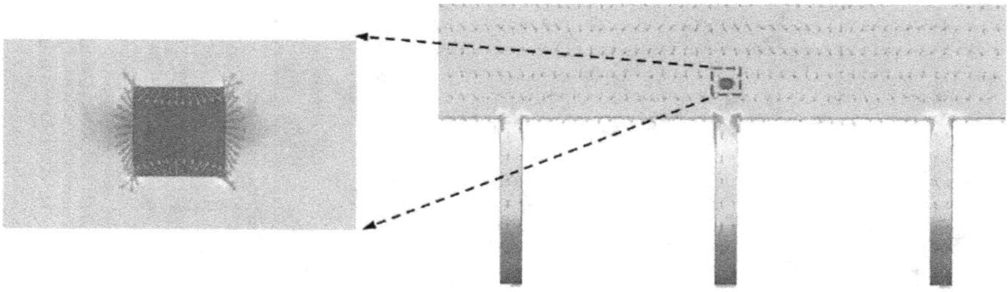

Figure 2.31 Electric field and normalized force distribution inside the SSPP waveguide in the presence of the dielectric, when the operating frequency is 0.93 THz and the field concentration is high near the groove opening.

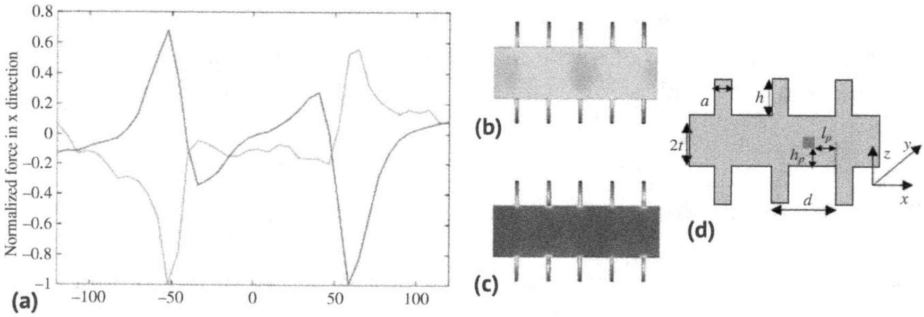

Figure 2.32 Electric field and normalized force distribution inside the SSPP waveguide in the presence of the dielectric, when the operating frequency is 0.93 THz and the field concentration is high near the groove opening.

grooves and, as a result, to alter the force acting on a particle inside the SSPP waveguide. Figure 2.32 shows the variation of the x component of the force (along the axis of the waveguide) with the particle position along the x-axis for two different frequencies. The particle position is depicted in Figure 2.32d. When frequency changes from 0.93 THz (which is close to the resonant frequency of the grooves and lies in region A) to 0.75 THz (which is far from region A, and the waveguide in it acts as a conventional rectangular waveguide), the field concentration at the opening of the grooves decreases, and, at the same time, the intensity of the field at the space between the groove increases, this change in field intensity results in changing the force direction.

2.7 Recent Advances in Spoof Plasmonics

In recent years, researchers have been exploring innovative ways to mimic plasmonic phenomena and properties at lower frequencies, particularly when metals behave like perfect electrical conductors (PECs). The key principle to achieving this is by constituting negative effective permittivity in the Drude model through structural dispersion. Cui et al. successfully realized spoof surface plasmons (SSPs) and spoof LSPs using ultrathin metallic patterns on printed circuit boards (PCBs) in 2013 and 2014, respectively, and reviewed the recent progress in the field of spoof plasmonics [50, 51]. These ultrathin corrugated metallic disks exhibited the behavior of high-quality resonators, paving the way for exciting applications in various domains.

The excitation of spoof LSPs can be achieved through different schemes, each offering unique advantages and applications. Probe excitation, although presenting a low excitation efficiency, allows for flexible selection of the electric components of the source and detector. As a result, it has been used extensively in principal investigations of basic modes, coupling mechanisms, and plasmonic topological photonics. Free-space excitation, on the other hand, is wireless and often utilizes periodic arrays to improve the excitation efficiency. Excitation of spoof LSPs via spoof SP transmission lines has also been reported, where the asymmetric electromagnetic distributions in the spoof SPs efficiently excite the spoof LSPs and may even trigger orbital angular momentum modes with proper geometries. Furthermore, microstrip line excitation has been favored due to its efficiency, stability, convenience, and compatibility with PCBs, leading to compacted package sizes. This makes it valuable for engineering applications, especially the development of sensors, filters, etc.

In 2014, Cui et al. reported observations of multipole resonances in an ultrathin spoof LSP resonator [53]. The resonances were intrinsically interpreted as standing waves of propagating spoof SPs around the perimeter of the resonator. Interestingly, when the resonator's perimeter equaled integer multiples of the effective wavelength of the propagating spoof SP modes, the spoof LSP resonances appeared as the dipole, quadrupole, hexapole, and higher-order Mie modes, refer to Figure 2.33. This standing wave interpretation bore a resemblance to that observed in optical whispering gallery modes. Furthermore, owing to the exponential asymptotic dispersion curve of spoof SPs, the resonant modes were found to be distributed closer to the microstrip line ring of the same diameter.

Spoof LSPs have also been successfully demonstrated in the THz frequencies, as shown in Figure 2.34, serving as a bridge between optics and microwaves. The THz sensing is particularly important for biomolecules, as their vibrational and rotational energy levels are present in THz bands. Fabrication of THz samples requires higher precision, and lithography is commonly employed along with PCBs or quartz substrates for CMOS ICs. While THz sources and detectors are still under development, THz time-domain spectroscopy is a widely used technique to measure the transmittance and reflectance of THz spoof LSP arrays in free space. For THz excitation, array structures are used to enhance efficiency. Coupling wires introduced alongside spoof LSP resonators trigger electromagnetically

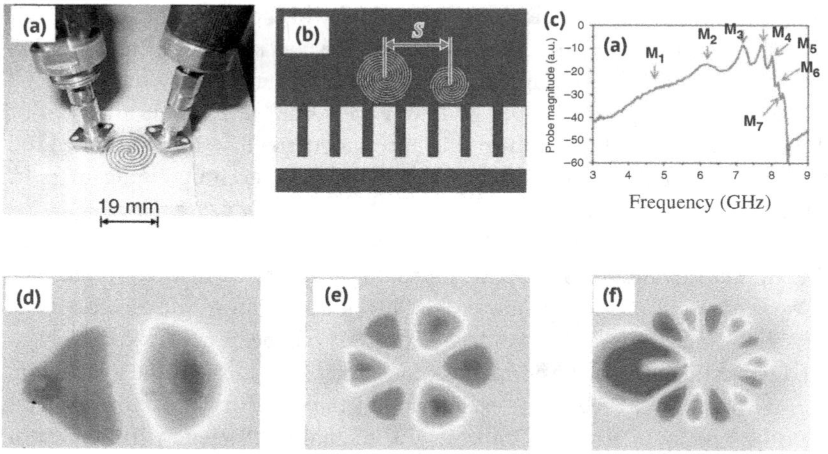

Figure 2.33 Design and measurement of localized SSPP. (a) Measurement setup (Zhang et al. [52]/John Wiley & Sons); (b) design of coupled localized SSPP; (c) measured scattering magnitude; and (d–f) simulated eigenmodes at different orders in localized SSPP.

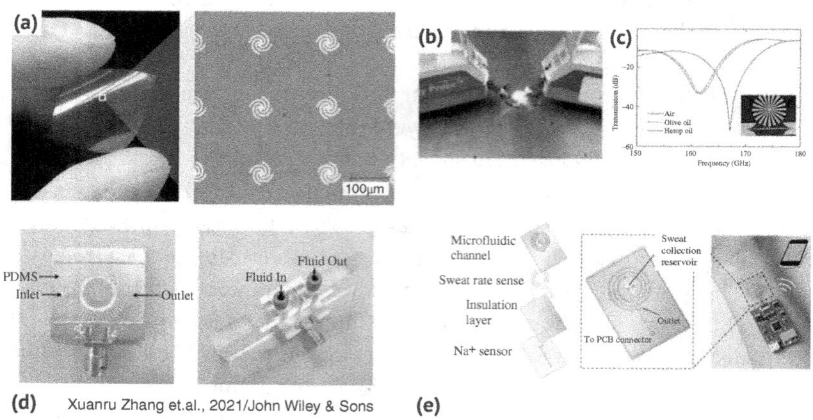

Figure 2.34 Spoof plasmonic structure designed at sub THz frequencies: (a) demonstration of microstructures at micron scale; (b) THz probing and (c) measurement of SSPP signal; (d) SSPP-based microfluidic sensor design; and (e) SSPP-based human health monitoring system (Xuanru Zhang et.al., 2021/John Wiley & Sons).

induced transparency, and spatial symmetry-breaking slits construct Fano resonances, improving the performance for sensing applications. To enable on-chip sensing in ICs, spoof LSPs have been excited via transmission lines, utilizing vector network analyzers with frequency extenders to generate THz signals and probe pins for signal measurement. On-chip permittivity sensing has been successfully demonstrated, distinguishing between olive oil and hemp oil.

Spoof LSP sensing has found applications in liquid sensing within microfluidic channels, offering contactless advantages similar to microwave sensing [52]. The microfluidic channel, when placed on the resonator, separates the target liquid from the sensor surface, making both the sensor and microfluidic channel reusable and conveniently processed via photolithography. High-resolution glucose sensing has been achieved using spoof LSP resonators, making it promising for wearable health monitoring devices, which have gained attention due to their integration with computation and communication systems, especially in the context of IoT development. A microwave resonator on a flexible substrate integrated with a flexible microfluidic channel is an example of wearable health monitoring. Another wireless and miniaturized sensing system using RFID tags has been reported to monitor human pulse signals. RFID techniques offer the advantage of wireless detection even when the RFID tag is inserted into the human body.

2.8 Conclusion

The field of spoof plasmonics has a rich and multifaceted history, drawing inspiration from two distinct but interconnected areas of engineering: slow wave structures in microwave amplifiers and SPR in noble metals. Researchers around the world, including the Mazumder Laboratory at the University of Michigan, led to the fusion of these two domains, aided by theoretical advancements and nanofabrication techniques.

SSPs and spoof LSPs have emerged as powerful tools for controlling electromagnetic waves on patterned metal surfaces. These versatile structures can be excited through various methods, including probes, free-space radiation, spoof surface plasmon transmission lines, and microstrip lines. Their applications span a wide spectrum, from revolutionizing wireless communication to enabling high-resolution sensors and wearable health monitoring devices. As spoof plasmonics extended its reach to THz frequencies, it opened new avenues in sensing, spectroscopy, and imaging. Integrated circuits and RFID technology have further contributed to the development of wireless and miniaturized sensing systems, facilitating health monitoring and remote sensing.

Looking ahead, the field of spoof plasmonics can have a versatile impact, which will leverage enhanced wave control capabilities in telecommunications, healthcare, and beyond. As researchers continue to push the boundaries in this field, we can anticipate new developments that will shape the way we interact with and harness electromagnetic waves in the near future.

References

1 Mohamad Farhat, Bal Singh Virdee, and Nader H Ghareeb. Microstrip bandpass filter with wide stopband and high out-of-band rejection. *American Journal of Engineering and Applied Sciences*, 12(4):543–548, 2019.

2 Jin Tao, Xu Guang Huang, Xianshi Lin, Jihuan Chen, Qin Zhang, and Xiaopin Jin. Systematical research on characteristics of double-sided teeth-shaped nanoplasmonic waveguide filters. *JOSA B*, 27(2):323–327, 2010.

3 Zhong-Tao Tian, Ze-Peng Zhuang, Zhi-Bin Fan, Xiao-Dong Chen, and Jian-Wen Dong. High-efficiency grating couplers for pixel-level flat-top beam generation. In *Photonics*, volume 9, page 207. MDPI, 2022.

4 JB Pendry, L Martin-Moreno, and FJ Garcia-Vidal. Mimicking surface plasmons with structured surfaces. *Science*, 305(5685):847–848, 2004.

5 LC Steinhauer and WD Kimura. Slow waves in microchannel metal waveguides and application to particle acceleration. *Physical Review Special Topics—Accelerators and Beams*, 6(6):061302, 2003.

6 Minghao Zhang, Yanyu Wei, Xianbao Shi, Lingna Yue, Wanghe Wei, Jin Xu, Guoqing Zhao, Minzhi Huang, Zhanliang Wang, Yubin Gong, et al. A modified slow-wave structure for backward-wave oscillator design in THz band. *IEEE Transactions on Terahertz Science and Technology*, 4(6):741–748, 2014.

7 Robert Williams Wood. Xlii. on a remarkable case of uneven distribution of light in a diffraction grating spectrum. *The London, Edinburgh, and Dublin Philosophical Magazine and Journal of Science*, 4(21):396–402, 1902.

8 Robert Williams Wood. Xxvii. diffraction gratings with controlled groove form and abnormal distribution of intensity. *The London, Edinburgh, and Dublin Philosophical Magazine and Journal of Science*, 23(134):310–317, 1912.

9 Lord Rayleigh. On the dynamical theory of gratings. *Proceedings of the Royal Society of London. Series A, Containing Papers of a Mathematical and Physical Character*, 79(532):399–416, 1907.

10 Daniel Maystre. Theory of wood's anomalies. In *Plasmonics: From Basics to Advanced Topics*, pages 39–83. Springer, 2012.

11 Ugo Fano. The theory of anomalous diffraction gratings and of quasi-stationary waves on metallic surfaces (Sommerfeld's waves). *JOSA*, 31(3):213–222, 1941.

12 Heinz Raether. *Excitation of Plasmons and Interband Transitions by Electrons*, volume 88. Springer, 2006.

13 CJ Powell and JB Swan. Effect of oxidation on the characteristic loss spectra of aluminum and magnesium. *Physical Review*, 118(3):640, 1960.

14 W Lukosz. Principles and sensitivities of integrated optical and surface plasmon sensors for direct affinity sensing and immunosensing. *Biosensors and Bioelectronics*, 6(3):215–225, 1991.

15 Bo Liedberg, Claes Nylander, and Ingemar Lundström. Biosensing with surface plasmon resonance—how it all started. *Biosensors and Bioelectronics*, 10(8):i–ix, 1995.

16 Daniel J O'Shannessy, Michael Brigham-Burke, and Kim Peck. Immobilization chemistries suitable for use in the biacore surface plasmon resonance detector. *Analytical Biochemistry*, 205(1):132–136, 1992.

17 James S Richey. Biacore's little brother: Bialite. *Biosensors and Bioelectronics*, 8(7–8):xxxviii–xl, 1993.

18 Walter Huber and Francis Mueller. Biomolecular interaction analysis in drug discovery using surface plasmon resonance technology. *Current Pharmaceutical Design*, 12(31):3999–4021, 2006.

19 Limin Lin. *An SPR-based Sensor to Measure DNA-protein Interaction*. Irvine: University of California, 2005.

20 J Nkoma, R Loudon, and DR Tilley. Elementary properties of surface polaritons. *Journal of Physics C: Solid State Physics*, 7(19):3547, 1974.

21 Kevin Welford. Surface plasmon-polaritons and their uses. *Optical and Quantum Electronics*, 23:1–27, 1991.

22 Chongchong Chen. A new kind of spoof surface plasmon polaritons structure with periodic loading of t-shape grooves. *AIP Advances*, 6(10):105003, 2016.

23 AI Fernández-Domínguez, Esteban Moreno, L Martin-Moreno, and FJ Garcia-Vidal. Guiding terahertz waves along subwavelength channels. *Physical Review B—Condensed Matter and Materials Physics*, 79(23):233104, 2009a.

24 AI Fernández-Domínguez, Esteban Moreno, L Martín-Moreno, and FJ García-Vidal. Terahertz wedge plasmon polaritons. *Optics Letters*, 34(13):2063–2065, 2009b.

25 Liangliang Liu, Zhuo Li, Changqing Gu, Bingzheng Xu, Pingping Ning, Chen Chen, Jian Yan, Zhenyi Niu, and Yongjiu Zhao. Smooth bridge between guided waves and spoof surface plasmon polaritons. *Optics Letters*, 40(8):1810–1813, 2015.

26 F de León-Pérez, G Brucoli, FJ García-Vidal, and L Martín-Moreno. Theory on the scattering of light and surface plasmon polaritons by arrays of holes and dimples in a metal film. *New Journal of Physics*, 10(10):105017, 2008.

27 J-J Wu, T-J Yang, and LF Shen. Subwavelength microwave guiding by a periodically corrugated metal wire. *Journal of Electromagnetic Waves and Applications*, 23(1):11–19, 2009.

28 Hua Lu, Chao Zeng, Qiming Zhang, Xueming Liu, Md Muntasir Hossain, Philipp Reineck, and Min Gu. Graphene-based active slow surface plasmon polaritons. *Scientific Reports*, 5(1):1–7, 2015.

29 Kyungjun Song and Pinaki Mazumder. Active terahertz spoof surface plasmon polariton switch comprising the perfect conductor metamaterial. *IEEE Transactions on Electron Devices*, 56(11):2792–2799, 2009.

30 Christopher R Williams, Steven R Andrews, SA Maier, AI Fernández-Domínguez, L Martín-Moreno, and FJ García-Vidal. Highly confined guiding of terahertz surface plasmon polaritons on structured metal surfaces. *Nature Photonics*, 2(3):175–179, 2008.

31 ML Nesterov, Diego Martín-Cano, AI Fernandez-Dominguez, Esteban Moreno, Luis Martin-Moreno, and FJ Garcia-Vidal. Geometrically induced modification of surface plasmons in the optical and telecom regimes. *Optics Letters*, 35(3):423–425, 2010.

32 Zhao Xu and Pinaki Mazumder. Bio-sensing by Mach–Zehnder interferometer comprising doubly-corrugated spoofed surface plasmon polariton (DC-SSPP) waveguide. *IEEE Transactions on Terahertz Science and Technology*, 2(4):460–466, 2012a.

33 Kyungjun Song and Pinaki Mazumder. Dynamic terahertz spoof surface plasmon–polariton switch based on resonance and absorption. *IEEE Transactions on Electron Devices*, 58(7):2172–2176, 2011.

34 Mahdi Aghadjani and Pinaki Mazumder. Terahertz switch based on waveguide-cavity-waveguide comprising cylindrical spoof surface plasmon polariton. *IEEE Transactions on Electron Devices*, 62(4):1312–1318, 2015a.

35 Kyungjun Song and Pinaki Mazumder. Nonlinear spoof surface plasmon polariton phenomena based on conductor metamaterials. *Photonics and Nanostructures-Fundamentals and Applications*, 10(4):674–679, 2012.

36 Zhao Xu and Pinaki Mazumder. Terahertz beam steering with doped GaAs phase modulator and a design of spatial-resolved high-speed terahertz analog-to-digital converter. *IEEE Transactions on Electron Devices*, 61(6):2195–2202, 2014.

37 Fariborz Parandin, Mohamadreza Jomour, Reza Kamarian, Morteza Gholipour, Naser Mahtabi, and Asghar Askarian. Recent advances in photonic crystal based all-optical analog to digital converter: A review. *Journal of Optical Communications*, 45(s1):s2421–s2437, 2023.

38 Mahdi Aghadjani and Pinaki Mazumder. Thz analog to digital converter using single sided spoof surface plasmon polariton waveguide. In *2016 IEEE 16th International Conference on Nanotechnology (IEEE-NANO)*, pages 845–848. IEEE, 2016.

39 Mahdi Aghadjani, Mikhail Erementchouk, and Pinaki Mazumder. Spatial accumulation of phase difference in spoof plasmon based Mach–Zehnder interferometer. *Optics Communications*, 410:248–253, 2018b.

40 Mahdi Aghadjani and Pinaki Mazumder. Thz polarizer controller based on cylindrical spoof surface plasmon polariton (C-SSPP). *IEEE Transactions on Terahertz Science and Technology*, 5(4):556–563, 2015b.

41 Mahdi Aghadjani, Mikhail Erementchouk, and Pinaki Mazumder. Spoof surface plasmon polariton beam splitter. *IEEE Transactions on Terahertz Science and Technology*, 6(6):832–839, 2016.

42 Zhao Xu and Pinaki Mazumder. Spoofed surface plasmon Mch-Zehnder interferometer (MZI) structure for THz bio-sensing applications. In *2012 12th IEEE International Conference on Nanotechnology (IEEE-NANO)*, pages 1–5. IEEE, 2012b.

43 Zhao Xu, Kyungjun Song, and Pinaki Mazumder. Spoofed surface plasmon polariton (SSPP) gap structure for high sensitivity bio-sensing in THz. In *2013 13th IEEE International Conference on Nanotechnology (IEEE-NANO 2013)*, pages 311–314. IEEE, 2013.

44 WHP Pernice, Mo Li, and HX Tang. Theoretical investigation of the transverse optical force between a silicon nanowire waveguide and a substrate. *Optics Express*, 17(3):1806–1816, 2009.

45 Eduardo A Coronado, Ezequiel R Encina, and Fernando D Stefani. Optical properties of metallic nanoparticles: Manipulating light, heat and forces at the nanoscale. *Nanoscale*, 3(10):4042–4059, 2011.

46 Michelle L Povinelli, Steven G Johnson, Marko Lončar, Mihai Ibanescu, Elizabeth J Smythe, Federico Capasso, and JD Joannopoulos. High-Q enhancement of attractive and repulsive optical forces between coupled whispering-gallery-mode resonators. *Optics Express*, 13(20):8286–8295, 2005.

47 Ji Xia, Qifeng Qiao, Guangcan Zhou, Fook Siong Chau, and Guangya Zhou. Opto-mechanical photonic crystal cavities for sensing application. *Applied Sciences*, 10(20):7080, 2020.

48 Xiaodong Yang, Yongmin Liu, Rupert F Oulton, Xiaobo Yin, and Xiang Zhang. Optical forces in hybrid plasmonic waveguides. *Nano Letters*, 11(2):321–328, 2011.

49 Mahdi Aghadjani, Mikhali Erementchouk, and Pinaki Mazumder. Force distribution inside the spoof surface plasmon polariton waveguide. *JOSA B*, 35(5):1113–1118, 2018a.

50 Wen Xuan Tang, Hao Chi Zhang, Hui Feng Ma, Wei Xiang Jiang, and Tie Jun Cui. Concept, theory, design, and applications of spoof surface plasmon polaritons at microwave frequencies. *Advanced Optical Materials*, 7(1):1800421, 2019.

51 Francisco J Garcia-Vidal, Antonio I Fernández-Domínguez, Luis Martin-Moreno, Hao Chi Zhang, Wenxuan Tang, Ruwen Peng, and Tie Jun Cui. Spoof surface plasmon photonics. *Reviews of Modern Physics*, 94(2):025004, 2022.

52 Xuanru Zhang, Wen Yi Cui, Yi Lei, Xin Zheng, Jingjing Zhang, and Tie Jun Cui. Spoof localized surface plasmons for sensing applications. *Advanced Materials Technologies*, 6(4):2000863, 2021.

53 Xiaopeng Shen and Tie Jun Cui. Ultrathin plasmonic metamaterial for spoof localized surface plasmons. *Laser & Photonics Reviews*, 8(1):137–145, 2014.

3

Fundamental Electrodynamics of Spoof Plasmonic Mode

3.1 Baleen Whales: What They Teach Us on Novel Communication

Nature frequently offers insights that aid in solving intricate technological dilemmas, often stemming from diverse and unexpected domains of study. A prime illustration can be drawn from the realm of bioacoustics. Sound waves, like other mechanical and electromagnetic waves, adhere to the inverse square relationship with distance, meaning their intensity diminishes as they extend from the source at a rate of $\frac{1}{r^2}$. Interestingly, "Baleen whales," a specific whale species, are able to communicate across extensive distances of hundreds or even thousands of kilometers, which defies the conventional constraints of the inverse square law. This ability is rooted in their substantial size and utilization of low-frequency, long-wavelength sounds, refer to Figure 3.1.

The exceptional communication prowess of baleen whales can be attributed to their exploitation of a distinct phenomenon in acoustic wave propagation within the ocean—the Sound Fixing and Ranging (SOFAR) channel. This channel centers around a depth where the combined influence of temperature and water pressure creates an area of minimum sound speed in the water column. As one descends in the ocean, pressure increases linearly while temperature peaks at the surface and then rapidly drops through the thermocline until it stabilizes at approximately 4 °C around a depth of 1,000 m. The swift temperature decrease near the surface prompts a reduction in sound speed, yielding a negative sound speed gradient. Conversely, greater depths experience escalating pressure, resulting in increased sound speed and a positive gradient. Consequently, a sound wave navigating through the thermocline bends downward due to the decreasing water temperature, only to be refracted upward as it encounters amplified sound speed owing to augmented depth and pressure. This cyclical bending pattern

Artificial Plasmonics for VLSI Interconnects: Bridging the Gap between Electronics and Optics, First Edition. Soumitra R. Joy and Pinaki Mazumder.

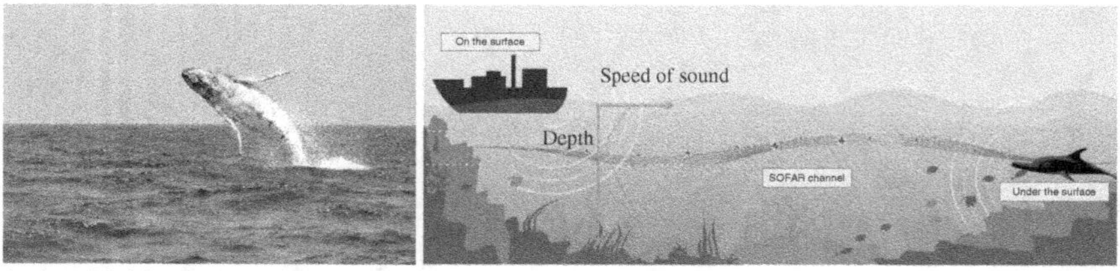

Andrea / Adobe Systems Incorporated

Figure 3.1 Baleen whale can transmit information over astonishingly large distance. The underlying physics of their information transfer mechanism may inspire us to solve a long-standing problem of VLSI communication.

permits low-frequency sound waves to be confined within a channel, enabling them to traverse thousands of meters or even kilometers without substantial energy loss. Baleen whales capitalize on this SOFAR channel to transmit songs over vast distances.

Taking inspiration from the distinct world of bioacoustics, we found a novel approach to tackle the communication challenges of very large-scale integration (VLSI) chips. This approach involves harnessing a different facet of surface wave behavior through the utilization of a specially engineered electromagnetic medium, known as plasmonics, and its evolutionary counterpart, spoof plasmonics. Despite being rooted in Maxwell's electromagnetic equations, which fundamentally differ from the principles governing acoustic waves, the propagation characteristics of plasmonic waves remarkably mirror those of surface waves in oceans. This parallelism becomes evident when examining the mathematical descriptions of acoustic and electromagnetic wave propagation:

$$\frac{\partial^2 p}{\partial t^2} = v_s^2 \nabla^2 p \text{ : acoustic wave equation} \tag{3.1}$$

$$\frac{\partial^2 E}{\partial t^2} = \frac{1}{\mu\epsilon} \nabla^2 E \text{ : electromagnetic wave equation} \tag{3.2}$$

This mathematical resemblance between acoustic and electromagnetic waves provides us with a valuable clue for designing electromagnetic channels in a manner that can defy the constraints imposed by the inverse square law. This innovative approach aims to enhance signal integrity, curbing signal dispersion within its immediate surroundings and paving the way for more resilient and effective communication in VLSI chips.

3.2 Plasmonics Aided High Speed VLSI Communication

Initiated in 2010, the Mazumder Laboratory began a research initiative in response to the burgeoning need for plasmonic circuitries—a demand driven by the anticipated convergence of computing and optical communications within future-generation VLSI chips. These chips are envisioned to accommodate photonic signals racing at speeds of hundreds of terahertz, which will be introduced through fiber optic cables. Subsequently, the optical data is slated to undergo down-conversion and processing within computing circuitry operating at frequencies in the tens of gigahertz range.

An immediate challenge surfaces due to the disparity in scale between optical waveguides used for routing optical signals, typically spanning micrometer dimensions (10–100 μm), and the sub-50 nm complementary metal-oxide semiconductor technology employed by electronic components. This incongruity stems from the fact that optical pulses struggle to propagate along conventional waveguides when the waveguide size is smaller than the wavelength of the signal pulse.

In response, plasmonics emerge as a potential solution, aiming to fabricate nanoscale photonic components that mitigate the coupling issue between optical and electronic signals. Metal's negative dielectric permittivity within the optical spectrum enables light confinement within subwavelength structures, allowing light to be guided over metallic structures. Consequently, plasmonic devices hold promise for integrating optical components with electronic devices, given their nanoscale dimensions, which can operate at optical frequencies while mirroring the dimensions of electronic components.

However, fundamental limitations cast a shadow over the potential of plasmonics. While plasmonics enable the concentration and guidance of light through subwavelength structures, the high attenuation of plasmonic signals restricts their use of local elements within chip-scale applications. Propagation loss along the plasmonic architecture remains a formidable challenge, especially as dimension contraction amplifies damping coefficients due to increased momentum wave. Energy attenuation further complicates the utilization of plasmonics within on-chip nanoscale waveguides and other near-field optical applications.

To address these limitations, efforts have been directed toward subwavelength localization methods at a frequency lower than the natural plasmonic resonance, so as to reduce metallic loss. Surface topology engineering, including the incorporation of holes, grooves, and dimples, mimics realistic surface plasmon polariton (SPP) behavior typically observed in visible or ultraviolet spectrums at THz or sub-THz range. This leads to the emergence of spoof SPP (SSPP) modes, fostering highly localized surface-bound modes on perfect conductors. These SSPP modes have paved the way for the development of efficient passive and active elements by capitalizing on their mimicking properties.

By mimicking realistic SPPs, SSPP modes offer intriguing attributes. The SSPP dispersion curves and mode profiles within sandwiched structures confer efficient confinement methods with reduced damping coefficients. Consequently, strong localization and a minimized damping mechanism are achievable within the radio frequency to terahertz (RF–THz) frequency domain. This endeavor has led to the exploration of new avenues in subwavelength localization methods, potentially overcoming the inherent challenges of plasmonics.

The Mazumder Laboratory's work in the field of plasmonics and spoof plasmonics has yielded a sequence of accomplishments. The core objective of these research endeavors was to pinpoint pragmatic applications for the transmission of plasmonic mode information within conventional VLSI systems. This objective was achieved through a methodology of theoretical modeling and the utilization of simulation tools, and verification through GHz and sub-THz experiments.

The Mazumder Laboratory developed the Plasmonic Interconnect Simulator Tool (PIST), an analytical framework that effectively illustrates the propagation behavior of surface plasmon (SP) pulses along intricate metallic nanoparticle (MNP) arrays and metallic nanowires (MNWs). Subsequent research efforts focused on elucidating the energy transfer mechanisms within these structures, leading to the formulation of comprehensive theoretical frameworks. These frameworks served as the foundation for creating SPICE-compatible equivalent models for MNPs and MNWs, facilitating high-fidelity circuit-level simulations.

Further progress was achieved through the design of passive circuit elements (resistors, inductors, capacitors, and conductance) tailored to MNP arrays and MNWs. This development enabled the realization of rapid system-level simulations comparable to the SPICE simulation environment. Coupled with circuit-level simulations, this led to the implementation of plasmonic wave propagation and the integration of photonics and electronic signal modeling.

Besides the integration of plasmonic waveguides into on-chip applications, the Mazumder Laboratory also focused on the design of an active plasmonic switch or modulator in order to expand the device's capabilities. This device offers intricate signal routing and the execution of arbitrary Boolean logic, enhancing the potential functionality of plasmonic systems.

Another key outcome was the development of an ultrafast, highly sensitive nanophotodetector, strategically embedded within a compact plasmonic architecture. This work enables the conversion of optical signals into electronic signals, opening avenues for high-speed and high-sensitivity applications.

3.2.1 Innovative Spice Simulation Tool Development for Plasmonics

Song and Mazumder [1] conducted work concerning the study of a chain of MNPs as a conduit for propagating SPPs at near-optical frequencies. They developed a methodology that leveraged the electric dipole moment (EDM) model, accounting for the oscillation of free electrons within individual MNP. This approach aimed to determine the equivalent circuit components associated with an MNP for the passive circuit elements of the structure. Within this framework, they calculated electromagnetic power flows, encompassing factors such as relaxation damping, radiation outflow, host matrix electromagnetic coupling, and interactions with applied signals, to derive the pertinent passive circuit elements for the MNPs. Employing the principle of energy conservation, they computed an extended equation of motion, describing the dynamics of a spherical MNP.

Furthermore, the resonant attributes of an individual MNP were effectively captured using a lumped resonant circuit model, with circuit parameters (RLC) derived from the equation of motion of the EDM and the electromagnetic (near-field energy enveloping the MNP. The work described the E-field received by the mth nanoparticle in the chain as:

$$E_{p,m} = \frac{\mu_i}{4\pi\epsilon_0 n^3 d^3} \left(p_{m+1}(t) + p_{m-1}(t)\right) \tag{3.3}$$

The propagating wave solution through the MNP chain has the following form:

$$p_{i,m} = p_{i,m-1} \exp\left(-\alpha d + j\left(\omega t \pm kd\right)\right) \tag{3.4}$$

The work developed an equivalent circuit for a linearly equidistant array of MNPs functioning as a plasmonic wire. This was portrayed as a voltage-controlled voltage source (VCVS), leveraging the influence of the nearest SP interactions to facilitate the understanding of plasmonic behavior within this complex system. Figure 3.2 captures the extracted model parameters for modeling a single nanoparticle and a chain of MNPs using the RLCG model (Figure 3.3) developed using the EDM electrodynamic modeling technique.

Subsequently, Song and Mazumder [2] explored the use of the MNW as a conduit for propagating SPPs at near-optical frequencies. Their investigation centered on establishing an equivalent circuit model for the nonradiative transport of SP energy along the MNW. By deriving the SP dispersion and damping relation

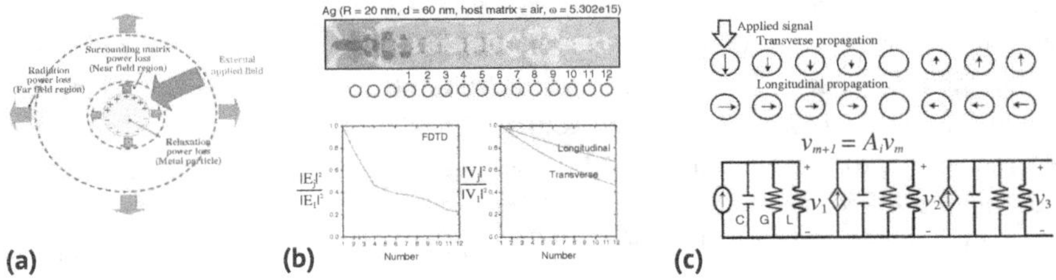

Figure 3.2 (a) Energy dissipation mechanisms in different regions for the oscillatory modes in MNP. (b) Scattered electric field in the Ag plasmon wire. (c) Two fundamental coupled oscillations of a plasmon wire, and the equivalent circuit modeling for MNPS plasmon wire as a VCVS.

	Cu	Au	Ag
eV(resonance)	2.11	2.36	3.49
ω_0 (rad/s)	3.206e15	3.585e15	5.302e15
ϵ	−7.262 + j3.460	−4.261 + j2.435	−2.068 + j0.279
γ (rad/s)	1.439e14	1.075e14	3.226e13
C	7.439af	3.124af	0.880af
L	13.08fH	24.91fH	40.43fH
G	1.078mS	0.339mS	0.0283mS

Figure 3.3 Nanocircuit elements of three noble MNPs (r = 20 nm) that is surrounded by a free space.

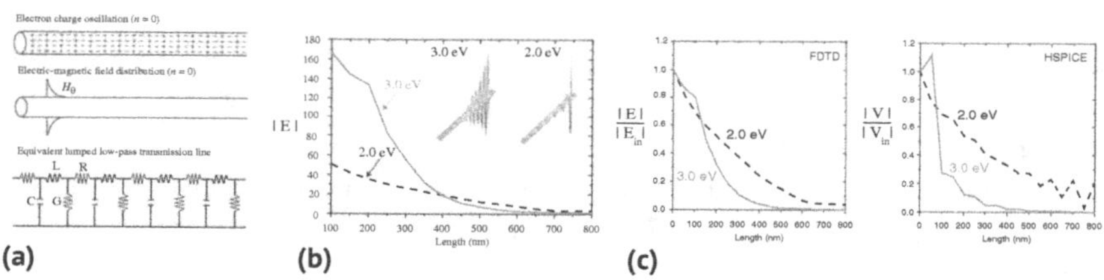

Figure 3.4 (a) Equivalent models of nonradiative SP wave along the MNW with fundamental mode n = 0 with no surface charge variation in the direction. (b) Electric field distribution in Ag nanowire across its axis of signal propagation. (c) The average E-field density in the Ag plasmon wire (r = 10 nm) calculated by using FDTD simulation and HSPICE simulation in commercial tool.

through a modified Bessel function electromagnetic field expansion, they introduced a low-pass transmission line (TL) model to illuminate SP dynamics. Particularly noteworthy was their calculation of parameters intrinsic to the low-pass TL model, including series impedance (Z) and shunt admittance (Y), using a lumped element model and harmonic voltage (current) distribution. Employing the finite difference (FD) discretization method—specifically T-cell RGLC networks—Song and Mazumder extracted equivalent circuit parameters such as resistance, inductance, capacitance, and conductance. These parameters provided an understanding of MNW behavior.

The derived equivalent circuit elements were verified through HSPICE circuit simulation and the three-dimensional (3D)-scattered finite time-domain method (FDTD), which confirmed the accuracy and effectiveness of the proposed model, as depicted in Figure 3.4.

3.3 A Universal Theoretical Framework for Spoof Plasmonics

The physics of spoof plasmonics has previously been understood through an analogy involving a negative permittivity effective medium. This approach, however, often confines the analysis to simple cases of spoof plasmonic structure variation. To overcome this limitation and establish a broader theoretical framework capable of examining diverse shapes, geometries, and configurations of spoof plasmonic waveguides, Erementchouk et al. [3] developed a comprehensive electrodynamics theory for spoof plasmonics. In their article titled "Electrodynamics of spoof plasmons in corrugated metallic waveguides" [3], the authors systematically explored the electromagnetic field states near periodically corrugated surfaces of perfect conductors, known as SSPPs. They approached this problem using an adapted mode matching technique within the transfer matrix formalism.

This approach enables the investigation of systems with arbitrary transversal structures, particularly those with narrow grooves, thereby transcending the limitations of effective medium descriptions and conventional mode matching implementations. A concise expression for the SSPP coupling parameter, accounting for higher Bloch modes, was derived. The broader analysis was then applied to examine the impact of a dielectric environment on SSPP spectra. Interestingly, the effective SSPP plasma frequency remained unaffected by the dielectric constant of the medium outside the grooves. For sufficiently wide dielectric slabs covering the corrugated surface, the main effect was characterized by a straightforward rescaling of the maximal Bloch wavenumber and coupling parameter.

The developed formalism based on a transversal transfer matrix is directly applied for finding a dispersion law governing states of the field confined to the corrugated surface, SSPPs. The dispersion law is obtained by requiring that the transfer matrix must map a state vector corresponding to the boundary condition at one point, say, $|\alpha L\rangle$ at the lower end, into a state vector corresponding to the condition at the opposite point, say $|\alpha U\rangle$. Explicitly, this condition is written as

$$D(\omega, \beta) \equiv \langle \alpha\bot_U | T | \alpha_L \rangle = 0 \tag{3.5}$$

where $\langle \alpha\bot_U|$ is an element of space defined by $\langle \alpha\bot_U|\alpha U\rangle = 0$. It should be emphasized that since for any chosen state vector $|\alpha L\rangle$ satisfying the boundary condition, the above equation must hold for an arbitrary $\langle \alpha\bot_U|$. The above equation, in fact, defines a system of homogeneous equations with respect to amplitudes entering $|\alpha L\rangle$.

When the width of the grooves is small compared to the period of the structure, the above equation can be analyzed using an approximation developed similarly to the Rayleigh–Schrodinger perturbation theory. This approach is illustrated by considering the SSPP dispersion equation in a half-closed waveguide (Figure 3.5),

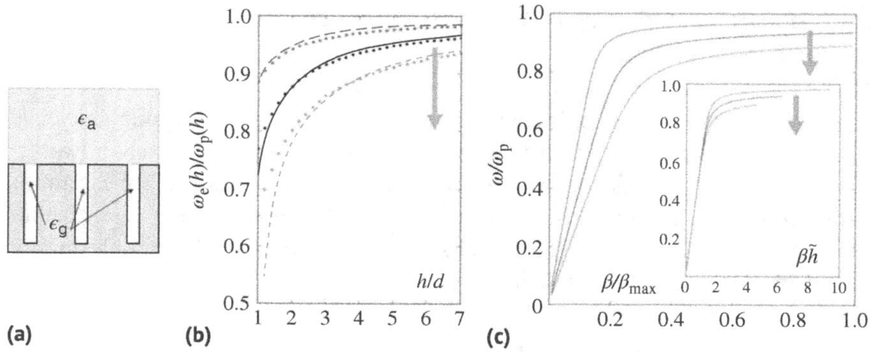

Figure 3.5 (a) The cross-section of a half-closed waveguide filled by material with the dielectric function ϵ_g bounded by half-space with the dielectric function ϵ_a. (b) The normalized position of the edge of the fundamental band $\frac{\omega_e(h)}{\omega_p(h)}$ as a function of the grooves height for different values of ϵ_a (the arrows show the variation with increasing $\epsilon_a = 1, 2.25, 4$): solid lines are obtained from analytical expression developed in the Mazumder Laboratory, dotted lines show the results of full-wave numerical simulations. (c) Variation of the SSPP dispersion diagram with the refractive index of the surrounding medium. The inset shows the dispersion diagrams in rescaled parametrization $\omega(\beta h)$.

when $|\alpha_L\rangle = |-x; \alpha_l\rangle$ and $\langle \alpha\bot_U| = \langle -1; \alpha\bot_m$. Using a representation of the action of the transfer matrix, the dispersion equation can be written in the form

$$\cos(P_l h)\,\alpha_l - i\sum_{l'} \eta_{l,l'} \sin(P_{l'} h)\,\alpha_{l'} = 0 \tag{3.6}$$

Furthermore, in the scenario of a thin dielectric layer, it was revealed that SSPPs demonstrate sensitivity to subwavelength scale variations in the layer's thickness.

Since $\eta_{l,l'} \sim \frac{a}{d}$, in structures with narrow grooves, the above equation can be approached with the help of a perturbation theory. The most significant effect is of the first order. It yields a dispersion equation of SSPP formed by modes of different orders along the x-axis inside the grooves,

$$D_{OS,l}(\omega, \beta) \equiv \cos(P_l h) - \lambda_l \frac{P_l}{\kappa 0} \sin(P_l h) = 0 \tag{3.7}$$

where we have defined $\kappa_m^2 = \beta^2 - P^2$, and $\lambda_l \equiv \frac{\eta_{l,l'}\,\kappa 0}{P_l}$ so that,

$$\lambda_l = \frac{\kappa_0}{ds_l} \int_{-a}^{a} dx \left[\left(1 - \frac{|x|}{a}\right) \cos\left(\frac{\pi l x}{a}\right) + \frac{(-1)^l}{\pi l} \sin\left(\frac{\pi l}{a}(a - |x|)\right)\right] e^{i\beta x} F(x) \tag{3.8}$$

with

$$F(x) = \sum_{m} \frac{1}{\kappa_m} e^{i2\pi mx/d} \tag{3.9}$$

Function $F(x)$ has logarithmic singularities at points $x = nd$ with integral n, which can be investigated by considering the vicinity of these points. Approximating the series components of $F(x)$, we obtain

$$F(x) = \frac{1}{\kappa_0} - \frac{d}{\pi} \ln\left(\frac{2\pi |x|}{d}\right) \tag{3.10}$$

This approximation eventually helps obtain the analytical expression for the coupling term as

$$\lambda_0 \approx \frac{a}{d}\left[\text{sinc}^2\left(\frac{a\beta}{2}\right) + \frac{\kappa_0 d}{\pi}\left(\frac{3}{2} - ln\left(\frac{2\pi a}{d}\right)\right)\right] \tag{3.11}$$

The second term in the braces, deviating from the approximation routinely used in studies of SSPP in periodically corrugated structures, accounts for the effect of multiple Bloch modes. This correction becomes important in the regime of well-formed SSPP when the states of the electromagnetic field are characterized by strong confinement to the surface.

Extension of this approach to more complex structures is straightforward owing to the fact that, as long as at one end, the state of the field is given by $| -x; \{\alpha_l\}\rangle$, its transfer through the structure can always be presented as:

$$|\alpha_f| = \left|x; \hat{T}^{(a)} \{\alpha_l\}\right| + \left|-x; \hat{T}^{(b)} \{\alpha_l\}\right| \tag{3.12}$$

with appropriate matrices $\hat{T}^{(a,b)}$.

3.3.1 Dispersion Law of SSPP Waveguides

3.3.1.1 Half-closed Waveguide

The simplest SSPP structure is a one-sided waveguide, which is essentially just a conducting corrugated surface. The SSPP dispersion law for such structure has the form

$$\cos(\text{Ph}) - \lambda \frac{P}{\kappa} \sin(\text{Ph}) = 0 \tag{3.13}$$

This equation can be utilized to clarify the analogy with true plasmon. The dispersion equation for the true SPPs at the interface between air and metal with the dielectric constant ϵ_m

$$\sqrt{\left\{\frac{\beta^2 - \left(\frac{\omega}{c}\right)^2}{\beta^2 - \epsilon_m\left(\frac{\omega}{c}\right)^2}\right\}} = -\frac{1}{\epsilon_m} \tag{3.14}$$

By defining ϵ_m in such a way that both of the above equations represent the same mathematical form, it can be found that

$$\epsilon_m = \frac{1}{\lambda^2}\left(1 - \lambda - \frac{1}{\sin^2(\omega h/c)}\right) \sim \frac{1}{\lambda^2}\left(1 - \frac{\omega_p^2}{\omega^2}\right) \tag{3.15}$$

where $\omega_p = \frac{\pi c}{2h}$ is the effective plasma frequency for the SSPP defined by the condition of vanishing effective dielectric function $\epsilon_m(\omega = \omega_p) = 0$. This shows that, indeed, from the spectral point of view, the electromagnetic waves confined near the corrugated surface look similar to SPPs.

In order to outline the main effect of the geometry of the structure, note that the spectral and geometrical parameters enter the dispersion equation of SSPP through three variables: Ph, βh, and η. Therefore, SSPP dispersion diagrams of all half-closed structures have the same overall form shown in Figure 3.5c. In wide structures, where $\frac{\pi c}{w} << \omega_p$, the transition from weakly attenuated to SSPP regime occurs near $\beta = \beta_c = \frac{\omega_p}{c}$, where the light line intersects the effective plasma frequency. Below the transition, $\beta < \beta_c$, the dispersion curve only slightly deviates from the light line, and for $\beta > \beta_c$, the dispersion curve gradually approaches ω_p. Thus, varying the geometry of the structure will only change the "smoothness" of the transition between two regimes on the $(Ph, \beta h)$-plane and, if the period is changed, will limit the dispersion diagram at different $(\beta h)_{\max} = \frac{\pi h}{d}$.

The form of the dispersion curve is determined by a set of characteristic frequencies. First, these are frequencies (ω_p and $\omega_\kappa(\beta)$ at which $P(\omega)$ and $\kappa(\omega, \beta)$ vanish. The curve $\omega = \omega_\kappa(\beta)$ separates regions corresponding to propagating and attenuated modes outside the grooves and thus defines the light line. For example, in this case, $w \to \infty$, we have $\omega_p = 0$ and $\omega_\kappa(\beta) = \beta_c$. Another set of characteristic frequencies, $\omega_p^{(n)}$, is given by the positions of zeros of $\cos(Ph)$, so that $P\left(\omega = \omega_p^{(n)}\right) = \frac{\pi\left(\frac{1}{2}+n\right)}{h}$ with integer n.

From the relationships between the characteristic frequencies, it can be seen that for all β's, Eq. (3.13) has at least one solution, which can be labeled as the fundamental branch. Its character strongly depends on the relationship between $\omega_Q(\beta)$ and $\omega_p^{(n)}$, that is, between the light line and SSPP plasma frequencies. Indeed, in the case of small β, when $\omega_Q(\beta) \ll \omega_p^{(n)}$, it can be written as:

$$P^2 \approx \beta^2(1 - \lambda^2\beta^2 h^2) \tag{3.16}$$

Taking into account the relation $\omega^2 = (\pi c/w)^2 + P^2c^2$, such a represented solution is valid for the arbitrary width of the structure and the dielectric function of the medium. This solution only slightly deviates from the light line and thus corresponds to a weakly attenuated field outside the grooves. In the opposite limit, $\omega_Q(\beta) \gg \omega_p(0)$, the solution is close to $\omega_p(0)$, so that,

$$P \approx \frac{\pi}{2h} - \frac{\pi\lambda}{2\beta h^2} \tag{3.17}$$

Thus, there is significant deviation from the light line, which corresponds to strong attenuation of the field outside the grooves. The transition between regimes of weak and strong confinement occurs near the crossing of the light line and the SSPP plasma frequency, i.e., $\omega_Q(\beta_c) = \omega_p(0)$. This consideration suggests identifying the strong confinement regime with the formation of SSPP, because it corresponds to the emergence of the effective Drude model as expressed by the second half of Eq. (3.15). As β is limited by the Brillouin zone, $\beta \le \beta_{\max} = \pi/d$, this yields the criterion that the corrugated surface must satisfy in order to support a well-defined SSPP, $\beta_c < \pi/d$ or

$$h > \frac{d}{2} \tag{3.18}$$

Thus, the effect of SSPP can be expected to be well developed in structures with sufficiently long grooves. With increasing the length of the groove, the characteristic frequencies $\omega_P^{(n)}$ decrease, which may lead to the appearance of additional bands. It can be seen that the number of branches at given β equals to $1 + n_{\max}$, where $n_{\max}$ is maximal n such that $\omega_P(n) < \omega_Q(\beta)$ with $\omega_P^{(n)}$ defined as zeros $\sin(Ph)$ in Eq. (3.13), so that $P(\omega = \omega_P^{(n)}) = \pi n$. The analysis of the nth higher branch can be performed using the same arguments as above with substitutions $\omega_P \to \omega_P^{(n)}$ and $\omega_p(0) \to \omega_P^{(n)}$. An interesting feature of the higher branches can be noticed: in contrast to the fundamental branch, they exist only when β is large enough (Figure 3.6) $\beta > \beta_0^{(n)}$, where $\beta_0^{(n)}$ is found from $\omega_Q(\beta_0^{(n)}) = \omega_P^{(n)}$, which yields the condition for the nth-order band to exist. For example, in wide waveguides with $w > h$, the nth higher band appears when $h > nd$.

3.3.1.2 Open SSPP Structure

As we have seen above, in structures with open grooves, there are two classes of SSPP modes corresponding to different transformation properties under reflection of the (x, y)-plane passing through the center of the grooves. Even modes are characterized by vanishing E_x at $z = z_C$ and, as a result, the dispersion equation of even modes reproduces that of half-closed structures with renormalized grooves height $h \to h/2$, as illustrated by Figure 3.7. Thus, the consideration provided above directly applies to even modes in structures with open grooves. The spectrum of odd modes can be studied by introducing the phase parameter $\zeta = Ph - \pi/2$ or $\tilde{P} = P - \pi/2h$. Equation relating $\tilde{P}$ and β has a structure similar to Eq. (3.13) and

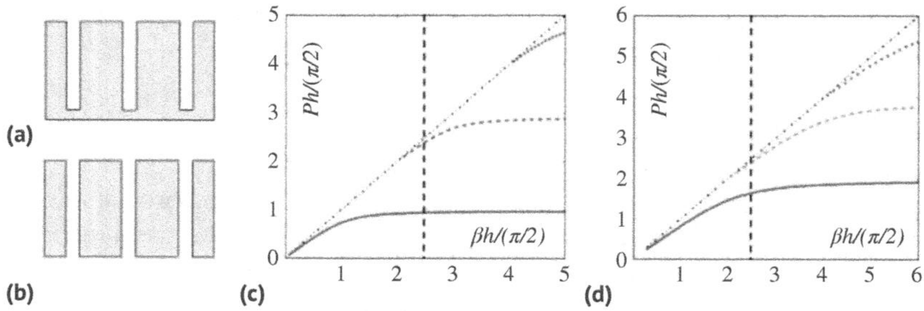

Figure 3.6 Half-closed and open SSPP structures. (a,b) The cross-sections of half-closed and open waveguides, respectively. (c,d) Dispersion diagrams on the $(Ph, \beta h)$-plane of SSPP in half-closed and open waveguides, respectively. The vertical and horizontal axes are additionally rescaled by $\pi/2$, so that the coordinates of the characteristic points are expressed in integer numbers. The dash-dotted line shows the light line, $\omega = \beta c$. The vertical, dashed line shows the right boundary of the dispersion diagrams determined by $\beta_{\max} = \frac{\pi}{d}$ in a structure with $\frac{h}{d} = 5$.

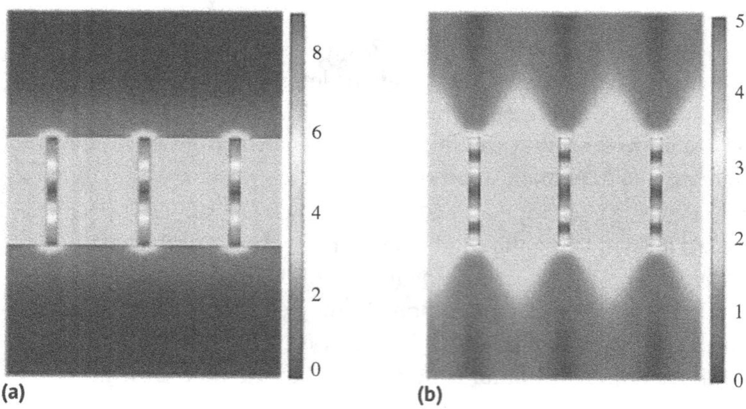

Figure 3.7 Electromagnetic field energy distribution in structures with open grooves with $\frac{h}{d} = 1.6$ at $\beta = \beta_{\max}$. (a) Even modes with $Ex = 0$ at the line passing through the center of the grooves are analogous to states in half-closed waveguides with the grooves of halved height. (b) Odd modes bands reside in the gaps left by the even modes. Their distinctive feature is the symmetry of the field distribution. The difference between panels (a) and (b) is due to reduced attenuation and increased contrast of beatings in higher bands compared to the fundamental band.

can be analyzed in a similar manner by considering relationships between characteristic frequencies. It follows from this analysis that odd modes are similar to higher order bands: they appear in structures with sufficiently long grooves, $h > d$, start from nonzero $\beta_o^{(n)} = \pi(1 + 2n)/h$, and occupy bands approaching effective plasma frequencies $\omega_o^{(n)} = (1 + n)2\pi c/h$ and are situated between the bands of even modes. The general structure of the spectrum of structures with open grooves is shown in Figure 3.6d. While it looks similar to the spectrum of the half-closed structure, an important difference should be noted. Odd branches occupy the space between even branches, which leads to a significant reduction of the bandgaps. The starting point of odd bands is the spoof plasma frequency of the previous even band. As a result, the bandgaps are determined solely by the distance between the edges of the bands and the respective plasma frequencies.

3.3.1.3 Dispersion in Double-sided SSPP Structure

SSPPs in double-sided structures (Figure 3.8a) are known for their ability to prevent radiation loss [4]. Because of this property and in view of the similarity of the dispersion equations governing modes of double-sided corrugated waveguides and half-closed structures, so that the same analysis can be performed, we limit ourselves to discussion of some general properties only. As the structure is closed, solutions of the dispersion equation must be sought for in a class of nonattenuated states outside the grooves (i.e., with real Q) as well. Let the transfer matrices through the lower and the upper groove be $\mathcal{T}_{OS}^{(1,2)}$, respectively, and the transfer across the arm is described by $\mathcal{T}_a$. Then, the dispersion equation can be presented schematically as:

$$D(\omega, \text{beta}) = D_{OS}^{(1)} \langle -1|\mathcal{T}_a| - 1\rangle \; D_{OS}^{(2)} - \Lambda(\omega, \beta) \tag{3.19}$$

where $D_{OS}^{(1)} = \langle x \mid \mathcal{T}_{OS}^{(2)} \mid -1 \rangle$ can be seen to yield the dispersion equation for the second half-closed waveguide, $D_{OS}^{(2)} = \langle -1|\mathcal{T}_{OS}^{(1)}| - x\rangle$, and $\lambda(\omega, \beta) = \langle x|\mathcal{T}_{OS}^{(2)}|1\rangle\langle 1|\mathcal{T}_a|1\rangle\langle 1|\mathcal{T}_{OS}^{(1)}| - x\rangle$. Expression for $D(\omega, \beta)$ and $\Lambda(\omega, \beta)$ containing diagonal elements of the transfer matrix through the arm. Depending on whether the states inside the arm are attenuated or not (i.e., $\beta > \beta_c$ or $\beta \sim 0$), the effect of the coupling between opposite sides of the waveguide is drastically different.

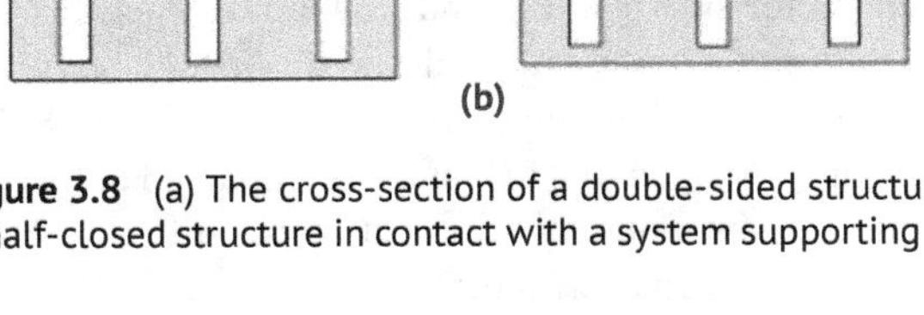

(a) (b)

Figure 3.8 (a) The cross-section of a double-sided structure; and (b) a schematic view of a half-closed structure in contact with a system supporting nontrivial propagating modes.

When $\beta >> \beta_c$, the coupling parameter $e^{-2\kappa_0 t}$ is exponentially small. As a result, in this region, we have weakly coupled modes at the opposite sides of the waveguide. In symmetric structure, this leads to slightly repelled symmetric and antisymmetric modes. It can be shown that the separation between the edges of even and odd modes is expressed as:

$$\Delta = \omega_p \frac{a}{h} \frac{\text{sinc}^2(a\beta/2)}{\sinh(2\kappa t)} \tag{3.20}$$

where $\kappa t = \pi t d^{-1}\sqrt{(1-(d/2h)^2)}$. In the opposite case, $\beta \to 0$, the matrix elements of $\mathcal{T}_a$ are pure phase factors and, as a result, we have strong coupling of modes confined to opposite sides. As a result, at $\beta = 0$, the even branch resides at $P \approx 0$, while the odd branch is at $P = \pi/2h(1-\delta)$ with

$$\delta \approx \frac{a}{d}\left[\tan\left(\frac{\pi t}{2h}\right) + \frac{d}{2h}\left(\frac{3}{2} - ln\left(\frac{2\pi a}{d}\right)\right)\right] \tag{3.21}$$

3.4 Electrodynamics of Spoof Plasmon in Finite Structure

A drawback in the previous theoretical analyses of spoof plasmonics was that they operated under the assumption that the spoof plasmonic structure possessed an infinitely large cross-section in the transverse direction of field propagation. Although this assumption simplified the treatment of spoof plasmonics' electrodynamics due to reduced dimensionality, it posed challenges from an engineering standpoint. Creating such a structure, particularly on a chip-scale, was impractical. To bridge this significant theoretical gap and address the feasibility of a realizable spoof plasmonic waveguide with a finite cross-section, Joy et al. [5] formulated an electromagnetic theory aimed at describing the characteristics of such waveguides.

Based on the argument of conservation of field polarization, it was hypothesized that [6], even in thin structures, the main characteristics of SSPP metamaterial with infinite thickness are retained, and hence the standard theory for spoof plasmon [7–11], originally derived for 3D structure, requires no revision. Because of the apparent resemblance between the dispersion behavior of a 3D SSPP structure and that of its two-dimensional (2D) planar counterpart, it was deemed redundant to perform an exclusive analysis of the planar SSPP structure. The objective of the authors in the subsequent sections of this chapter is to develop a complete theory explaining the optical characteristics of SSPP mode in planar structure coupled with its environment, including the substrate underneath. It may be noted that Liu et al. [12] performed numerical simulations to study the effect of SSPP thickness on dispersion as shown by Yang et al. [13], and the impact of substrate on SSPP mode was characterized by introducing an effective permittivity term. However, a quantitative analysis of a planar structure is needed to explain the counter-intuitive redshift occurrence in the SSPP waveguide with the reduction of the thickness of SSPP at a given propagation constant (β). Wan and Cui [14] presented a theoretical analysis of spoof plasmon in thin structures. Their analysis, however, does not predict any frequency shift or field confinement in the lower region of the SSPP band, although in practice the whole band undergoes redshift with reduction of SSPP thickness as established through experimental measurement [6]; as well as COMSOL simulation. In addition to explaining this anomaly, our analysis also shows that field confinement is achieved at a nonzero frequency on the lateral wall of SSPP even at a frequency away from the spoof plasma resonance. In other words, subwavelength confinement in planar SSPP can be a broadband phenomenon contrary to the common notion of resonance-based narrowband confinement [15, 16]. In this chapter, based on the work of the authors (Joy and Mazumder), the idea of the effective thickness of SSPP mode in an ultrathin structure is introduced, which offers an intuitive view of why spoof plasma mode exists even in the limit of vanishing metal thickness. We shall also provide a quantitative analysis of the ultrathin structure using the proposed effective thickness model, which agrees with 3D simulation results.

This work endeavors to make both intuitive and quantitative analysis of spoof plasmon mode in planar structures. The following sections discuss a theory, which shows that, for an SSPP waveguide surrounded by dielectrics on each of its lateral sides, an imaginary wave vector $P = i\delta_P$ describes the mode profile along lateral direction. This unusual cross-sectional mode with hyperbolic profile explains why the operating frequency at a given propagation constant makes redshift with the thickness reduction, which is quite contrary to a conventional structure where frequency scales inversely with size. I also discuss an interesting concept of the effective thickness t_{eff} of SSPP groove to explain why we retain a solution of SSPP mode even in the limit of vanishing metal thickness. A useful contribution of this work is the analysis of the effective refractive index $n_{g,\mathrm{eff}}$ of the groove, which takes into account both the geometry of the waveguide and the environment outside of it in governing its dispersion relation. These parameters simplify other analyses relevant to the design of bandwidth-efficient SSPP data buses, and the estimation of bandwidth modulation rate with the change in the external environment.

3.5 Modal Analysis of Spoof Plasmon

3.5.1 Dispersion Relation in Wide Structures

The distribution of the field across a periodic structure is conveniently described within the formalism of the transversal transfer matrix when the field is represented as a superposition of upward and downward propagating waves with transfer matrices relating amplitudes in different regions. The developed formalism is

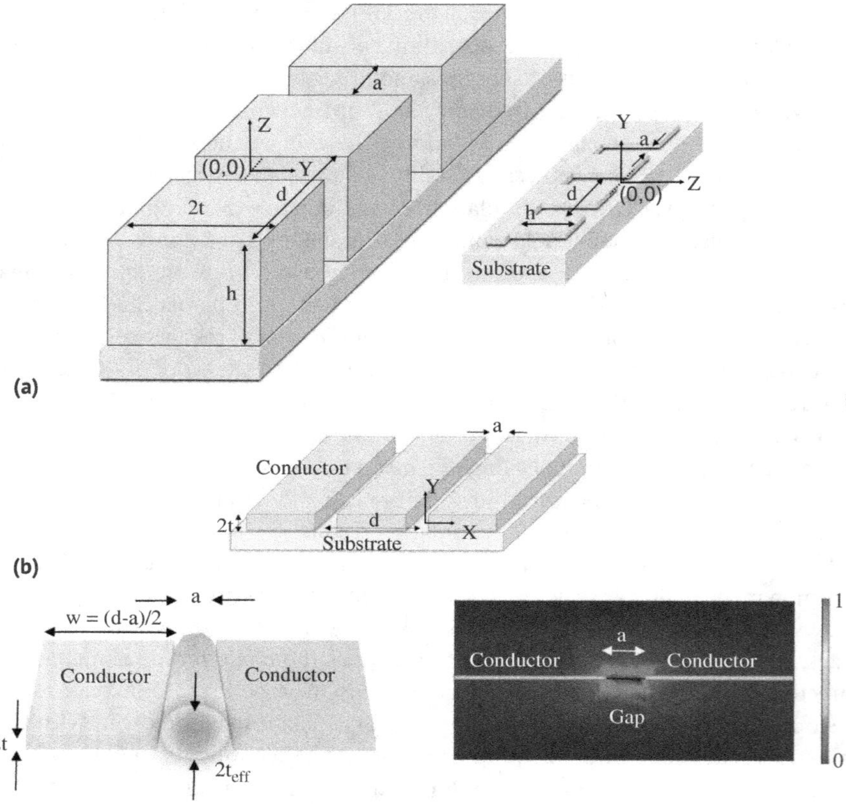

Figure 3.9 (a) Schematics of a 3D SSPP structure standing alone and a 2D SSPP structure on substrate. (b) Planar SSPP seen from a lateral viewpoint. The groove is very long along *z* direction. (c) (left) Illustration of effective thickness of SSPP groove. The optical mode formed between the pair of conductors extends beyond the physical thickness of metal. (right) 3D simulation result of the mode formation inside the groove, justifying the concept of effective thickness.

directly applied for finding dispersion law governing states of the field confined to the corrugated surface, SSPPs. The dispersion law is obtained by requiring that the transfer matrix must map a state vector corresponding to the boundary condition at one point, say, $|\alpha_L\rangle$ at the lower end, into a state vector corresponding to the condition at the opposite point, say, $|\alpha_U\rangle$. Explicitly, this condition is written as:

$$D(\omega, \beta) \equiv \langle \alpha_U^{\perp} | \mathcal{T} | \alpha_L \rangle = 0 \tag{3.22}$$

where $|\alpha_U^{\perp}\rangle$ is an element of space defined by $\langle\alpha_U^{\perp}|\alpha_U\rangle = 0$. It can be shown that the dispersion relation, when written in terms of the spoof plasmon waveguide geometry with one-dimensional (1D) periodic grooves of quarter-wavelength height (h), yields the following:

$$D(\omega, \beta) = \cos(Q_0 h) - \eta \frac{Q_0}{\kappa_0} \sin(Q_0 h) = 0 \tag{3.23}$$

where Q_0 is the zeroth order wave vector inside the groove along the direction of its height, $\kappa_0^2 = \beta^2 - Q_0^2$, where β is the propagation wave vector of spoof plasmon mode. Taking higher order Bloch modes into account, the coupling co-efficient η can be shown to be

$$\eta \approx \frac{a}{d}\left[\text{sinc}^2\left(\frac{a\beta}{2}\right) + \frac{\kappa_0 d}{\pi}\left(\frac{3}{2} - \ln\left(\frac{2\pi a}{d}\right)\right)\right] \tag{3.24}$$

The second term in the braces, deviating from the approximation routinely used in studies of SSPP in periodically corrugated structures, $\eta_{SMA} = \frac{a}{d}\text{sinc}^2\left(\frac{a\beta}{2}\right)$, accounts for the effect of multiple Bloch modes.

3.5.2 Impact of Dielectric Half-space on Dispersion

We start by considering the case when the interior of the grooves of a half-closed waveguide and the outside space are filled with materials with different dielectric constants, ϵ_g and ϵ_a, respectively (Figure 3.10a). The mismatch between ϵ_g and ϵ_a leads to modification of the interface transfer matrix. It has been shown by our group that, in this case, the SSPP dispersion equation has the same general form as for a dielectrically homogeneous environment.

$$\cos(Qh) - \eta(\epsilon_a)\frac{Q^2(\epsilon_a)}{Q^2(\epsilon_g)}\frac{Q(\epsilon_g)}{\kappa(\epsilon_a)}\sin(Qh) = 0 \tag{3.25}$$

Changes induced by the dielectric environment in this case are quite straightforward. We discuss them in the limit $\omega \to \infty$, when after introducing $\tilde{Q} = \sqrt{\epsilon_a}\omega/c$ and $\tilde{h} = h\sqrt{\epsilon_g/\epsilon_a}$, we arrive at the same equation that governs the electromagnetic dispersion relation in an all-homogeneous-dielectric (i.e., all-air) SSPP waveguide structure with rescaled coupling parameter $\eta \to \eta\sqrt{\epsilon_a/\epsilon_g}$. Noteworthy, due to the strong attenuation of the electromagnetic field exterior to the groove in the regime of $\beta d >> \frac{\omega}{c}h$, the spoof plasmon resonance frequency $\tilde{\omega}_p$ is independent of the dielectric environment outside the groove and is only a function of the dielectric constant of the material interior to the groove.

$$\tilde{\omega}_p = \frac{\omega_p}{\sqrt{\epsilon_g}} \tag{3.26}$$

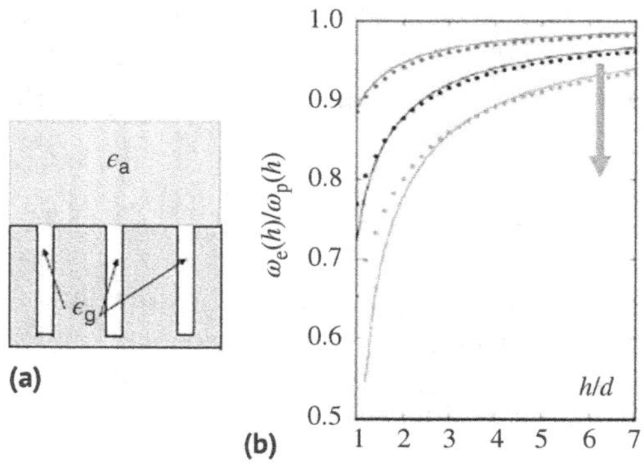

Figure 3.10 (a) The cross-section of a half-closed waveguide filled by material with the dielectric function ϵ_g bounded by halfspace with the dielectric function ϵ_a. (b) The normalized position of the edge of the fundamental band $\omega_e(h)/\tilde{\omega}_p(h)$ as a function of the grooves height for different values of ϵ_a (the arrows show the variation with increasing $\epsilon_a = 1, 2.25, 4$): solid lines are obtained from Eq. (3.27), dotted lines show the results of full-wave numerical simulations.

In the case of a dielectric half-space, the dispersion relation can be characterized by the band-edge frequency $\omega_e = \tilde{\omega}_p - \Delta\omega_e(\epsilon_a, \epsilon_g)$ of the fundamental band, where

$$\Delta\omega_e(\epsilon_a, \epsilon_g) = \tilde{\omega}_p \sqrt{\frac{\epsilon_a}{\epsilon_g}} \frac{2\eta(\epsilon_a)}{\pi\sqrt{(2h/d)^2 \epsilon_g/\epsilon_a - 1}} \tag{3.27}$$

Its dependence on the height of the groove is shown in Figure 3.10 together with the results of full-wave numerical simulations. It demonstrates, in particular, that as long as the system is in the SSPP regime, $\beta\tilde{h} > \pi/2$, Eq. (3.27) agrees very well with numerical simulations.

3.5.3 Dispersion Relation Correction in Structures of Finite Thickness

For the sake of simplicity, we will start our analysis of a planar (with finite thickness) corrugated metal structure with groove of infinite length, as depicted in Figure 3.9b. This will help us isolate analysis of the mode profile in lateral plane (*XY* plane) and longitudinal plane (*ZX* plane). By the term of "lateral" side of SSPP, we refer to the half-space extended along the Y direction exterior to the groove. This half-space is defined by $|y| > t$, where $2t$ is the thickness of the structure, the grooves are oriented along the *Z* direction and $(x, y) = (0, 0)$ line longitudinally bisects a groove into equal halves. Finite cross-section of SSPP will induce a nonzero wave vector P along the lateral direction of the groove. We will make two alternative assumptions to analyze what sort of modification is needed to study the dispersion characteristics of an SSPP waveguide with finite thickness ($2t$).

3.5.3.1 Assumption I: Absence of Ez Field

The perfect electric conductor (PEC) approximation of metal at terahertz frequencies and below, along with the fact that the operating wavelength is much larger than twice the groove width, justifies the assumption that the electric field component E_z, because of being tangential to every point on the surface of patterned metal (including the interior wall of the grooves) vanishes on the waveguide surface and is necessarily nominal compared to the other components of electric field of SSPP mode in space. Let $E_{g,x}$ and $E_{a,x}$ are x-components of the electric field interior and exterior to the groove, respectively, and we will first consider the even mode of $E_{g,x}$ (even/odd mode is defined according to the symmetry of $E_{g,x}$ component along Y axis with its center at the middle of the groove).

$$E_{g,x} = \sum_m G_m \cos\left(\frac{m\pi}{a}\left(x - \frac{a}{2}\right)\right)\cos(P_m y)e^{iQz} \tag{3.28a}$$

$$E_{a,x} = \sum_n A_n e^{i\beta_n x} e^{-\kappa_n(y-t)} e^{iQz} \tag{3.28b}$$

where Q and β are propagation wave vector components along the groove length and waveguide's periodicity, respectively; P_m and $K_n = i\kappa_n$ are the transverse wave vector components along Y axis describing modes at the interior and exterior to the groove, respectively; $\epsilon_g = n_g^2\epsilon_0$ is the dielectric constant of the groove, G_m and A_n are the amplitudes of the field components in the groove and air, respectively; m, n are integers corresponding to optical modes. κ_n must have to be of real value in order to sustain guided mode. An assumption of $E_z = 0$ in addition to the divergence relation of the electric field in the source free region, i.e., $\nabla.\mathbf{E} = 0$ leads to the following simple expression of other field components.

$$E_y = -\int \frac{\partial E_x}{\partial x} dy \tag{3.28c}$$

$$H_z = -\frac{i}{\omega\mu}\left(\frac{\partial E_y}{\partial x} - \frac{\partial E_x}{\partial y}\right) \tag{3.28d}$$

Hence, the other field components inside (outside) the groove can be expressed in terms of the amplitude $G_m(A_n)$.

Now, matching the electric field boundary condition $E_{a,x} = E_{g,x}|_{y=t,-a/2<x<a/2}$ at the interface and taking the fact that $E_x = 0$ outside the window of $-a/2 < x < a/2$ on the interface, and also keeping the continuity condition of the tangential magnetic field, it can be shown that

$$\tan(Pt) = \frac{-P}{\kappa}\frac{a}{d}\text{sinc}^2(a\beta/2) \tag{3.29}$$

where we assumed the prevalence of $n = m = 0$ mode, and used the relation $\beta^2 - \kappa^2 + Q^2 = P^2 + Q^2 = \left(\frac{\omega n_g}{c}\right)^2$; n_g being the refractive index of the material that fills the groove. It can be seen that, irrespective of the metal thickness $2t$, Eq. (3.29) is satisfied for $P = 0$, which, in turn, leads to the prediction of $Q = \frac{\omega n_g}{c}$. Thus, assumption I predicts the existence of SSPP mode in structures with arbitrary thickness; however, it falls short in explaining the small shift in frequency with change of metal thickness.

3.5.3.2 Assumption II: TM Mode Prevalence

For a structure with thickness ($2t$), standing free in air environment, with the wave propagating along the groove's direction of periodicity (x direction), we can set $H_x = 0$ at the lower frequency regime inside the narrow groove for a PEC boundary. Then, the field components for TM mode inside the groove (E_g, H_g) and outside of it (E_a, H_a) can be written in terms of the geometric features (such as groove length, period, thickness, etc.) of spoof plasmon waveguides, assuming $e^{-i\omega t}$ time dependence.

Then, from the condition of tangential field (i.e., E_x and H_z) continuity at the interface ($y = t$), it can be shown that,

$$P\tan(Pt) = -\left(\frac{\omega}{c}\right)^2 \frac{a}{d}\text{sinc}^2(a\beta/2)\frac{\kappa}{\beta^2 - (\omega/c)^2} \tag{3.30}$$

As we can see, because the right-hand side (RHS) of Eq. (3.30) is a negative number for a guided mode, Pt must have to be greater than $\pi/2$ to make the left-hand side (LHS) negative as well. This sets a lower limit on the SSPP waveguide thickness in turn. This prediction is harmonious to the result of some of the earlier attempts, i.e., Martin-Cano et al. [17] for the size reduction of SSPP waveguide where a structure with a thickness comparable to its periodicity was considered to be the best possible solution. Interestingly, the LHS of Eq. (3.30) can also be negative for arbitrary thickness $2t$ if P assumes a small, purely imaginary value, i.e., $P = 0 + i\delta_P$, since the imaginary P yields $LHS = P\tan(Pt) \approx P^2 t = -\delta_P^2 t$ for small Pt. Then, using $(\omega/c)^2 = \frac{1}{n_g^2}(P^2 + Q^2) \approx \frac{Q^2}{n_g^2}$, we get:

$$(\delta_P)^2 t \approx \frac{1}{n_g^2}\frac{a}{d}\text{sinc}^2(a\beta/2)\frac{Q^2\sqrt{\beta^2 + Q^2(1 - \frac{1}{n_g^2})}}{\beta^2 - Q^2} \tag{3.31}$$

where $\tan h(x) \approx x$ is used, and we kept our interest limited to a thin groove. The corresponding frequency will be $\omega/c = \frac{1}{n_g}\sqrt{Q^2 - (\delta_P)^2} \approx \frac{1}{n_g}Q(1 - \frac{1}{2}\frac{\delta_P}{Q})$. So, the frequency detuning due to the finite size of SSPP waveguide standing free in air is $\Delta = -\frac{1}{2}\delta_P$.

For odd mode with $\sin(P_m y)$ variation of $E_{g,x}$ field, Eq. (3.30) should be replaced with the following

$$P\cot(Pt) = \left(\frac{\omega}{c}\right)^2 \frac{a}{d}\mathrm{sinc}^2(a\beta/2)\frac{\kappa_0}{\beta^2 - (\omega/c)^2}$$

To determine the value of the wave vector P, we need to know the phase factor along the groove length (i.e., wave vector Q) for a given propagation constant β. The relation between β and Q follows from the standard theory of SSPP mode [3, 9] in 1D-corrugated metal structure.

$$\sqrt{\beta^2 - (\omega/c)^2} = Q\frac{a}{d}\mathrm{sinc}^2(a\beta/2)\tan(Qh) \tag{3.32}$$

For a structure of thickness $2t$ bound with metallic walls on the lateral sides, the lowest order lateral wave vector would be $P_0 = \frac{\pi}{2t}$. This indicates that reduction of thickness $2t$ in such a structure would increase P, which would in turn cause an increase of frequency ω (i.e., blueshift) at a given β, as $P^2 + Q^2(\beta) = \left(\frac{\omega}{c}\right)^2$. On the contrary, SSPP structures whose lateral sides have an interface with dielectric show the opposite response, i.e., for a given β, the operating frequency demonstrates redshift with the reduction of its thickness, as the wave vector P takes an imaginary value following Eq. (3.31).

3.5.4 Effect of Substrate

Usually, SSPP waveguides are grown on top of a substrate such as silicon, which introduces asymmetry to the cross-section profile of the waveguide. Let us assume that the grooves are filled with dielectric constants $\epsilon_r = \epsilon_g$, and are fabricated on substrate with $\epsilon_r = \epsilon_s$.

Following the same strategy stated above, we obtain the following dispersion relation, taking into account the presence of a substrate:

$$\left(\cos(2Pt) + S^2\frac{\kappa_a}{k_0^2 - \beta^2}\sin(2Pt)\right)\left(\cos(2Pt) + \epsilon_r S^2\frac{\kappa_s}{k_s^2 - \beta^2}\sin(2Pt)\right) = 1 \tag{3.33}$$

where $S^2 = \frac{k_0^2}{P}\frac{a}{d}\mathrm{sinc}^2(a\beta/2)$, and $\epsilon_r = \frac{\epsilon_s}{\epsilon_0}$.

Also $\beta^2 + Q^2 - \kappa_a^2 = \frac{1}{n_g^2}(P^2 + Q^2) = (\omega/c)^2 = \frac{1}{n_s^2}(\beta^2 + Q^2 - \kappa_s^2)$.

3.5.5 Effective Refractive Index of a Planar SSPP

In an SSPP medium with infinite lateral dimension, the propagation constant inside the groove is, $Q = \frac{\omega}{c}$. From the above analysis, we can also determine an effective propagation constant for an SSPP medium with finite thickness as follows:

$$\begin{aligned} Q^2 &= \left(\frac{\omega}{c}n_g\right)^2 - P^2 \\ &= \left(\frac{\omega}{c}n_g\right)^2 + \left(\frac{\omega}{c}\right)^2 \frac{1+n_s^2}{2}\frac{a}{d}\mathrm{sinc}^2(a\beta/2)\frac{1}{\beta t} = \left(\frac{\omega}{c}n_{g,\mathrm{eff}}\right)^2 \end{aligned} \tag{3.34a}$$

Thus, we can state that the impact of reducing the thickness of the SSPP structure from an infinite value to some finite one is equivalent to filling up the interior of the groove of the SSPP waveguide (with infinite thickness) with some material of refractive index $n_{g,\mathrm{eff}}$.

$$n_{g,\mathrm{eff}}^2 = n_g^2 + \frac{1+n_s^2}{2}\frac{a}{d}\mathrm{sinc}^2(a\beta/2)\frac{1}{\beta t} \tag{3.34b}$$

The above analysis implies that the refractive index of the material lying outside a planar SSPP on its lateral side can redefine the spoof plasma frequency of the medium. Consequently, the lateral surface of a planar SSPP structure is the most sensitive interface among all. This knowledge will help us to design an efficient SSPP signal modulation scheme by placing an electro-optic material on the lateral SSPP interface. This will obviate the need to perform uniform deposition of material into the walls of narrow grooves of SSPP waveguides.

3.6 Thin-film of SSPP

3.6.1 The Concept of Effective Thickness

The single-mode approximation method, while providing a simple way to analyze the behavior of subwavelength mode in periodic SSPP structure, has one drawback. As we exclude the higher order modes in our calculation, ignoring the field fringing effect at the edges, the extrapolation from the analysis begins to depart from the exact results as the thickness of the metallic structure becomes comparable to the dimension of the surface emanating from the fringe field, refer to Figure 3.11. Roy et al. [5], the authors of this book, proposed a different strategy of finding an analytical solution for SSPP mode in an ultrathin waveguide. We introduce a simple yet powerful concept of the effective thickness of the SSPP groove in order to incorporate the effect of field fringing. Figure 3.9c gives a visualization

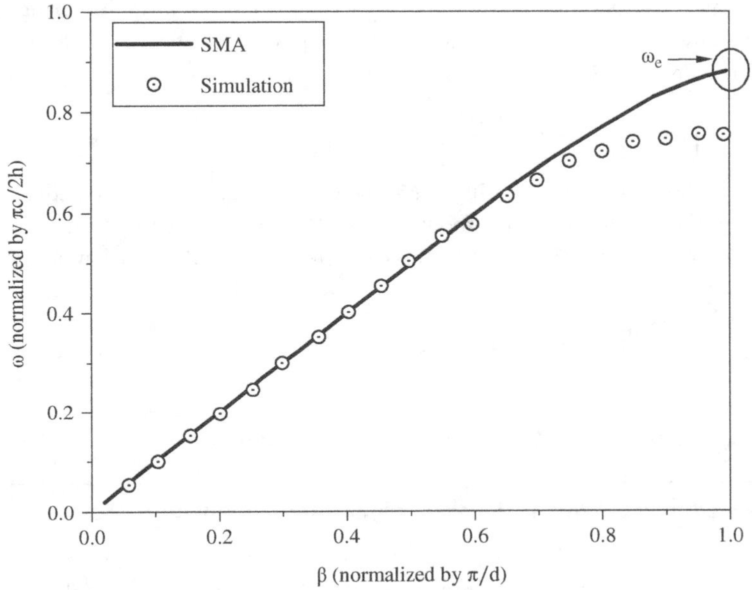

Figure 3.11 Dispersion relation in SSPP structure with infinite width. Solid lines are generated from analytical results with single-mode approximation, while the dots are from 2D simulation in COMSOL Multiphysics.

of the idea of effective thickness and corresponding simulation. The idea of defining the effective thickness of SSPP is owing to the observation that the grooves of SSPP structures essentially behave as capacitors. The capacitance per unit length can be estimated as the following:

$$C_g = \frac{Q_{g,s}}{V_g} = \epsilon_0 \frac{2tE_{g,x}}{aE_{g,x}} = \epsilon_0 \frac{2t}{a}$$

where the surface charge per unit length $Q_{g,s}$ is estimated from the boundary condition $\epsilon_0 E_g.\hat{x} = \rho_s = \frac{Q_{g,s}}{2t}$.

As we know, the electric field can be expressed as a contribution from both a scalar potential and a magnetic vector potential, i.e., $E_{g,x} = -\nabla_x \phi - i\omega A_x = -\frac{\partial \phi}{\partial x} - i\omega A_x$. In our case, $\nabla_x \phi$ is on the order of $\frac{1}{a}$, while the normalized frequency $Q \sim \omega/c$ that we are interested in is much less than π/a. This is why $|\frac{\omega A_x}{\nabla_x \phi}| << 1$, and we can approximate the state inside the groove as a quasi-static field, where $E_{g,x} \approx -\nabla_x \phi$.

A slightly better formula for parallel plate capacitance incorporating field fringing effect up to the limit of $t > a$, provided by Parker [18, 19] enables us to define the effective thickness of SSPP groove as:

$$t_{\text{eff}}|_{t \geq a} = t\left[1 + \frac{a}{2\pi t}\left(1 + ln\left(\frac{4\pi t}{a}\right)\right)\right]$$

We capitalize the very idea of finding the static capacitance between the edges of neighboring metal strips forming grooves of SSPP waveguide (see Section 3.6.2 for details), which yields an effective thickness for SSPP waveguide with vanishing thickness (i.e., $t \to 0$), as:

$$t_{\text{eff}}|_{t \to 0} = t_{TF} = \frac{a\pi}{4}\frac{1}{\ln\left(\frac{w+a+\sqrt{a^2+2wa}}{w}\right)} \tag{3.35}$$

Here, $w = (d - a)/2$, and d is the periodicity of the structure.

3.6.2 Capacitance Between Edges of Neighboring Conductor Plates

A metal strip of width w with surface charge density σ can be thought of as consisting of many infinitely long thin wires, which have a known standard expression for an electric field. The x component of the electric field for the metal strip is calculated as:

$$\begin{aligned} E_{x1} &= \frac{1}{2\pi\epsilon_0}\int_0^w \frac{\sigma}{\sqrt{1-(x/w)^2}}\frac{1}{w+r-x}dx \\ &= \frac{\sigma}{2\pi\epsilon_0} w \frac{\pi - \cos^{-1}\left(\frac{w}{w+r}\right)}{\sqrt{r(r+2w)}} \approx \frac{\sigma}{2\pi\epsilon_0}\frac{w\pi}{\sqrt{r^2+2wr}} \end{aligned}$$

The x component of the electric field E_{x2} produced by the neighboring strip that carries an equal amount of (opposite polarity) charge density σ is also found by replacing r by $(a - r)$ in the above expression. Since E field is conservative, we can find the potential difference V_g between the two strips by integrating $(E_{x1} - E_{x2})$ over a convenient path.

$$\begin{aligned} V_g|_{t\to 0} &= \int_0^a (E_{x1} - E_{x2})dr \\ &= \frac{\sigma w\pi}{2\pi\epsilon_0}\int_0^a \left[\frac{1}{\sqrt{r^2+2wr}} + \frac{1}{\sqrt{(a-r)^2+2w(a-r)}}\right]dr \\ &= \frac{\sigma w}{\epsilon_0}\ln\left(\frac{w+a+\sqrt{a^2+2wa}}{w}\right) \end{aligned}$$

From this result, we can obtain the effective capacitance per unit length $C_g = \frac{1}{V_g}\int_0^w \frac{\sigma}{\sqrt{1-(x/w)^2}}dx = \frac{\pi w\sigma}{2V_g}$. Comparing the result with the standard formula for parallel plate capacitance $C = \epsilon_0 \frac{2t_{TF}}{a}$, we obtain the effective thickness of a thin-film SSPP structure as

$$t_{TF}|_{t\to 0} = \frac{a\pi}{4}\frac{1}{\ln\left(\frac{w+a+\sqrt{a^2+2wa}}{w}\right)}$$

Here, $w = (d - a)/2$, and d is the period of the structure.

3.6.3 Effect of a Substrate on Thin SSPP

In the case of SSPP of nominal thickness on a substrate, we will first find the gross refractive index that an infinitely long and thin conducting strip will experience close to the interface of substrate by applying image theory for dielectric [20] as $n_{\text{strip}} = \frac{\sqrt{n_s^2+1}}{\sqrt{2}}$. The newly defined renormalized refractive index for the hypothetical uniform space around a thin conductor strip n_{strip} basically takes the effect of the substrate into account. In contrast, the effective refractive index $n_{g,\text{eff}}$ of the SSPP structure takes care of phase variation inside a patterned metal, described by Eq. (3.34b). In particular, we see a stronger impact of the environment on thin-film SSPP reflected through its effective index $n_{g,TF_{\text{eff}}}$.

$$n^2_{g,TF_{\text{eff}}} = \frac{n_s^2+1}{2}\left(1 + \frac{a}{d}\text{sinc}^2(a\beta/2)\frac{1}{t_{TF}}\frac{1}{\beta}\right) \tag{3.36}$$

A visual comparison of the band-edge frequencies $\omega_e = \omega|_{\beta=\pi/d}$ (marked with a circle in Figure 3.12a,b) normalized with respect to the quarter-wavelength frequency (corresponding to $\frac{\pi}{4}$ phase change along the groove length) for SSPP waveguides of different thicknesses is given in Figure 3.12a. It shows that, for an SSPP waveguide with geometric parameters taken as $h = d \approx 10a$, the modification of the band due to the change of waveguide thickness is nominal, the maximum detuning of ω_e for thin-film SSPP waveguide being not more than 20% (i.e., $\frac{\frac{\pi}{2}-\frac{\omega_e h}{c}}{\frac{\pi}{2}} \approx 0.2$). On the other hand, as depicted in Figure 3.12, the impact of the substrate on detuning of the band-edge frequency of a thin-film SSPP waveguide from quarter-wavelength frequency is remarkable, reaching as much as 70% for silicon substrate, whose refractive index is ≈ 3.5 at 1 THz. This spectacular change in the operating frequency of planar SSPP structure owes to the fact of

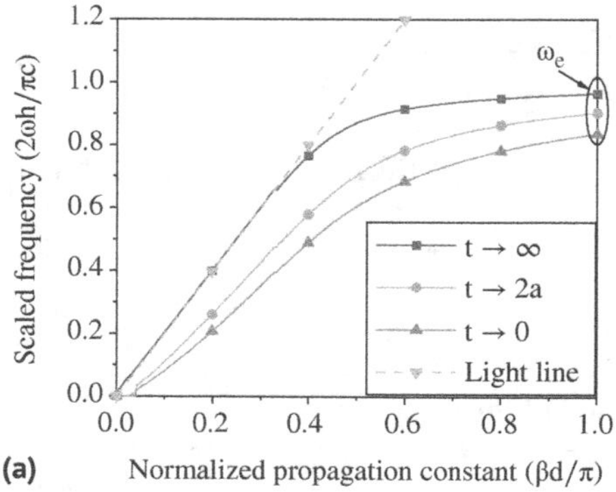

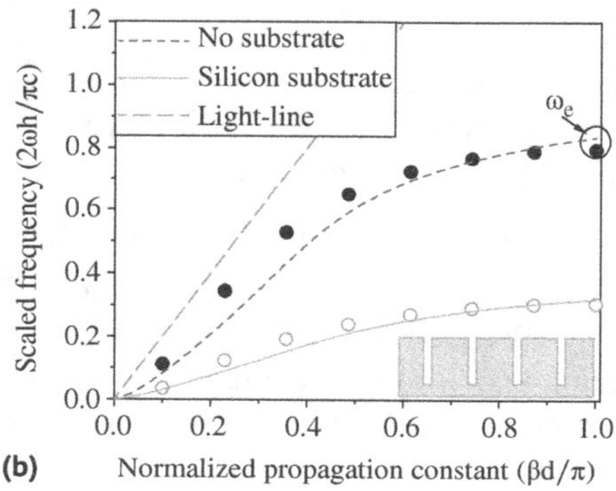

Figure 3.12 (a) Comparison between dispersion diagram of three different SSPP structure: 3D SSPP (with large thickness), planar 2D SSPP with finite thickness ($t = 2a$), and thin-film SSPP (with $t \to 0$). The other geometric parameters are: $h = d \approx 10a$. (b) Comparison between dispersion diagram of thin-film SSPP structure ($t \to 0$) for two different cases: waveguide placed in air, and placed on a silicon substrate. The other geometric parameters are: $h = d \approx 10a$. The solid lines are from the developed theoretical model in this paper, while the discrete dots (•) and (○) are obtained by FDTD numerical simulation in COMSOL Multiphysics, Version 5.2; Comsol, Inc.

modified plasma frequency $\omega_p = \frac{\pi}{2n_{g,\text{eff}}h}$, where $n_{g,\text{eff}}$ is described by Eq. (3.34b) for SSPP with thickness $t > t_{TF}$ and by Eq. (3.36) for $t \leq t_{TF}$. Figure 3.13b shows the simulation result by COMSOL Multiphysics of the profile of an electric field component (E_x) taken along a line passing through an m thick SSPP unit cell. While the evanescent field profile exterior to the groove appears as expected, we emphasize on the nature of field interior of the groove in particular, as it shows a peculiar profile of hyperbolic spatial distribution with its center in the middle of the groove.

This profile justifies our description of the interior E_x field by the function $\cos(i\delta_P t)$, δ_P being a small positive number with unit of wave vector.

3.7 Properties of Confined Modes

3.7.1 Degree of Confinement

While the dispersion relation of spoof plasmon remains approximately the same in both thin and wide structures (up to the small frequency shift), the field confinement along two different directions on the plane transverse to the direction of propagation is subject to different degrees of localization at a given frequency. Figure 3.13a gives a visual illustration of the profile of confined E_x field in a unit cell of SSPP waveguide along two different transverse directions, namely lateral and vertical. In order to quantify the degree of localization in these two directions, let us attribute the following condition on the transverse wave vector κ of evanescent mode exterior to the groove for distinguishing the spectrum of modal confinement: $\kappa \geq \frac{s\omega}{c}$, where $s > 0$ is a dimensionless parameter specifying the threshold degree of modal confinement in the system of interest. Then, using the relation between β and ω, and leveraging the fact that $\beta \approx \kappa$ on the lateral side of a planar SSPP structure, we arrive at the following inequality condition for the spectra of modal confinement along the lateral direction.

$$s^2 \leq 1 + \left(\frac{a}{d}\right)^2 \tan^2\left(\frac{\omega}{h}c\right) \tag{3.37}$$

As the RHS of the inequality, Eq. (3.37) is always greater than 1, the inequality for modal confinement is always satisfied for all frequencies, as long as $s < 1$ suffices for the system, and we will have modal confinement for all possible frequencies up to the spoof plasma resonance, i.e., the entire band. For an arbitrary positive value of s, the fraction of the band over which we will achieve threshold confinement along lateral direction can be shown to be

$$\Delta_{\text{c,lat}} = \frac{\omega_p - \omega_{\min} hc}{\omega_p} = 1 - \frac{2}{\pi} \tan^{-1}\left(\frac{d}{a}\sqrt{(s^2 - 1)u_{s-1}}\right) \tag{3.38}$$

where u_{s-1} is a unit step function of s with value equal to 1 for $s > 1$, ω_p is the normalized spoof plasma frequency of the meta-structure, and $\omega_{\min}$ is the frequency corresponding to the threshold degree of confinement.

On the contrary, for vertical confinement along Z direction, the bandwidth over which the confinement can exceed the threshold value can be shown to be

$$\Delta_{\text{c,vert}} = 1 - \frac{2}{\pi} \tan^{-1}\left(\frac{d}{a}s\right) \tag{3.39}$$

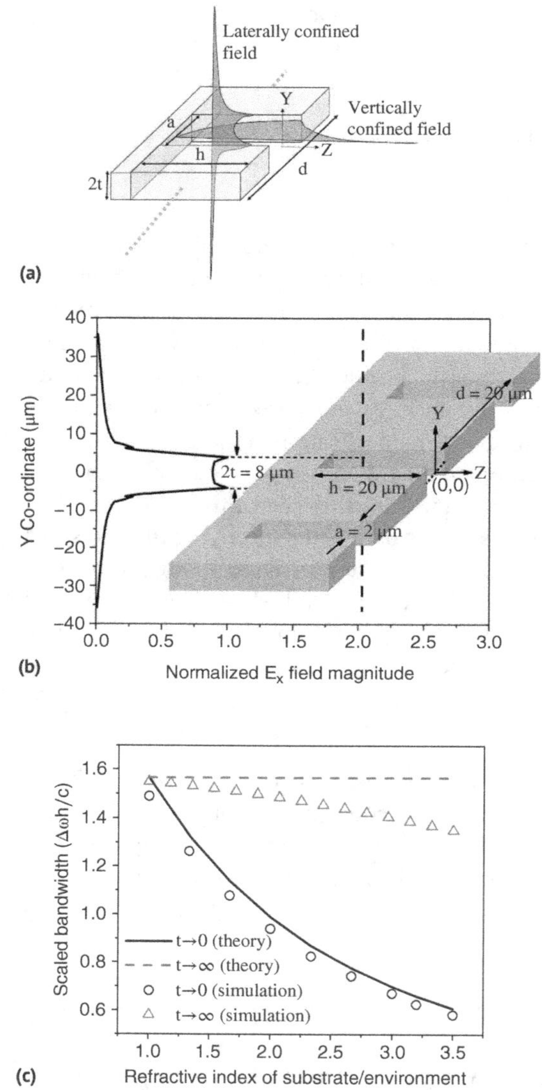

Figure 3.13 (a) Illustration of confined E_x field profile in 2D transverse *YZ* plane for a unit cell of SSPP waveguide, while the mode propagates along *X* direction. Field profile along *Y* axis is labeled as laterally confined field, and that along *Z* direction as vertically confined field. (b) Simulation result of the profile of electric field component (E_x) taken along a line crossing through an SSPP unit cell of 8 μm thickness at a frequency $\omega_p/2$, where $\omega_p = \frac{\pi c}{2h}$. (c) Change of bandwidth of a thin-film SSPP structure and an infinitely thick SSPP structure with the change of the refractive index of the substrate/external environment. Groove length $h = 10d$, and groove width $a = \frac{d}{10}$. Solid lines are obtained by our developed theory, while the discrete circles (*o*) and triangles (Δ) are obtained via numerical simulation in COMSOL Multiphysics [21].

Comparing Eqs. (3.38 and 3.39), by taking, for instance, $s \to 1$, we can see that the fraction of the band of modal confinement $\Delta_{c,\text{vert}}$ along vertical (Z) direction is much smaller than unity for SSPP structures with narrow grooves. Whereas, for the same value of s, $\Delta_{c,\text{lat}}$ approaches unity, which means the degree of confinement along the lateral (Y) direction, can exceed the threshold limit almost over the entire band of thin SSPP waveguide, irrespective of the value of a/d. For $s >> 1$, i.e., for stringent requirement of deep subwavelength confinement, the two equations become similar, as expected.

3.7.2 Bandwidth Modulation of Confined Mode

In order to find the bandwidth (BW_{sp}) of SSPP waveguide of the fundamental band, it suffices to know the spectral position of the band-edge frequency $\omega(\beta_{\max}) = \omega_e$, as $\text{BW}_{sp} = \frac{1}{2\pi}\omega_e$. Band-edge frequency also indicates the frequency region where the sub-diffraction characteristics of SSPP mode are the strongest. For 3D SSPP structure, with groove length of h, refractive index of n_g, and a dielectric half-space of index n_d residing on top of it, we can estimate the difference between the plasma frequency (asymptotic frequency) and the band-edge frequency. Let δ_ω be the deviation of the band-edge frequency from normalized plasma frequency $\tilde{\omega}_p = \frac{\omega_p}{c} n_g h = \frac{\pi}{2}$. We shall let $\tilde{\omega}_e = \frac{\omega_e}{c} h = (\frac{\pi}{2} - \delta_\omega)\frac{1}{n_g}$ to be the normalized band-edge frequency for SSPP mode.

The sensitivity of the band-edge frequency on dielectric half-space can be found as follows:

$$\left.\frac{\partial \tilde{\omega}_e}{\partial n_d}\right|_{\text{3D-SSPP}} = -\frac{n_d}{8n_g^2}\frac{a}{h}\left(\frac{d}{h}\right)^2 \text{sinc}^2\left(\frac{\pi a}{2d}\right) \tag{3.40}$$

The approximation is valid for SSPP structures with groove length $h >> d/2$.

On the other hand, for SSPP mode in planar structure, the sensitivity of $\tilde{\omega}_e$ on the refractive index of the substrate can be shown to be

$$\left.\frac{\partial \tilde{\omega}_e}{\partial n_s}\right|_{\text{2D-SSPP}} = -\frac{\pi}{2n_{\text{g,eff}}^2}\frac{\partial n_{\text{g,eff}}}{\partial n_s} \tag{3.41a}$$

where

$$\frac{\partial n_{\text{g,eff}}}{\partial n_s} = n_s \frac{a}{d}\text{sinc}^2(a\beta/2)\frac{1}{\beta t}\frac{1}{2n_{\text{g,eff}}} \tag{3.41b}$$

For thin-film SSPP structures, the sensitivity of band-edge frequency on the substrate index is even more pronounced due to the following term:

$$\frac{\partial n_{g,TF_{\text{eff}}}}{\partial n_s} = n_s\left(1 + \frac{a}{d}\,\text{sinc}^2(a\beta/2)\frac{1}{\beta t_{TF}}\right)\frac{1}{2n_{g,TF_{\text{eff}}}} \tag{3.42}$$

It is easy to see from Eq. (3.40) that, while in the limit of $\frac{a}{h} \to 0$ for a given groove width a, the dependence of band-edge frequency of a 3D SSPP structure over the dielectric half-space simply vanishes. The dependency of band-edge frequency of a planar SSPP structure (i.e., structure with small thickness) over substrate index is given by Eq. (3.41a).

$$\left|\frac{\partial \tilde{\omega}_e}{\partial n_s}\right|_{2\text{D-SSPP},a/h\to 0} \approx n_s \frac{1}{n^3_{g,\text{eff}}} \frac{a}{4t} \operatorname{sinc}^2\left(\frac{\pi a}{2d}\right)$$

where $n_{g,\text{eff}}$ is always greater than n_g. We see that, as $t \to \infty$, the band-edge frequency ceases to vary with the refractive index of the substrate.

The sensitivity of the band-edge frequency on the optical properties of the environment can be more significant for the planar SSPP structure than for its 3D counterpart. In the case of strongly confined SSPP mode, which requires $d << h$, the band edge of the 3D SSPP structure will have scant regard to its environment, while in planar structure the sensitivity of band edge can remain significant, independent of the ratio of d/h.

For thin-film SSPP, the dependency of band-edge frequency over substrate index is the strongest of all, since the groove mode expands into the substrate itself. Our analysis suggests the sensitivity of band-edge frequency of thin-film SSPP to be, for $n_s >> 1$

$$\left|\frac{\partial \tilde{\omega}_e}{\partial n_s}\right|_{TF,a/h\to 0} \approx n_s \frac{\pi}{4} \frac{1}{n^3_{g,TF_{\text{eff}}}}$$

Figure 3.13c illustrates a comparative view of the change in SSPP bandwidth with the change in the refractive index of the external environment, for a 2D and 3D SSPP structure. The groove length h is taken five times the periodicity d, so as to single out the sole contribution of the change of plasma frequency ω_p in governing the change of bandwidth. For 3D structure, the change of bandwidth is nominal, as its plasma frequency is independent of the condition of the environment. However, for planar thin-film SSPP structure, the change of bandwidth is tremendous, since the plasma frequency in this case is indirectly dictated by the environment.

3.8 Summary of the Chapter

The chapter presents an exploration into the electrodynamic analysis of SPPs and their predecessor, SSPPs, crucial for accelerating ultrafast data transfer within VLSI chip scales. With a focus on the Mazumder Laboratory's decade-long theoretical advancements, encompassing a spectrum of plasmonic and spoof plasmonic structures, the chapter sheds light on the evolution of SPPs to SSPPs. The work extends to the development of a simulation tool, incorporating plasmonic waveguides with conventional circuit components via spice simulations. Engineered metal patterns serve to reduce signal radiation loss and increase modal confinement. The theoretical framework explores interactions between spoof plasmonic modes and the material.

The chapter discusses a theoretical framework elucidating the properties of spoof surface plasmon modes in corrugated conductors. This theory incorporates the interplay of adjustable geometric parameters, coupling the SSPP structure with the substrate environment. Noteworthy new parameters, such as the effective refractive index and thickness of SSPP waveguides, streamline analyses within advanced networks. A quantitative derivation of field profiles successfully explains the propagation mode for a wide range of geometry of Spoof plasmon waveguides, subsequently validated through simulations. The chapter also discusses the utilization of the lateral surface of SSPP waveguides, enabling expansive modal confinement and heightened sensitivity. Ultimately, this collective effort lays the foundation for subwavelength engineering techniques and applications rooted in spoof plasmonics, with the potential to reshape electromagnetic wave propagation in modern VLSI chip technology.

References

1 Kyungjun Song and Pinaki Mazumder. An equivalent circuit modeling of an equispaced metallic nanoparticles (MNPs) plasmon wire. *IEEE Transactions on Nanotechnology*, 8(3):412–418, 2008.

2 Kyungjun Song and Pinaki Mazumder. Equivalent-circuit modeling of nonradiative surface plasmon energy transfer along the metallic nanowire. *IEEE Transactions on Nanotechnology*, 10(1):111–120, 2009b.

3 Mikhail Erementchouk, Soumitra Roy Joy, and Pinaki Mazumder. Electrodynamics of spoof plasmons in periodically corrugated waveguides. *Proceedings of the Royal Society A*, 472(2195):20160616, 2016. https://doi.org/10.1098/rspa.2016.0616.

4 Hao Chi Zhang, Tie Jun Cui, Qian Zhang, Yifeng Fan, and Xiaojian Fu. Breaking the challenge of signal integrity using time-domain spoof surface plasmon polaritons. *ACS Photonics*, 2(9):1333–1340, 2015.

5 Soumitra Roy Joy, Yu Hao, and Mazumder Pinaki. Properties of spoof plasmon in thin structures. *Proceedings of the Royal Society A: Mathematical, Physical and Engineering Sciences*, 2018.474(2220):20180205, https://doi.org/10.1098/rspa.2018.0205.

6 Xiaopeng Shen, Tie Jun Cui, Diego Martin-Cano, and Francisco J Garcia-Vidal. Conformal surface plasmons propagating on ultrathin and flexible films.

Proceedings of the National Academy of Sciences, 110(1):40–45, 2013. https://doi.org/10.1073/pnas.1210417110.

7 Mikhail A Kats, David Woolf, Romain Blanchard, Nanfang Yu, and Federico Capasso. Spoof plasmon analogue of metal-insulator-metal waveguides. *Optics Express*, 19(16):14860–14870, 2011. https://doi.org/10.1364/OE.19.014860.

8 David Woolf, Mikhail A Kats, and Federico Capasso. Spoof surface plasmon waveguide forces. *Optics Letters*, 39(3):517–520, 2014. https://doi.org/10.1364/OL.39.000517.

9 Kyungjun Song and Pinaki Mazumder. Active terahertz spoof surface plasmon polariton switch comprising the perfect conductor metamaterial. *IEEE Transactions on Electron Devices*, 56(11):2792–2799, 2009a. https://doi.org/10.1109/TED.2009.2030838.

10 Zhao Xu, Kyungjun Song, and Pinaki Mazumder. Analysis of doubly corrugated spoof surface plasmon polariton (DC-SSPP) structure with sub-wavelength transmission at THz frequencies. *IEEE Transactions on Terahertz Science and Technology*, 2(3):345–354, 2012. https://doi.org/10.1109/TTHZ.2012.2188558.

11 Chongchong Chen. A new kind of spoof surface plasmon polaritons structure with periodic loading of T-shape grooves. *AIP Advances*, 6(10):105003, 2016. https://doi.org/10.1063/1.4964619.

12 Xiaoyong Liu, Yijun Feng, Bo Zhu, Junming Zhao, and Tian Jiang. High-order modes of spoof surface plasmonic wave transmission on thin metal film structure. *Optics Express*, 21(25):31155–31165, 2013. https://doi.org/10.1364/OE.21.031155.

13 Jie Yang, Maoxiong Zhao, Liang Liu, Hong Xiang, and Dezhuan Han. Analysis of the symmetric and anti-symmetric modes in spoof–insulator–spoof waveguides. *Journal of the Physical Society of Japan*, 86(6):064401, 2017. https://doi.org/10.7566/JPSJ.86.064401.

14 Xiang Wan and Tie Jun Cui. Guiding spoof surface plasmon polaritons by infinitely thin grooved metal strip. *AIP Advances*, 4(4):047137, 2014. https://doi.org/10.1063/1.4874307.

15 Shiwen Luo, Jun Zhao, Duluo Zuo, and Xinbing Wang. Perfect narrow band absorber for sensing applications. *Optics Express*, 24(9):9288–9294, 2016. https://doi.org/10.1364/OE.24.009288.

16 Jingjing Liu, Tengfei Wang, and Fuqiang Nie. Narrow band resonance in the UV light region of a plasmonic nanotextured surface used as a refractive index sensor. *RSC Advances*, 7(57):35957–35961, 2017. https://doi.org/10.1039/C7RA01642E.

17 Diego Martin-Cano, Maxim L Nesterov, Antonio I Fernandez-Dominguez, Francisco J Garcia-Vidal, Luis Martin-Moreno, and Esteban Moreno. Plasmons for subwavelength terahertz circuitry. *Optics Express*, 18(2):754–764, 2010. https://doi.org/10.1364/OE.18.000754.

18 George W Parker. What is the capacitance of parallel plates? *Computers in Physics*, 5(5):534–540, 1991.

19 George W Parker. Electric field outside a parallel plate capacitor. *American Journal of Physics*, 70(5):502–507, 2002. https://doi.org/10.1119/1.1463738.

20 Ismo V Lindell, Keijo I Nikoskinen, and Ari Viljanen. Electrostatic image method for the anisotropic half space. *IEE Proceedings - Science, Measurement and Technology*, 1997.144(4):156–162,https://doi.org/10.1049/ip-smt:19971317.

21 COMSOL Multiphysics. Comsol multiphysics reference manual, version 5.2a. 2016. www.comsol.com.

4

Information Capacity of Spoof Plasmonic Interconnect

4.1 The 1858 Transatlantic Telegraph: Lessons from a Failed Project

Understanding the intricacies of information transmission, particularly in the face of natural impediments like noise, signal distortion, and attenuation, required considerable time and a series of innovations. The ill-fated 1858 Transatlantic data transfer project serves as a stark reminder of the challenges encountered before we comprehended the nuances of information transmission.

This historic project marked the first recorded attempt at communication across the vast 2,000-mile expanse of the Atlantic Ocean, spanning two continents, depicted in Figure 4.1. This system was founded upon Samuel Morse and his renowned code of dots and dashes, which had already gained widespread usage in many nations, including the United Kingdom and the United States. To achieve this feat of connecting the "old" world with the "new" one, a staggering 2,500 miles of wire were needed underwater. After enduring multiple unsuccessful attempts to lay the cable, efforts that nearly led to the financial collapse of Cyrus Field's company, the ships Niagara and Agamemnon embarked on their historic journey on 10 June 1858. At the helm of this ambitious endeavor was "Wildman Whitehouse," a surgeon by profession and an electrical enthusiast by passion. His role as Electrician to the Atlantic Telegraph Company in 1856 entrusted him with the critical tasks of testing the 1857/1858 cables and developing the equipment for transmitting telegraph signals between Ireland and Newfoundland.

However, despite the expertise of individuals like William Thompson, a renowned authority in wired telephony of that era, who advocated for the use of high-quality copper wire with a substantial cross-section and highly sensitive receivers to combat signal attenuation and distortion, Whitehouse took a different approach. He adamantly insisted on augmenting the signal's power at the

Artificial Plasmonics for VLSI Interconnects: Bridging the Gap between Electronics and Optics, First Edition. Soumitra R. Joy and Pinaki Mazumder.

(a) Mike Peel / Wikimedia Commons / Public domain

(b) Frederick (attr. to) Barnard / Bridgeman Images

Figure 4.1 (a) The installation of trans-atlantic cable in Agamemnon ship in 1858. (b) Wildman Whitehouse, a retired physician and an amateur electrician of the Agamemnon voyage. Many historians are of the opinion that Whitehouse's idea of brute-forcing the signal into the cable caused the failure of the system within a month of its installation.

transmitting end, employing what could be described as a "brute-force" method. Regrettably, this decision led to the generation of a 2,000-V signal by the induction coil, ultimately resulting in insulation breakdown and irrevocable damage to the underwater cable.

This grandiose project, as ambitious as it may have sounded, met a tragic demise, serving as a stark reminder that we, as humans, had not yet amassed the knowledge required to accomplish such endeavors.

4.2 Data Transfer Through Noisy Channel: Condition the Signal, Don't "Brute-Force"

Transmitting signals underwater, particularly in saline conditions, poses significant challenges due to heightened signal attenuation and leakage in electrically conductive saltwater. To compound matters, increasing the power of the transmitted signal also escalates thermal noise, adversely affecting message quality. It would take nearly a century of research, spearheaded by pioneers like Harry Nyquist, Ralph Hartley, and Claude Shannon, to systematically unravel the principles of information transmission and the prerequisites for achieving error-free communication over noisy channels.

While the initial groundwork for relating thermodynamic entropy to information entropy had been laid by Hartley, Shannon (refer to Figure 4.2) played a pivotal role in highlighting that not all bits of information carry

The Regents of the University of Michigan

Figure 4.2 The sculpture of Claude Shannon in the University of Michigan, Ann Arbor, where he was an alumnus (B.S.E., 1936). For many researchers, including the authors of the book, Shannon has remained an iconic source of inspiration, driving their exploration and expansion of his concept of information.

equal probabilities. He demonstrated the immense potential in leveraging this nonuniform distribution of probability by introducing data redundancy into the information transmission process over noisy channels, ensuring reliable error-free retrieval. Shannon's information theory underscores the strategic use of redundancy within the intended information to combat channel noise effectively.

In contemporary interconnect technologies, redundancy is strategically deployed to bolster signal resilience against noise and eliminate improbable bit configurations. It is widely acknowledged that as we move toward higher frequencies and bandwidths, interference-induced noise escalates, necessitating additional error correction mechanisms.

Traditionally, the communication channel itself has been considered a passive element, playing no active role in error correction or noise mitigation. However, in the chapters ahead, the authors of this book will demonstrate that the seemingly inert nature of the channel in addressing noise in very large scale integration (VLSI) chip communication can be overcome through the introduction of a disruptive technology based on spoof plasmonics.

4.3 Challenges in Millimeter-scale Communication: A Call for Innovation Beyond Shannon's Paradigm

The groundbreaking work of Nyquist, Hartley, and Shannon paved the way for addressing the data transfer conundrum by introducing redundancy into the message. The question is whether this approach can serve as a universal remedy for all communication challenges. Can the concept of signal conditioning, encompassing source and channel coding, effectively extend to extreme scenarios? These scenarios may involve the need for exceptionally high data density, environments rife with crosstalk, thermal noise, and interference, and the stark constraint of mere millimeters of space available between transmitter and receiver chips, leaving little room for conventional design innovations to ensure reliable and high-speed communication.

In such extreme contexts, it becomes evident that the standard signal conditioning solutions may not be feasible due to their substantial power consumption and area footprint overhead. Consider, for instance, VLSI communication within a multichip module. Data must traverse between chiplets with minimal latency, yet there is limited design area on the substrate to accommodate separate sophisticated circuits for error correction and signal amplification. The challenge with the conventional approach of signal compression and data redundancy lies in the fact that the chip responsible for data compression or source coding would itself occupy a substantial area, introducing additional latency. For example, a data compressor/decompressor chip for a power quality monitoring system necessitates a die size of and houses 400,000 equivalent gates (2-input NAND gates).

Hence, there arises a compelling need for an entirely novel and innovative solution, one rooted in a fundamentally different science of signal propagation. What if we could empower a channel to inherently safeguard itself as noise levels rise, without the need for additional power-hungry circuitry? Is this even theoretically feasible?

Before delving into the quest for an answer to this question, and to gain a deeper understanding of the problem at hand, it is essential to explore why millimeter-scale information transfer is increasingly pertinent in modern semiconductor technology. In the ensuing sections, we will further explore the importance of this question and investigate potential approaches to tackle the challenges posed by millimeter-scale communication.

4.4 Millimeters-scale Communication: Its Growing Relevance in Data-driven World

Over the last half-century, semiconductor technology has been driven by Moore's Law, which forecasted an exponential increase in transistor density. However, we are now approaching the fundamental limits of transistor scaling. Notably, not all semiconductor components scale equally. For instance, input/output (I/O) components, which may include analog circuitry, tend to scale at about half the rate of logic components. Even with the recent transition to TSMC's 3 nm technology, SRAM cell sizes hardly scale at all. It is evident that we must explore new avenues for scaling technology beyond the traditional notion of transistor scaling.

Interestingly, Gordon Moore himself alluded to an alternative path in the same seminal paper that gave birth to Moore's Law. He suggested that it might be more cost-effective to construct large systems from smaller, separately packaged and interconnected functions.

As semiconductor technology advances beyond the era of Moore's Law, we may have to relinquish the idea of relying solely on transistor scaling and monolithic integration. Instead, we should embrace the concept of a complex architecture composed of chiplets through heterogeneous integration. This shift is necessitated by several factors. Transistor scaling is approaching the molecular limit, resulting in increased wafer production costs due to complex lithography and additional processing steps. Furthermore, enlarging chip sizes beyond a few centimeters raises the likelihood of critical defects, significantly impacting device yield. Given these constraints, it becomes increasingly clear that, to keep pace with Moore's Law, we should adopt multichip module technology through heterogeneous integration. This approach involves designing various chips customized for specific functions (such as power management, memory, processors, and I/O), each capitalizing on different process nodes and material systems suited to their respective roles. Figure 4.3 illustrates how the adoption of multiple chiplets can significantly increase yield compared to monolithic integration, assuming a 0.1 defect per square centimeter scenario. Notably, a 360 mm^2 monolithic die yields only 15%, while a 4-chiplet design (each 99 mm^2) more than doubles the yield to 37%.

With the adoption of chiplet-based architecture, high-performance packaging with millimeter-scale data transfer has become a critical area of research and investment for leading industries. Data center networking, high-performance computing, and autonomous vehicles are driving the adoption of high-end performance packaging, further evolving this technology. Presently, the trend is to harness larger computing resources at the cloud, edge computing, and device levels. Consequently, a growing demand is fueling the adoption of high-end performance packaging.

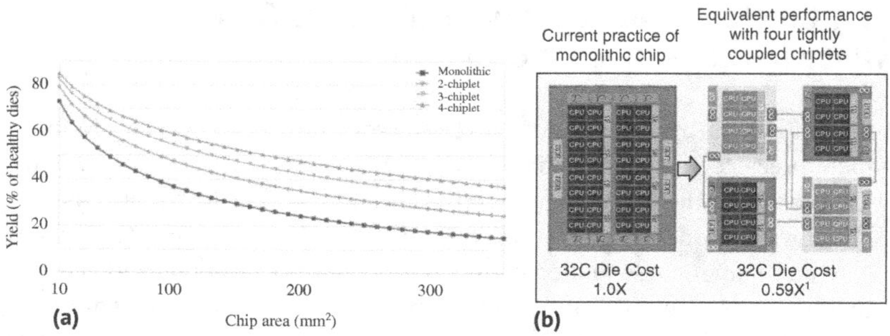

Figure 4.3 (a) Yield improvement brought by multichip design strategy over monolithic growth of die as size increases; and (b) the shift of semiconductor chip design principle from monolithic growth to chiplet design coupled with millimeter scale interconnects for more economic manufacturing.

4.5 A Growing Industry Investment in Millimeter-scale Chip Packaging

The High-end Performance Packaging market is poised for significant growth, with revenue projected to soar to 7.87 billion by 2027, a substantial increase from the 2.74 billion recorded in 2021. In the same year, top industry players made substantial capital expenditures totaling approximately 11.6 billion for packaging activities, underscoring their awareness of its pivotal role in combating the deceleration of Moore's Law. Figure 4.4 illustrates the significant efforts and substantial billion-dollar investments made by leading semiconductor companies in recent years to achieve high-performance packaging with millimeter-scale length of interconnects.

At the forefront of these investments is Intel, the leading investor, allocating 3.5 billion. Intel's three-dimensional (3D) chip stacking technology, known as Foveros, involves stacking a die on an active silicon interposer. Notably, Intel's embedded multi-die interconnect bridge (EMIB) technology offers a 2.5D packaging solution with a 55-micron bump pitch. The synergy of Foveros and EMIB has given rise to Co-EMIB, a key component used in the Ponte Vecchio GPU. Furthermore, Intel has strategic plans to adopt hybrid bonding technology for Foveros Direct.

TSMC, another significant contributor in the field, is committed to investments with a 3.05 billion Capital expenditure. While expanding its business with Ultra High-density Fan-out (UHD FO) solutions, TSMC is actively shaping new system-level roadmaps and technologies for 3D system-on-chip (SoC) integration. TSMC's CoWoS platform offers diverse solutions, including redistribution layers (RDL) and silicon interposers, while its large-scale integration platform competes directly

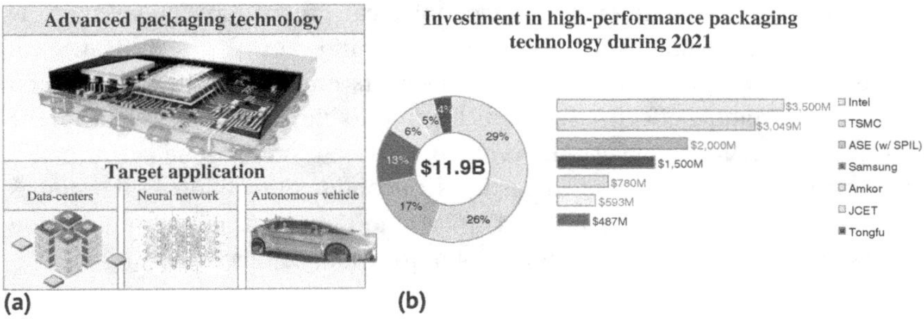

Figure 4.4 (a) Advanced packaging technology is getting significant interest on the context of growing demand in high-speed data centers, AI-driven chips, and autonomous vehicle that requires beyond the Moore's law innovation; and (b) the share of investment in high-performance packaging technology by the leading semiconductor industries during 2021.

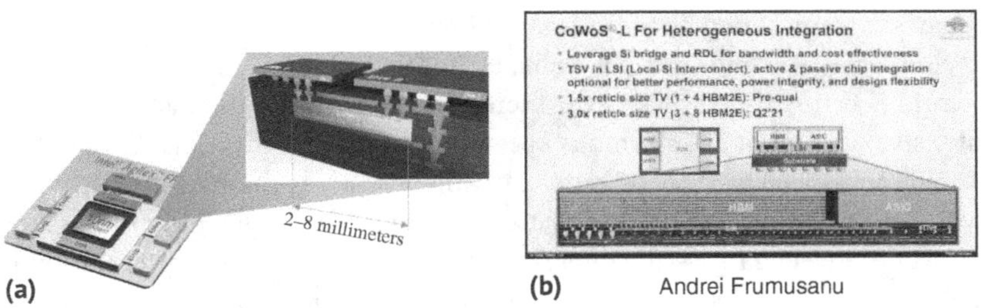

Figure 4.5 Two competing 2.5D package integration technologies: (a) Intel's EMIB technology; and (b) TSMC's LSI technology. The size of EMIB and LSI is in the order of few millimeters, corresponding to the reticle size in a 300 mm wafer.

with EMIB. TSMC's robust capabilities in advanced front-end (FE) nodes position it as a powerhouse in high-end packaging, enabling it to dominate next-generation system-level packaging.

To meet the growing demand for high-density package integration and millimeter-scale data transfer, various technological design innovations have emerged. Take the example of Intel's EMIB. EMIB stands out as a unique packaging technology, offering localized high-density wiring, as visually depicted in Figure 4.5a. While the majority of package interconnects continue to rely on traditional organic package interconnects, EMIB's fundamental concept involves employing thin silicon pieces with multilayer back-end-of-line (BEOL) interconnects, embedded within organic substrates, to facilitate localized dense interconnections. TSMC, on the other hand, introduced a comparable 2.5D package integration technology, in 2020, based on local semiconductor interconnects (LSI), as portrayed in Figure 4.5b.

The remarkable investments and technological innovations undertaken by industry leaders vividly illustrate their recognition of the significance of millimeter-scale communication. These endeavors underscore the urgency and competition in finding effective solutions for chip-to-chip data transfer within the millimeter scale. As we navigate the evolving landscape of high-end performance packaging and the challenges of advancing semiconductor technology, it becomes increasingly evident that the pursuit of efficient millimeter-scale communication is central to unlocking new frontiers in computing and connectivity.

4.6 Quest for a Fundamentally Different Propagation Mode for Millimeter-scale Packaging

The preceding sections have underscored the critical importance of millimeter-scale information transfer in the modern semiconductor-chip design strategy. This urgency is exemplified by the substantial investments and research efforts of leading companies striving to find a viable technological solution.

The world is being driven by an ever-increasing demand for the distribution of data-intensive content at an enormous speed of information transfer. Large data-centers and high-speed broadband network came into being in order to cope with this vive of information explosion. For example, the bandwidth of chip-to-chip interconnects between large-scale integrated circuits will double every 2 years and has reached 10 Tbit/s by 2020 [1]. It is also surmised that by the late 2020s, conventional electrical wiring will not be able to provide any practical solutions [2]. As explicated by Meindl [3], the aggressive scaling of devices leads to a Beyond Moore's Law era, marked by "tyranny of interconnects", where interconnects supplant logic devices in the role of the main factor determining circuit performance. It is predicted that the decisive role of interconnects in circuit characteristics would only strengthen with every future technology node. Therefore, it is imperative for us to investigate new kinds of interconnect technology that can overcome the limitations of the current ones.

In response to this pressing challenge, the authors of this book present a visionary approach to address the dilemma of high-performance chip packaging that involves interconnects spanning a few millimeters. The innovative method harnesses the potential of surface-confined electromagnetic waves in a spoof plasmonic channel. As we delve into these chapters and the following ones, we will unveil the remarkable properties of metamaterials that challenge conventional wisdom. Importantly, we will demonstrate how, even without signal encoding or additional bit redundancy, the proposed spoof plasmonic interconnect emerges as a self-resilient solution against crosstalk noise. Moreover, we will showcase its capability to facilitate information transfer at speeds exceeding hundreds of gigabits per second across distances of tens of millimeters within VLSI systems.

4.7 Limitations of Standard Interconnect Technologies

The shortcomings of conventional electrical data bus are that their bandwidth, even with the assumption of the lossless conductor, is limited by the crosstalk between the adjacent wires. Higher frequencies are more susceptible to crosstalk and therefore compromise signal integrity.

There also exists an alternative interconnect implementation based on an optical communication link [4], renowned for its gigantic bandwidth of tens of Tbps [5, 6]. However, optical interconnects come with a price. Conventional photonic technologies are quite different from those of electronics in terms of materials, device sizes, and fabrication processes and thus lack high-density photonic device integration [7–10]. Because of the large thermo-optic effect on silicon compared to the electro-optic effect, the projected advantage of power economization in chip-level optical interconnects is almost nullified once we account for the power consumed by the thermal stabilization circuit [11]. In addition, the high overhead energy cost associated with electro-optic signal conversion, and large (hundreds of wavelengths) foot-print of optical modulato [12] render optical interconnect almost prohibitive for data bus smaller than 10 cm length [13, 14]. According to Cho et al. [15], the critical length below which an optical interconnect is energetically less-favorable for data transfer than its electrical counterpart is 15 cm for a typical coupling loss of 3 dB at 6 Gbps data rate, even with the assumption of an ideal modulator. Thus, we are left with a range of distance of 0.1–10 cm, dubbed as the "last centimeter barrier," which falls into the category of inter-chip distance and looks insurmountable with any available cost-effective means [16].

4.8 Authors' Contribution to the Field of Interconnect Design

In the introductory chapter, we discussed in detail about the complexity of choosing between electrical and optical interconnects for short-distance communication, given that the rate of data traffic varies widely over time. We argued that the dilemma of favoring one interconnect technology over the other under the above circumstances can be circumvented by designing a reconfigurable interconnect technology, where the physical interconnect can alter its mechanism of guiding wave to minimize energy consumption per bit of transferred data at all possible traffic rates. With this quest in mind, we investigate the potential of spoof surface plasmon polariton (SSPP) interconnects. With feature sizes of a few tens of microns, which coincide with the dimensions of the wire pitches on the electrical backplane limited by the geometric size of the chip pins and pads, this alternative technology may become a new paradigm in inter-chip communications. Since the analysis of electrical means of information transfer via conductive SSPP interconnects associated with a metallic ground is well understood, we rather aim to discuss how to harness the potential of strong field confinement in the SSPP channel, unaided by any ground wires. Coined by the seminal work of Pendry et al. [17] and Garcia-Vidal et al. [18], SSPP is a special electromagnetic mode where the artificial plasma frequencies are design specific rather than material specific. In addition, the modal volume of SSPP can be squeezed far below the Abbe diffraction limit without requiring the electromagnetic field penetrate into the metal. Thus, we can, via SSPP, guide a signal through a sub-diffraction space.

The demonstration of the existence of SSPP mode in ultrathin structure by Shen et al. [19], Shen and Jun Cui [20] broadened the scope of the application of spoof plasmon waveguide [21]. In 2016, Hao Yu and his group demonstrated subwavelength-sized standalone interconnect operating at THz range implemented in thin, planar CMOS-compatible SSPP metamaterial [22–24]. This revealed the potential of the SSPP interconnect to become a novel way of data transfer media between chips at THz speed and beyond.

Because of the nature of field confinement around an SSPP channel, the channels in a network can be designed to remain essentially isolated from each other in a compact design scheme. The bandwidth limitation on SSPP channels derives from a quite different physics: it is the resonant frequency of the unit cell (which in turn depends on the geometry of the cell) of the metamaterial that defines the bandwidth. Spoof plasmonics can be a very rich field to explore for the future terascale data transfer between chips, as the spoof plasmon interconnect may amass the good sides of both electrical and optical interconnects. On the one hand, the SSPP interconnect possesses the great advantage of CMOS compatibility and its quasi-electrical mode can transfer data without a modulation scheme and thus liberates from the overhead energy cost of source driver, modulators' and detectors' power. On the other hand, like optical ones, it can suppress crosstalk at high frequencies where bandwidth per channel can be few hundreds of Gbps for a few tens of mm information propagation length, in addition to the possibility of ultra-low energy (subfemtojoule) [25] signal detection through facilitating "quantum impedance conversion" [26] of photon. We may switch between these two different types of information transfer mode in SSPP interconnect, namely electro-SSPP and opto-SSPP mode, which may offer an energy-economic data transfer solution between chips where the data traffic volume may vary by orders of magnitude over the course of time. In this chapter, I will show that, with my prescribed geometry, an SSPP-based data bus can provide ~ 1 Gbps/μm bandwidth density while the signal can propagate up to state-of-the-art chip edge length ~ 10 mm, where each channel would carry information at approximately 300 Gbps rate without invoking significant crosstalk.

The current chapter demonstrates the fundamental interplay between various figure-of-merits of information capacity of an inter-chip communication network comprising SSPP channels. In Section 4.9, we discussed the bandwidth density maximization strategy in an SSPP network. In Section 4.10, we defined the characteristic length of information propagation pertaining to SSPP channel and its relation to the geometry of the channel. Section 4.11 shows the relation between finite information propagation length and the resistivity of SSPP channels, which eventually led to the estimation of noise properties and Shannon information capacity in the limit of thermal noise contributed by the SSPP channels. Section 4.12 discusses the quantitative comparison between various alternatives of chip-to-chip interconnects in terms of crosstalk in the system.

4.9 Bandwidth in Crosstalk-mediated SSPP Channels

Figure 4.6a is a schematic of the data transfer system that we propose for inter-chip communication where the data bus consists of SSPP channels running in parallel. A pair of channels are formed by placing two SSPP waveguides back to back. Each of such channel-pair maintains a guard space between the nearest face-to-face SSPP waveguides to avoid crosstalk. An SSPP mode possesses a continuous band, starting from zero frequency—where there is no field confinement and hence the channel is subjected to significant crosstalk with the neighboring channels; upto the spoof plasma frequency—where the degree of field confinement is maximum, and the channels can be decoupled from each other (Figure 4.6b). Between these two extremities, the field localization property undergoes gradual changes. Figure 3.12 shows the dispersion diagram of an isolated SSPP channel. In order to generate a dispersion diagram independent of the geometric dimensions of interconnects, we, in Figure 3.12, instead of plotting the absolute value of frequency ω and propagation constant β, choose to plot a unit-less frequency parameter $\hat{\omega} = 2\omega h/\pi c$ against another unit-less propagation parameter $\hat{\beta} = \beta d/\pi$. At a given frequency ω, any change in the geometric dimension of the SSPP interconnect would cause a corresponding change in the absolute value of β in such a way that the relation between the normalized parameter $\hat{\omega}$ and $\hat{\beta}$ will remain unaltered, as they absorb the effect of geometric change into themselves. We may distinguish two different regions in the dispersion curve of SSPP channels. One of them is the strong SSPP regime, where the dispersion curve becomes almost flat and provides deep sub-wavelength field confinement, which can be leveraged for high-resolution imaging. The other region is the soft SSPP regime, where the field localization is moderate; however, the bandwidth of this region is appreciably large. As soft SSPP regime can offer both large bandwidth and field confinement together, this can be leveraged toward making an interconnect channel with a high information capacity which may defy RC limited bandwidth.

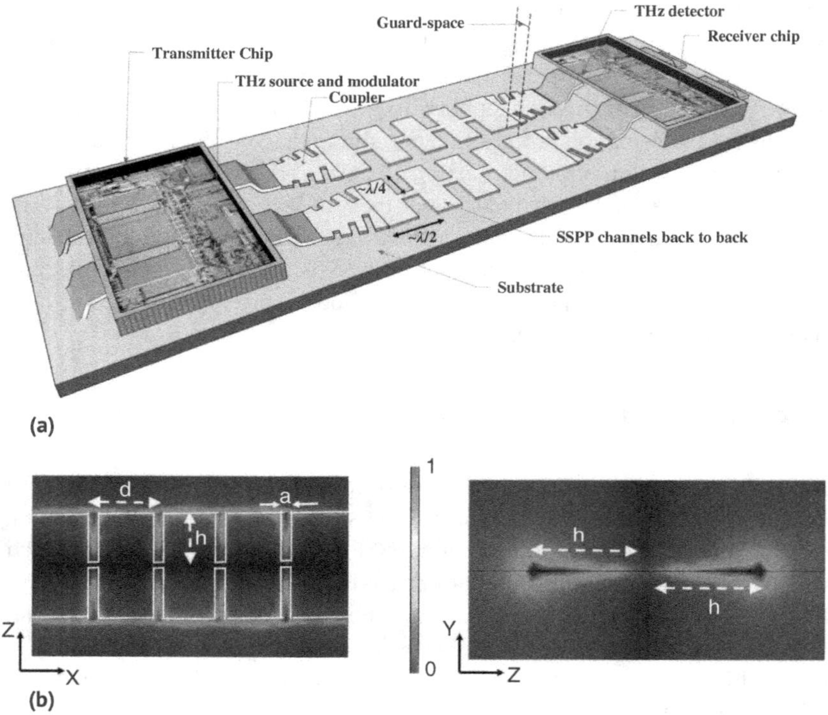

Figure 4.6 (a) An inter-chip communication network employing SSPP channels. Scales of interconnects are shown in terms of free-space wavelength (λ) corresponding to spoof plasmon resonance, a typical wavelength being 300 μm. (b) The left figure is the top-view of electric field distribution in SSPP channels, while the right figure is a cross-section view of the field, showing modal confinement. The simulation frequency is $\sim$ 1 THz, roughly the spoof plasmon resonance frequency of an SSPP interconnect with groove length 80 μm.

Let us begin our analysis of an SSPP interconnect system by designating the groove length, width, period, thickness and channel separation as h, a, d, $2b$, and $2t$, respectively. It is shown in other works that the thickness ($2b$) of SSPP waveguide plays a trivial role in determining the dispersion relation. This helps to divide the SSPP system into two independent three-dimensional sub-systems for the sake of simpler analysis. The first subsystem comprises of a periodic groove of length h and period d on zx-plane where the subsystem is infinitely extended along y-direction. The second subsystem consists of a periodic array of metal slabs of small thickness $2b$, separation a, and period d on xy-plane where the subsystem is considered infinite along z-direction. Taking the midpoint of the groove as the origin of co-ordinates, the field distribution on the surface of the first subsystem can be written as follows.

$$\begin{aligned}
E_{ax}^{(d)}(z) &= \sum_n A_{an} e^{-\kappa_{c,an}(z+t)} \\
E_{bx}^{(d)}(z) &= \sum_n A_{bn} e^{\kappa_{c,bn}(z-t)} \\
E_{az}^{(d)}(z) &= -\sum_n \frac{i\beta_{an}\kappa_{an}}{k^2-\beta_{an}^2} A_{an} e^{-\kappa_{c,an}(z+t)} \\
E_{bz}^{(d)}(y) &= \sum_n \frac{i\beta_{bn}\kappa_{bn}}{k^2-\beta_{bn}^2} A_{bn} e^{\kappa_{c,bn}(z-t)}
\end{aligned} \tag{4.1}$$

where A_{an} and A_{bn} are the complex amplitude of nth mode of face-to-face posing SSPP waveguides "a" and "b," respectively, β is the propagation constant, and κ_c is the attenuation constant of evanescent field along z direction. κ_c can provide a measure of modal confinement that we shall capitalize to suppress crosstalk, and we shall be discussing about it in detail shortly. The magnetic field components outside the groove can be expressed as:

$$\begin{aligned}
H_{ay}^{(d)}(z) &= -\sum_n \frac{i\omega\epsilon_0\kappa_{an}}{k^2-\beta_{an}^2} A_{an} e^{-\kappa_{c,an}(z+t)} \\
H_{by}^{(d)}(z) &= -\sum_n \frac{i\omega\epsilon_0\kappa_{bn}}{k^2-\beta_{bn}^2} A_{bn} e^{\kappa_{c,bn}(z-t)}
\end{aligned} \tag{4.2}$$

Each of the above components has $e^{i\beta x}$ dependence along the propagation direction x. The dispersion relation of the subsystem is [27],

$$\sqrt{\beta^2 - Q^2} = Q\frac{a}{d}\mathrm{sinc}^2\left(\frac{a\beta_0}{2}\right)\tan(Qh) \tag{4.3}$$

where Q is the wave vector inside the grooves along their length, and for infinitely extended subsystem ($2b \to \infty$) along y direction, $Q = k_0 = \frac{\omega}{c}$. The relation of $Q \approx k_0$ holds even at the limit of vanishing thickness b of SSPP interconnects. Therefore, analysis of surface mode for the second subsystem with finite thickness can be postponed until we begin Section 4.11 that discusses thermal noise in SSPP interconnects. The length of modal confinement (t_c) of SSPP modes can be estimated as $t_c = \frac{1}{2\kappa_c} = \frac{1}{2\sqrt{\beta^2-(\frac{\omega}{c})^2}}$, where ω is angular frequency, and c is the speed of light. In order to transfer data with suppressed inter-channel coupling (less than −3 dB), we may choose to utilize the upper f_c fraction of the bandwidth (BW_{sp}) of an SSPP channel where modal confinement is stronger.

Then, $\kappa_{c,\min} \approx \frac{a}{d}\frac{\omega_{\min}}{c} n_{g,\text{eff}} \tan\left(\frac{\omega_{\min}}{c} n_{g,\text{eff}} h\right)$, where $\omega_{\min} = 2\pi(1 - f_c)\text{BW}_{sp}$, and $n_{g,\text{eff}}$ is the effective refractive index of the medium adjacent to the channel. Hence, the minimum amount of space required to interleave between a pair of neighboring SSPP channels is

$$t_{c,\max}/h = \frac{d}{2a}\left(\frac{1}{1-f_c}\right)^2 \left(\frac{c}{2\pi n_{g,\text{eff}}\, h.\text{BW}_{sp}}\right)^2 \tag{4.4}$$

The bandwidth BW_{sp} is itself a function of the groove geometry, particularly governed by groove length, as [27]:

$$\begin{aligned}\text{BW}_{sp} &= \left(\frac{\pi}{2} - \delta_w\right) \frac{c}{2\pi n_{g,\text{eff}}\, h} \\ &= \frac{c}{2\pi n_{g,\text{eff}}\, h}\left[\frac{\pi}{2} - \frac{a}{d}\frac{d}{\sqrt{4h^2 - d^2}}\right]\end{aligned} \tag{4.5}$$

An important figure-of-merit for an interconnect network is the bandwidth density (ρ_{BW}), defined as the amount of bandwidth available for data transfer per unit cross-sectional length of the data bus at reduced cross-channel coupling. It has been experimentally demonstrated that, by placing two SSPP waveguide back-to-back with nominal separation, the crosstalk between them can be kept below −20 dB. So, the critical separation which would determine the channel packing density is the separation between two SSPP waveguide placed face-to-face. Bandwidth density of SSPP data bus of a design shown in Figure 4.6a can be expressed as

$$\begin{aligned}\rho_{\text{BW}} &= \frac{2f_c.\text{BW}_{sp}}{t_{c,\max} + 2h} \approx \frac{cf_c}{2n_{g,\text{eff}}\, h(t_{c,\max} + 2h)} \\ &= \frac{c}{4n_{g,\text{eff}}\, h^2} \frac{f_c(1-f_c)^2}{(1-f_c)^2 + D}\end{aligned} \tag{4.6}$$

where $D = \frac{d}{\pi^2 a}$. As evident from Eq. (4.6), choosing a substrate with low refractive index ($n_{g,\text{eff}}$ lower) is beneficial for the design of a closely packed SSPP network for a given bandwidth of data transfer. In addition, it shows that reducing groove length h leads to a greater bandwidth density. The lowest limit on h is dictated by the condition on the ratio of geometric length of the structure to support spoof plasmon mode, which is $2h > d$ [27]. The dependence of ρ_{BW} over f_c (i.e., fraction of total SSPP bandwidth BW_{sp} which is crosstalk suppressive) is also worth looking at. For small f_c, crosstalk, limited bandwidth linearly increases with f_c; however, ρ_{BW} quadratically decreases as f_c approaches its maximum value of 1. Its optimum value $f_{c,\text{opt}}$ can be determined by considering $\frac{\partial \rho_{\text{BW}}}{\partial f_c} = 0$ that yields:

$$f_{c,\text{opt}} = 1 - \frac{(D + \sqrt{D^2 + D^3})^{2/3} - D}{(D + \sqrt{D^2 + D^3})^{1/3}} \approx \begin{cases} \frac{2}{5} - \frac{3}{50}\delta, & \text{when } D = 1 + \delta \\ \frac{1}{3}\left(1 + \frac{2}{3\sqrt{D}}\right), & \text{when } D >> 1 \end{cases} \tag{4.7}$$

In this limit of large D, an interesting relation of ($f_c \geq \frac{1}{3}$) emerges, which indicates that we can have at least 33% of the total bandwidth of the SSPP channel available for crosstalk suppressive data transfer. If we plug in the nominal value of $f_c = \frac{1}{3}$ in Eq. (4.6), then the maximum attainable bandwidth density in SSPP network is found to be $\rho_{\text{BW}}|_{\max} \approx \frac{\pi^2 c}{n_{g,\text{eff}}\, h^2} \frac{1}{12\pi^2 + 27d/a}$.

Figure 4.7a gives a visual depiction of bandwidth optimization strategy. It shows that we may utilize the upper 33% of each of the SSPP channel bandwidth to maximize the bandwidth density (ρ_{BW}). As the condition of existence of SSPP warrants $h_{\min} = d/2$, we can increase bandwidth density ρ_{BW} further by reducing periodicity d. However, a reduction of d would increase the mode coupling factor $\frac{a}{d}$, which would adversely affect the BW_{sp}. In addition, reduction of h would also move the system to a higher operating frequency, where the Ohmic loss in conductor through the skin effect would creep in and limit signal propagation length. Section 4.3 deals with the question of how far the information can travel through a channel consisting of a real metal with finite conductivity.

One other interesting feature of SSPP channel is its nature of confinement along the vertical (y) direction, which is in fact stronger than its lateral confinement and can be shown to be

$$t'^{-1}_{c,\max} = 2\kappa'_{c,\min} \approx 2\beta_{\min} \approx \frac{\pi(1 - f_c)}{n_g h} \tag{4.8}$$

where κ'_c is the vertical evanescent wave vector along y direction. If we stack SSPP channels one on top of the other with interleaving space between them equal to the vertical confinement length ($t'_{c,\max}$) of the mode, then the bandwidth density ρ'_{BW} of the vertically stacked channels becomes,

$$\rho'_{\text{BW}} \approx \frac{\pi c}{4 n_g^2 h^2} f_c(1 - f_c) \tag{4.9}$$

Comparison of bandwidth density for planar SSPP channel cluster and that for vertical channel cluster written in Eqs. (4.6) and (4.9), respectively, reveals an interesting fact. Equation (4.9), unlike Eq. (4.6) is free from the term d/a. This adds relative flexibility in designing the groove width a without affecting the bandwidth density. In addition, Eq. (4.9) shows that the optimal fraction of SSPP bandwidth

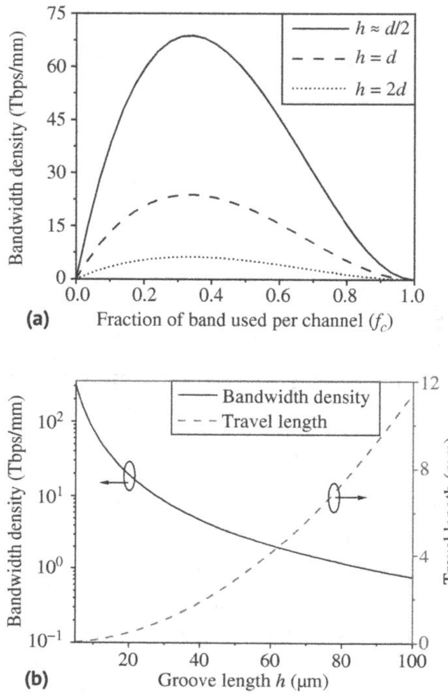

Figure 4.7 (a) Bandwidth density of SSPP data bus versus the fraction of the bandwidth utilized per SSPP channel. The graph is drawn for data bus consisting of crosstalk suppressive SSPP channels, each having a period, $d = 20$ μm,$a \sim \frac{1}{10}d$. (b) The left vertical axis shows optimum bandwidth density ($f_c = 1/3$) of the SSPP data bus. The groove length of the waveguide is varied, while the aspect ratios of its geometric parameters are kept fixed at $a/d = 0.1$, $h/d \approx \frac{1}{2}$. The right vertical axis shows the corresponding characteristic length of the SSPP channel made of gold (skin depth of gold at THz is ~ 60 nm) up to which half of the signal power retains.

in a crosstalk suppressive regime, in the case of vertical channel stacking, can be 50% significantly higher than that achievable in a planar arrangement of channels.

4.10 Traveling Length of SSPP Mode in Lossy Metal

Following Rusina et al. [28], we can incorporate metallic loss in our analysis of SSPP channel for metal with finite conductivity. We begin so by modifying the wave vector Q along the direction(z) of groove length as follows:

$$Q \approx Q_0 n_g \left(1 + i\frac{l_s}{2a}\right) \tag{4.10}$$

where $Q_0 = \frac{\omega}{c}$, l_s is the skin depth of the metal. Employing the characteristic dispersion relation of SSPP mode [27, 29], we determine the finite propagation length of SSPP mode l_p at a given frequency as:

$$\frac{1}{2}l_p^{-1} = Im[\beta] \approx \begin{cases} Q'', \text{for } \beta \sim Q_0 n_g \\ \frac{\pi}{2}\frac{d}{a}Q''\frac{4h^2 - d^2}{d^2}, \text{ for } \beta >> Q_0 n_g \end{cases} \tag{4.11}$$

where $Q'' = \mathrm{Im}[Q]$. The regime $\beta >> Q_0 n_g$ can be seen in an SSPP structure with groove length (h) significantly larger than half of the period (d). It should be emphasized that the structures considered in the present work are characterized by $h < d$, and therefore do not operate in the strong decay regime. While a knowledge of l_p at a single carrier frequency ω_c may suffice to estimate the propagation distance of a message signal with bandwidth $\Delta\omega_m$ as long as $\omega_c >> \Delta\omega_m$, it is not quite useful in case of information transfer over SSPP channel network, where the carrier frequency is supposed to be nearly comparable to the signal bandwidth. Under such circumstances, we find it worthwhile to introduce a slightly different characteristic length l_t which is defined as the length that a pulse with rectangular Fourier spectral profile of bandwidth $\Delta\omega_m$ can travel until the power distributed over its bandwidth reduces by half. This leads to the following relation of power loss in metal,

$$P_{\text{loss}} = \int_{\omega_{\text{min}}}^{\omega_{\text{max}}} A^2 e^{-2\beta_I(\omega)l_t} d\omega = \frac{A^2\Delta\omega_m}{2} \tag{4.12}$$

where $\beta_I = Im[\beta]$. If we operate in the weak SSPP regime for information transfer, it can be seen from Eqs. (4.10) and (4.11) that, for $\frac{a}{d} << 1$,

$$\beta_I \approx \frac{\omega}{c}\frac{l_s}{2a} \tag{4.13}$$

Replacing the variable ω with β_I in the equation of P_{loss}, we get

$$l_t = \frac{1}{2}\frac{a}{l_s}\frac{c}{\omega_{\text{mid}}} \tag{4.14}$$

where the approximation $e^x \approx 1 + x + \frac{1}{2}x^2$ is applied, and $\omega_{\text{mid}} = \frac{\omega_{\text{max}}+\omega_{\text{min}}}{2}$. In terms of the spoof plasma frequency $\omega_p = \frac{\pi c}{2h}$ of the SSPP channel, we can rewrite the expression for l_t as:

$$l_t = \frac{1}{2}\frac{a}{l_s}\frac{c}{\omega_p(1 - 0.5f_c)} = \frac{a}{l_s}\frac{h}{\pi(1 - 0.5f_c)} \tag{4.15}$$

where $\omega_{\max} \approx \omega_p$ is assumed, and $\Delta\omega_m = f_c\omega_p$ is substituted. It is obvious from the expression of l_t that, for a given information bandwidth $\Delta\omega_m$, the larger the length of the groove h of SSPP we choose, the further the signal can travel. The expression for bandwidth density and information propagation length reveals a fundamental trade-off about choosing groove length h. While choosing smaller h increases the crosstalk limited bandwidth density of SSPP data bus, it decreases signal traveling length through the channel. Figure 4.7b illustrates this trade-off, drawn for waveguide whose geometry is varied in a way so as to keep all relevant aspect ratios fixed at some optimal value (i.e., $h/d \sim \frac{1}{2}$, $a/d = 0.1$) and the groove length h is varied. The corresponding maximum bandwidth density of SSPP data bus and the traveling length of information at that maximum bandwidth density are plotted on the same graph. It can be seen from the diagram that, for interconnect length to be ~10 mm, which is roughly the state-of-the-art chip edge length, we can obtain ~ 1 Gbps/μm bandwidth density with an optimal geometry design of SSPP waveguide. Information carried by each channel can be determined as bandwidth density multiplied by an average space occupied by an individual channel including the guard space, which is found to be 300 Gbps. In order to determine the upper bound of the signal propagation limit, we choose gold while plotting the numerical value of Figure 4.7b, which has better conductivity over other commonly used metals. For a different choice of metal with conductivity σ_m, the numerical value of information transfer length can be found by scaling the value determined in Figure 4.7b by a factor of $\sqrt{\frac{\sigma_m}{\sigma_{Au}}}$, since skin depth $l_s \sim \frac{1}{\sqrt{\sigma}}$. If the optimal value of f_c is taken as $\frac{1}{3}$ as discussed previously, the corresponding value of information traveling length will be $l_{t,\mathrm{opt}} = \frac{6}{5\pi}\frac{a}{l_s}h$.

4.11 Information Capacity in the Limit of Thermal Noise

In the regime of low crosstalk, the main limitation on the Shannon information capacity of a channel originates from signal-to-noise ratio (SNR) in the channel. In the present work, we are particularly interested in the thermal noise characteristics of SSPP channels, which is expected to be the most dominant type of noise over others (such as flicker noise and shot noise) at higher operating frequency and low power transmission. If SNR of SSPP channel is limited by thermal noise, we may want to determine the equivalent electrical resistance of SSPP channel in order to estimate thermal noise. To this goal, we calculate the power ($P_{x,\mathrm{tot}}$) propagating in the longitudinal (x) direction on a planar SSPP channel on each of its sides and equate the fraction of the power lost in joule heating process with I^2R loss. Because of the boundary condition on the plane of the SSPP channel made of metal of good conductivity, it can be argued that the in-plane component E_z is negligible compared to the other component E_y of electric field. Hence, power spectral density $P_{x,\text{tot}}$ is mainly contributed by E_y and H_z and can be found as follows.

$$H_{z,a} = Ae^{i\beta x}e^{-\kappa(y-b)}e^{iQz}$$

$$E_{y,a} = \frac{\beta}{\omega\epsilon}Ae^{i\beta x}e^{-\kappa(y-b)}e^{iQz} \tag{4.16}$$

$$P_{x,\text{tot}}(\omega) = \frac{1}{2}\int_{z=0}^{h}\int_{y=b}^{\infty} Re[E_{y,a}H^{*}_{z,a}]dydz = \frac{1}{4}\frac{\beta}{\omega\epsilon\kappa}A^2h$$

The set of relations in Eq. (4.16) describes the second subsystem, discussed in Section 4.9, comprised of metal slabs of infinite length and $2b$ thickness that account for field distribution in the region defined by $y \geq b$. Here, A is the amplitude of the magnetic field with the SI unit of Amp/m/$\sqrt{\text{Hz}}$, and $\epsilon = n^2_{\text{g,eff}}$ is the effective homogeneous permittivity of the medium mediated by the SSPP channel. The surface current in the longitudinal direction is

$$I_x(\omega) = \int_0^h J_x dz \approx hH_{z,a}(x)|_{y=b} = Ahe^{i\beta x} \tag{4.17}$$

If a pulse of bandwidth $\Delta\omega_m$ is to be transferred through the SSPP channel, half of the power will be lost after traveling the characteristic length l_{tr}. Thus, the resistance (R_{SP}) of SSPP channel per unit length can be found from the following:

$$|I_x|^2 R_{SP} l_t \Delta\omega_m = \frac{1}{2}\int_{\omega_p-\Delta\omega_m}^{\omega_p} P_{x,\text{tot}}(\omega)d\omega \tag{4.18}$$

where we used the spoof-plasma frequency ω_p in the limit of integral as an approximation of BW_{sp}. If we denote the wave vector along the groove length as Q, then using the fact that $Q \approx \omega/c$ in the relation $\beta^2 - \kappa^2 + Q^2 = (\omega/c)^2$ describing the dispersion characteristics of the field over the SSPP surface, we can set $\beta \approx \kappa$. Thus, we obtain,

$$R_{sp} = \frac{1}{8l_t}\frac{1}{\epsilon h\Delta\omega_m}\ln\left(\frac{\omega_p}{\omega_p - \Delta\omega_m}\right) = \frac{1}{8l_t}\frac{1}{\epsilon h\Delta\omega_m}|\ln(1-f_c)| \tag{4.19}$$

where $\Delta\omega_m = f_c\omega_p$ is used. Then, the electronic noise generated by the thermal agitation of electrons on the patterned metal of the channel can be modeled as a noiseless resistor in series connection with a noise voltage source $\bar{v}_n$ as follows:

$$\bar{v}_n^2 = 4k_BTR_{sp}l_{ch}\Delta\omega_m/2\pi \tag{4.20}$$

where $l_{ch} = Nd$ is the channel length, N being the number of unit cells in the SSPP channel with the period d. In addition, k_B is the Boltzmann constant and T is the absolute temperature of the channel. Now, if the SSPP mode excites an rms voltage value of V_r at the signal receiving end, the Shannon information capacity (C_{info}) of SSPP channel in the limit of thermal noise can be written as:

$$\begin{aligned} C_{\text{info}} &= \rho_{\text{BW}} \log_2(1 + \text{SNR}) \\ &= \rho_{\text{BW}} \log_2 \left(1 + \frac{2\pi V_r^2(\Delta\omega_m)}{4k_B T R_{sp} l_{ch} \Delta\omega_m}\right) \end{aligned} \tag{4.21}$$

where $\text{SNR} = \frac{V_r^2}{\bar{v}_n^2}$ is the signal-to-noise ratio, ρ_{BW} is the SSPP bandwidth density limited by crosstalk among neighboring channels, and follows Eq. (4.6). Voltage drop V_r across the receiver resistance R_r is related to the transmitted information bandwidth $\Delta\omega_m$ as follows:

$$V_r^2 = P_r R_r \tag{4.22}$$

$$P_r = \frac{1}{4}\frac{1}{\epsilon}A^2 h \left| \ln\left(1 - \frac{\Delta\omega_m}{\omega_p}\right)\right| \left(1 - \frac{1}{2}\frac{l_{ch}}{l_t}\right) \tag{4.23}$$

where P_r is the power at the receiving end. Characterization of SSPP channel resistance R_{sp} facilitates determining the bit error rate (BER) due to channel-induced noise. Out of two types of noise, namely, thermal noise and shot noise, let us assume that thermal noise prevails over shot noise (i.e., thermal noise power is at least twice of shot noise power), and the receiver can be controlled in a way so as not to contribute to the thermal noise. In addition, let us also choose to design receiver input resistance (R_r) to be an order of magnitude larger than the channel resistance ($R_{ch} = R_{sp}Nd = \frac{1}{r}R_r$), so that we may have a large voltage drop (V_r) at the receiving end. Shot noise power can be written as $2qI_r\Delta f R_r$, where $I_r = \frac{V_r}{R_r}$. Thermal noise power can be written as $4k_B T R_{ch} \Delta f \frac{1}{R_r}$. Then, the above-stated condition leaves us with the following inequality condition:

$$\frac{4k_B T R_{ch} \Delta f}{R_r} \geq 2 \times 2q\frac{V_r}{R_r}\Delta f R_r \tag{4.24}$$

These conditions set an upper limit for the signal voltage (V_r) at the receiving end as $V_r \leq \frac{k_B T}{rq}$, where r is the ratio of resistance of the receiver to the channel, q is the charge of electron. Since SSPP channel is not suitable to carry DC current, as at $\omega = 0$, there would be no field confinement, we must choose a source encoding scheme that is devoid of any DC offset. Let us choose non-return-to-zero (NRZ) coding for digital signal transmission where bit-0 and bit-1 are represented by negative and positive pulses of equal amplitude. Since we also want to maximize the information capacity by choosing an optimal SNR, let us set $\pm V_{r,\max} = \pm \frac{k_B T}{rq}$ to denote the mean value for binary bits, and zero as the threshold value. Since NRZ encoding is symmetric for 0 and 1 bit, the probabilistic bit error rate can be calculated as:

$$\begin{aligned}\text{BER} &= \frac{1}{v_n\sqrt{2\pi}} \int_0^{\infty} \exp\left(-\frac{(V+V_r)^2}{2v_n^2}\right)dV \\ &= \frac{1}{2}\,\text{erfc}\left(\frac{V_r}{v_n\sqrt{2}}\right).\end{aligned} \tag{4.25}$$

where $V_r = V_{r,\max}$ is chosen for maximum SNR. Then, for the utilization of f_c fraction of SSPP bandwidth for data transmission with suppressed inter-channel coupling, the BER can be related with the geometry of the SSPP waveguide, as follows:

$$\text{BER}_{sp} = \frac{1}{2}\,\text{erfc}\left(\frac{1}{rq}\sqrt{\frac{\pi\epsilon k_B T}{|\ln(1-f_c)|}\frac{l_t}{l_{ch}}}\sqrt{h}\right) \tag{4.26}$$

If $f_c = \frac{1}{3}$ is chosen for the reason of bandwidth density maximization, as described in the previous section, and channel length is taken as $l_{ch} = l_t$, and $r = 10$ is considered, then the optimal BER turns out to be $\text{BER}_{sp,\text{opt}} = \frac{1}{2}\,\text{erfc}\,(\frac{1}{5q}\sqrt{\frac{\pi\epsilon K_B T}{2\ln(3/2)}}\sqrt{h})$. At the same optimal conditions for data transfer within the thermal noise limit, the optimal Shannon information capacity can be written as

$$\begin{aligned}C_{\text{info,opt}} &= \rho_{\text{BW,max}} \log_2(1 + \text{SNR}_{\max}) \\ &= \frac{\pi^2 c}{n_{\text{g,eff}} h^2} \frac{1}{12\pi^2 + 27\frac{d}{a}} \log_2\left(1 + \frac{8\pi\epsilon h K_B T}{25q^2 \ln(3/2)}\right)\end{aligned} \tag{4.27}$$

where $\text{SNR}_{\max}$ is the maximum SNR possible to obtain at the receiver while still operating in the limit of thermal noise generated by SSPP channel. Figure 4.8 shows how the Shannon information capacity and BER of an SSPP channel at room temperature vary with the groove length while SNR is maximum at the receiving end within the limit of thermal noise generated by SSPP channel.

4.12 SSPP Interconnect in Comparison with Others: The Benefit of Minimized Interference

In this section, I shall have a comparative discussion to facilitate choosing the means of interconnects depending on system requirements. I shall make an analytical estimation of the interference suppression in a parallel bus system made of

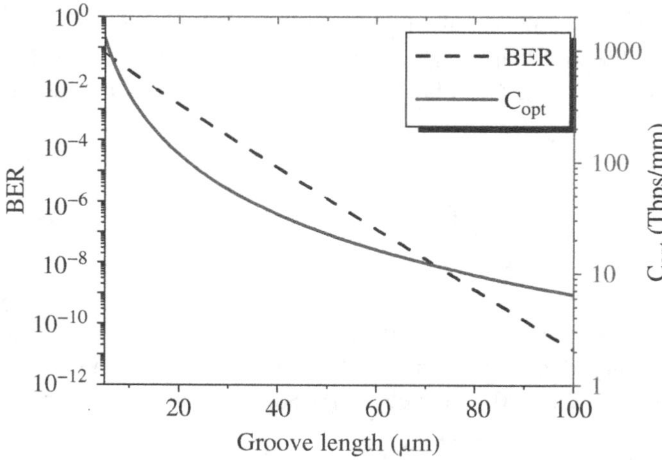

Figure 4.8 Variation of the optimized Shannon information capacity (C_{opt}) and the optimized BER in SSPP channel at $T = 300$ K with the variation of groove length. The channel length is taken as the characteristic traveling length (l_t) for the corresponding groove length. The data is plotted for maximum attainable SNR at the receiver end within thermal noise limit of the channel. The geometric ratio a/d of the SSPP channel is taken as 0.1.

conventional electrical interconnects, that of optical interconnects, and that of our proposed SSPP interconnects.[1]

4.12.1 Crosstalk in Optical Interconnect

A simple assumption of information transfer in the form of TE mode propagation in a dielectric slab may suffice for crosstalk estimation in a system of optical interconnects. The dispersion relation for the even TE mode in an isolated dielectric slab is

$$\kappa = k_z \tan(k_z w) \tag{4.28}$$

where $2w$ is the thickness of the slab, $k_z = \sqrt{k_0^2 n_d^2 - \beta^2}$ and $\kappa = \sqrt{\beta^2 - k_0^2 n_a^2}$ are the transverse wave vector inside and outside the dielectric waveguide, respectively, and n_d and n_a are refractive index of the slab and the environment,

1 We have not included plasmonic interconnects in our comparative discussion, as it is shown by others [33] that plasmonic interconnects are meant for communication within 10 μm distance, whereas we are concerned about centimeter-scale distance coverage, a typical separation between electrical chips on board.

respectively. Coupling between identical neighbor waveguide pairs, with the help of Eq. (4.28) is determined as follows [30, 31],

$$\begin{aligned} K_{op} &= \frac{\omega}{4}\epsilon_0(n_d^2 - n_a^2)E_{a0}E_{d0}\int_{w+2t}^{3w+2t}\big[e^{-\varkappa(z-w)} \\ &\quad\times \cos(k_z(z-2w-2t))\big]dz \\ &= \frac{\omega}{4}\epsilon_0\frac{E_{d0}^2}{k_0^2}\cos^2(k_z w)\big[\varkappa(1-e^{-2\varkappa w}) \\ &\quad+(1+e^{-2\varkappa w})k_z\tan(k_z w)\big]e^{-2\varkappa t} \\ &= \frac{\omega\epsilon_0}{2}\frac{E_{d0}^2}{k_0^2}\cos^2(k_z w)\varkappa e^{-2\varkappa t} \end{aligned} \tag{4.29}$$

where $2t$ is the separation between neighboring interconnects, $k_0 = \omega\sqrt{\epsilon_0\mu}$ is free space wave vector, and $E_{d0}(E_{a0})$ is the electric field magnitude inside (outside) the dielectric slab. In the RHS of the last relation, the continuity condition of tangential electric field at the interface ($E_{d0}\cos(k_z w) = E_{a0}$) is applied. We can obtain the behavior of the coupling parameter in the limit of high frequency as following:

$$\lim_{k_0 w\to\infty} K_{op} = \lim_{2K_z w\to\pi} K_{op} = \lim_{w\frac{\omega}{c}\to\infty}\frac{1}{w}\left(\frac{\pi c}{2\omega n_d w}\right)^2 e^{-2\omega n_d t/c} \tag{4.30}$$

The coupling constant K_{op} would enter as the nondiagonal terms of the operator H_{op}, where H_{op} describes the evolution of the modes along propagation direction in a system of optical interconnects, perturbed by the presence of neighbor waveguides. The eigenfunction analysis of H_{op} leads to the following relation of $a_1(x)$, where $a_1(x)$ is the mode amplitude of a waveguide due to unity amplitude signal injection into either one of its nearest neighbor interconnects.

$$a_1(x) = \frac{1}{2\sqrt{2}}(e^{ix\beta_-} - e^{ix\beta_+}) = \frac{-i}{\sqrt{2}}e^{i\beta_0 x}\sin(K_{op}\sqrt{2}x) \tag{4.31}$$

We can define crosstalk length l_x as the length at which $|a_1(l_x)| = \frac{1}{\sqrt{2}}$. This definition leads to the following expression of l_x.

$$l_{x,op} = \frac{\pi}{2\sqrt{2}}\frac{1}{K_{op}} \tag{4.32}$$

Combining Eqs. (4.30) and (4.32), we get the following expression of critical crosstalk length in optics at high-frequency regime.

$$\lim_{w\frac{\omega}{c}\to\infty} l_{x,op} = \lim_{w\frac{\omega}{c}\to\infty}\frac{w\sqrt{2}}{\pi}\left(\frac{\omega n_d w}{c}\right)e^{2\omega n_d t/c} \tag{4.33}$$

The above relation is expected to be reasonably correct for a frequency ω such that $\frac{\pi c}{2\omega n_d w} << 1$. It points out that optical interconnects are not favorable for carrying signals at low frequencies since crosstalk becomes most severe at low frequency, where modal confinement is poor.[2] Therefore, a baseband signal cannot be transmitted through an optical interconnect unless it is multiplied with a high-frequency carrier. This information will turn out to be crucial to compare with other alternatives of interconnects, as we are going to analyze them in a moment.

Figure 4.9 shows the critical length in an optical data bus limited by crosstalk, as calculated by Eq. (4.32). The geometry of the optical bus, for this specific plot, is taken as follows: each channel width is 20 μm, the guardspace between adjacent channel is 20 μm.

4.12.2 Crosstalk in Electrical Interconnects

For inter-chip communication implemented through purely electrical interconnects, we can analyze its bandwidth density by modeling the interconnect network as a distributed RLCG circuit, where R, L, C, G are defined as resistance, self-inductance, capacitance, and conductance of the dielectric, per unit length, respectively. Each interconnect is coupled to its nearest neighbor through mutual inductance L_m and mutual capacitance C_m, also defined per unit length. The dynamics of the system are governed by a set of equations discussed in Pozar [32] with a solution of the form $V_i(x) = V_0 e^{-\gamma x}$, where γ is the complex propagation constant, V_i is the voltage induced at ith electrical interconnect with initial complex amplitude of V_0.

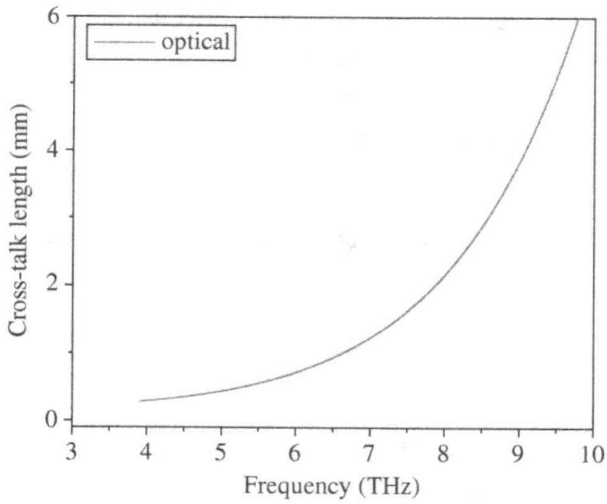

Figure 4.9 Critical crosstalk length is determined for optical interconnects. Optical interconnect width is taken as 20 μm, guardspace between adjacent interconnects is 20 μm.

2 The analysis is valid for only weakly perturbed mode.

Since we are concerned about interconnects which will be designed to operate in low interference conditions, an assumption of weak coupling among the neighbor interconnects is justified. Under this assumption,

$$\begin{aligned} \gamma_{\pm} &= \frac{1}{\sqrt{2}}(1+i)\sqrt{\omega R(C \pm C_m\sqrt{2})} \\ &\approx \frac{1}{\sqrt{2}}(1+i)\sqrt{\omega RC}\left(1 \pm \frac{C_m\sqrt{2}}{2C}\right) = \gamma_0 \pm \Delta\gamma \end{aligned} \tag{4.34}$$

where $\gamma_0 = \frac{1}{\sqrt{2}}(1+i)\sqrt{\omega RC}$, $\Delta\gamma = \frac{1}{\sqrt{2}}(1+i)\sqrt{\omega R}\frac{C_m}{\sqrt{2C}}$, and $C_m\sqrt{2} << C$ is assumed.

In order to estimate crosstalk, we should, as we did previously, inject a unit signal at the input terminal of one particular interconnect, and observe how the signal victimizes the neighboring channel. However, since the electrical system incorporates the parameter R (resistance per unit length) in the coupling term, which also leads to signal absorption, we should define the 3 dB crosstalk length $l_{x,el}$ in a way that satisfies the following relation:

$$\begin{aligned} &\left|\frac{a_1(l_x)}{a_2(l_x)}\right| = \frac{1}{\sqrt{2}}\left|\frac{e^{\Delta\gamma l_x} - e^{-\Delta\gamma l_x}}{e^{\Delta\gamma l_x} + e^{-\Delta\gamma l_x}}\right| = \frac{1}{\sqrt{2}} \\ &\implies |\tanh(\Delta\gamma l_x)| = 1 \end{aligned} \tag{4.35}$$

where $a_1(x)$ is the voltage amplitude of the crosstalk victim interconnect due to the unity amplitude of voltage injection (i.e., $a_2(0) = 1$) at the input terminal of one of its nearest neighbors. The last relation, after a little algebra, turns out to translate into the relation $\cos(2\mathrm{Re}[\Delta\gamma]l_x) = 0$, which results in the following:

$$l_{x,el} = \frac{\pi}{2}\frac{1}{\sqrt{\omega RC}}\frac{C}{C_m} \tag{4.36}$$

The above expression can also be written in terms of bandwidth ($2\pi \times \mathrm{BW}_{el} = \omega_c = \frac{1}{RCL^2}$) of the electrical channel of length L, as follows:

$$l_{x,el} = \frac{\pi}{2}\sqrt{\frac{\omega_c}{\omega}\frac{C}{C_m}}L \tag{4.37}$$

The expression of $l_{x,el}$ shows that signal contamination by crosstalk in an electrical interconnects network aggravates as the ratio $\frac{C_m}{C}$ increases. The ratio of mutual capacitance to self-capacitance increases if the spacing between the neighbor interconnects reduces. In addition, there exists an inverse square-root relation between critical crosstalk length $l_{x,el}$ and the operating frequency ω. This relation shows why electrical interconnects at higher frequencies have become more prohibitive for a system that requires interference suppression for signal fidelity. This relation between $l_{x,el}$ and ω in electrical interconnects is in stark contrast with that we demonstrated in optical interconnects.

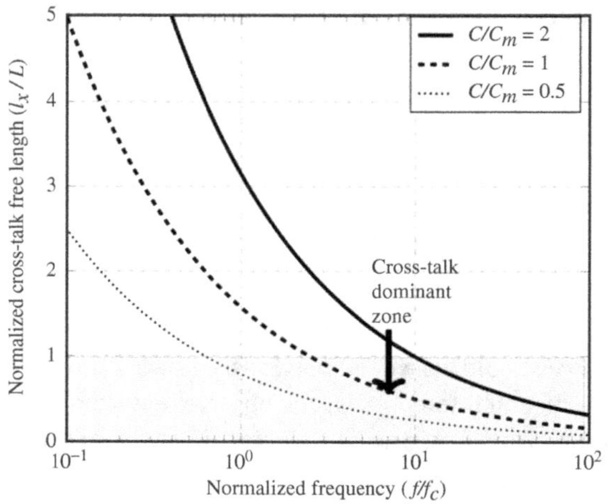

Figure 4.10 Normalized interconnect length that can remain free from crosstalk in a group of parallel electrical channels. The frequency scale is normalized with respect to the cut-off frequency of an isolated electrical channel. In the *y*-axis, the crosstalk free length is normalized with respect to the physical length of the channel.

Figure 4.10 shows the maximum allowable length for an electrical interconnect in order to avoid crosstalk. The diagram is drawn for a group of parallel data bus characterized by the ratio of mutual- and self-capacitance ($\frac{C_m}{C}$). In electrical channels, beyond a critical frequency, shown in the red circle in Figure 4.10, the crosstalk can become so severe that the undesired signal strength in the victim channel exceeds the signal strength in the intended channel. If we arrange the parallel data bus denser in an attempt to increase bandwidth density, this critical frequency will shift further left as the $\frac{C_m}{C}$ ratio increases.

4.12.3 Crosstalk in Spoof Plasmon Interconnects

Since SSPP interconnects can be appreciably coupled only if they are poised face-to-face (i.e., there is essentially no coupling between back-to-back SSPP interconnect pair), the operator $\mathbf{H}_{sp}$ in the relation $\mathbf{H}_{sp}\underline{\psi} = i\beta\underline{\psi}$ that describes the mode dynamics in an SSPP interconnects system is a block-diagonal matrix.

$$\mathbf{H}_{sp} = diag(\underline{\beta}, \underline{\beta}, \ldots) \tag{4.38}$$

Here, $\underline{\beta}$ is a 2×2 matrix with β_0 as its diagonal elements and K_{sp} as the off-diagonal ones, where β_0 is the propagation constant in an isolated SSPP waveguide that satisfies the dispersion relation in Eq. (4.3). Since it is unlikely that we would employ a weakly bound mode for information transfer, in our analysis we drop any self-coupling terms K_s and only incorporate cross-coupling co-efficient K_{sp} between two identical spoof plasmon waveguides "a" and "b," described as:

$$K_{sp} = -n_{\text{eff}}^2 \epsilon_0 \frac{\omega}{4} \int_{z=-t}^{t} (E_{az}^{(d)} E_{bz}^{(d)} - E_{ax}^{(d)} E_{bx}^{(d)}) dy + n_{\text{eff}}^2 \epsilon_0 \frac{\omega}{4} \frac{\partial}{\partial x} \left[\int_x \int_{z=-t-h}^{-t} E_{ax}^{(g)} . E_{bx}^{(d)} dxdy + \int_x \int_{z=t}^{t+h} E_{ax}^{(d)} . E_{bx}^{(g)} dxdy \right] \tag{4.39}$$

where the superscript (g) and (*d*) indicates field inside the groove and in the dielectric outside the groove, respectively, the field components follow Eq. (4.1) with subscript "a" and "b" for two different interconnects separated by a distance of 2*t*. Interestingly, if we introduce longitudinal misalignment between the grooves on two different SSPP interconnects facing each other, then the term in the third bracket in Eq. (4.39) of coupling strength would be of less importance due to the field localization effect in SSPP near grooves. We impose normalization conditions on the power propagation in a mode such that,

$$\frac{1}{2} \text{Re} \left[\int_{z=-t}^{\infty} E_{az}^{(d)} H_{ay}^{(d)*} dz \right] = 1 \tag{4.40}$$

If we assume $n = 0$ mode to be dominant, then from the above condition of normalization, we obtain

$$|A_{a0}|^2 = 4 \frac{(\beta_a^2 - \kappa_a^2)^2}{\omega \epsilon_0 \beta_a \kappa_a} \tag{4.41}$$

where κ_a is attenuation constant of the evanescent field in SSPP interconnect "a" along transverse direction z. We will obtain a similar expression for A_{b0} as the above one with the subscript "a" replaced by "b." Since we are considering interconnects of identical geometries, $\beta_a = \beta_b$, and $\kappa_a = \kappa_b$. Then, the coupling constant turns out to be

$$K_{sp} = 2tn_{\text{eff}}^2 \frac{\beta^4 + \kappa^4 - 3\beta^2\kappa^2}{\beta\kappa} e^{-2\kappa t} \tag{4.42}$$

Recall that, since we ignored any self-coupling terms, our obtained relation of K_{sp} is reliable as long as $2\kappa t >> 1$. The expression of K_{sp} can be further simplified if we are interested in the frequency regime, where $2\frac{\omega}{h}c \to \pi$ (i.e., $\frac{\omega}{h}c = \frac{\pi}{2} - \delta$, where δ is a small number). In that limiting case, following from the dispersion relation in Eq. (4.3), we obtain $\kappa = \lim_{\delta \to 0} \frac{\pi}{2h} \frac{a}{d} \frac{1}{\delta}$, and $\beta = \lim_{\delta \to 0} \kappa(1 + \frac{1}{2} \frac{d}{a} \delta) \approx \kappa$. Then, the coupling parameter K_{sp} becomes $\lim_{\frac{2\omega}{c} h \to \pi} |K_{sp}| = 2tn_{eff}^2 \kappa^2 \exp(-2\kappa t)$. The eigenvalues of the matrix $\bar{\beta}$ are $\beta_{\pm} = \beta_0 \pm K_{sp}$. It can be shown that, owing to unit signal injection at SSPP channel "a," its neighbor "b" will register a −3 dB interference at a length of $l_{x,sp}$ which abides by the following relation:

$$|b(l_{x,sp})| = \left| \frac{1}{2} e^{i\beta_0 l_x} (e^{iK_{sp} l_x} - e^{-iK_{sp} l_x}) \right| = \frac{1}{\sqrt{2}} \tag{4.43}$$

which eventually results in the following:

$$l_{x,sp} = \frac{\pi}{4} \frac{1}{|K_{sp}|} \tag{4.44}$$

In the limiting case of $\frac{2\omega}{h}c \to \pi$, critical crosstalk length in SSPP interconnects become

$$\lim_{2\omega h \to \pi c} l_{x,sp} = \lim_{2\omega h \to \pi c} \frac{h^2}{2\pi t n_{eff}^2} \left(\frac{d}{a}\right)^2 \left(\frac{\pi}{2} - \frac{\omega h}{c}\right)^2 \times \exp\left[\frac{\pi}{h}\frac{a}{d}\frac{2tc}{\pi c - 2\omega h}\right] \tag{4.45}$$

The above expression of critical crosstalk length in the SSPP interconnects system reveals that $l_{x,sp}$ diverges at a frequency $\omega = \frac{\pi c}{2h}$. Thus, we can achieve crosstalk suppression at some chosen high frequency by tailoring the geometry of SSPP structure, particularly the length h of the groove. Comparison of Eqs. (4.33) and (4.45) reveals that SSPP interconnects can mimic the behavior of optical interconnect upto some design-specific frequency in the sense that, the crosstalk suppression in both of the types of interconnects gets better with the choice of carrier of higher frequency.

Figure 4.11 illustrates the maximum length of interconnect that can remain necessarily unaffected by crosstalk at different frequencies in an arrangement of parallel SSPP channels. A typical geometry of the SSPP interconnects system is chosen with a groove length of $h = 250$ μm and period of $d = 375$ μm. The spacing

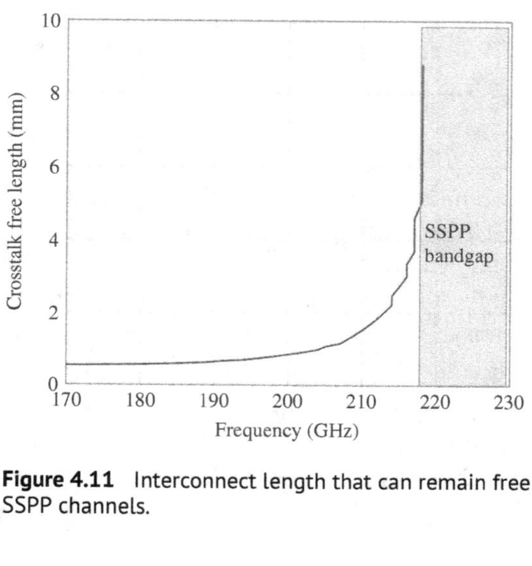

Figure 4.11 Interconnect length that can remain free of crosstalk in a group of parallel SSPP channels.

between neighboring interconnect is taken as $2h$ μm. It shows that SSPP interconnect demonstrates quite the opposite trend of crosstalk suppression compared to electrical interconnect: crosstalk in SSPP decreases with increasing frequencies, until we reach the band-edge frequency of SSPP channel. Thus, SSPP interconnect can be a superior choice over conventional electrical interconnect in circumstances where signal integrity is mainly limited by cross-interference among the nearby channels.

4.13 Dual Mode in Spoof Plasmon Waveguide

We figure out two different kinds of mode that can be supported by SSPP interconnect, namely electro-SSPP and opto-SSPP mode, as illustrated in Figure 4.12. Provided with a metallic ground plane, we can utilize SSPP interconnect in a way similar to pure electrical means of communication, where the signal is sent in the form of transverse electromagnetic (TEM) mode confined between the metallic wires and ground. The merit of a patterned metal wire (i.e., SSPP interconnect) in terms of bandwidth and crosstalk suppression doesn't change appreciably compared to that of an unpatterned, plain wire, since the product of resistance and capacitance per unit length $RC = \frac{\epsilon}{\sigma x_d}$ (x_d being the dielectric spacing between interconnect and the ground) is independent of the width of the wire. In addition to that, close to the resonant frequency (which can be designed to be at few THz), SSPP interconnects can transfer information in the form of surface confined spoof plasmon mode where metallic ground plane becomes redundant for waveguiding mechanism.

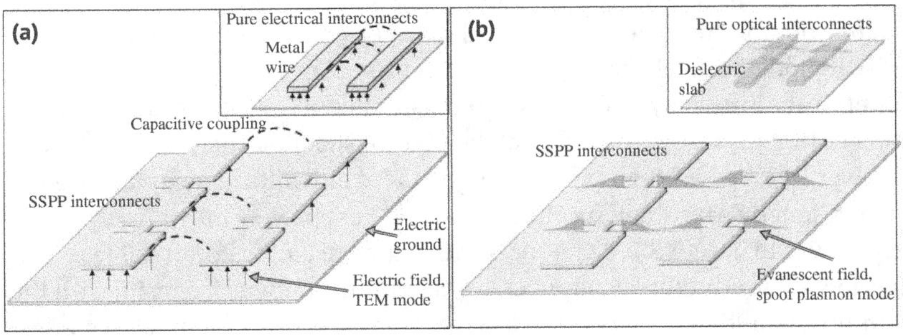

Figure 4.12 Illustration of two different kind of modes that can be supported in SSPP interconnects. (a) SSPP interconnects, in conjunction with a metallic ground plane are operating in electrical mode. The field distribution is analogous to the TEM mode that propagates in purely electrical interconnects (shown in inset). (b) SSPP interconnect operating in THz-optical mode. The transverse field distribution is analogous to that of optical interconnect (shown in inset).

As we showed, electrical and optical means of communication have a stark contrast of crosstalk behavior; electrical means are suitable for information transfer at low frequencies, while the optical ones are good for high frequency spectrum. In either case, we would prefer to minimize the energy spent for per bit of information transfer, E_b.

$$E_b = \frac{E_p}{\Delta t}\left(\frac{E_{ov}}{E_p N_p} + 1\right) \tag{4.46}$$

where $E_{\text{ov}} = E_{\text{source}} + E_{\text{modualate}} + E_{\text{detect}} + E_{\text{couple}}$ is the overhead energy associated with the carrier source (i.e., laser driver), modulator, detector and the coupling loss, N_p is the number of bit transferred within time Δt, and E_p is the energy associated with each bit. Clearly, an optical means of inter-chip communication will be energy-economic at a time when there is heavy traffic of information such that $N_p E_p >> E_{ov}$. Recall that we are concerned about centimeter-scale distance coverage between electrical chips, where it is not unlikely that the volume of information transfer between them will vary in a wide range from time to time. Thus, in a system with widely varying demands of data transfer rate, SSPP interconnects can be a potential alternative of energy-economic means of communication, as they allow us to switch between two different modes of communication, namely electro-SSPP and opto-SSPP mode.

4.14 Summary

In a nutshell, the present chapter comprises an information theory adopted for spoof plasmon channels which promises to be an energy-economic solution to traverse the broadly-discussed "last centimeter" distance between two chips. This is a unique solution for centimeter-scale data transfer that combines the advantages of both optical and electrical nature. The purpose of our analyses is to reveal and quantify the interplay between various parameters related to the information transfer process through a compact data bus of spoof plasmon channels. The authors' work addresses the optimization of the bandwidth density and signal propagation length. Our analysis shows that the best way of maximizing bandwidth density is to design SSPP channels with a groove length comparable to half of the period of the structure, leading to $\sim$ 1 Gbps/μm bandwidth density with the signal propagation length ~10 mm, and ~300 Gbps speed per channel. An effective resistivity of the quasi-optic SSPP channel has been introduced and used for the analysis of the noise characteristics of the channel at a given temperature. In the limit of thermal noise generated by the SSPP channel, we quantitatively demonstrated that an interplay exists between maximization of Shannon information capacity of the channel and the probability of bit error for digital data transfer via spoof plasmon channel. This work would potentially lay a foundation for the novel and prospective technique of spoof plasmon-based inter-chip communication to facilitate the next-generation fast and reliable data transfer process with high signal integrity.

References

1 Barry B Brey. *The Intel microprocessors*. Upper Saddle River, NJ, USA, Prentice Hall Press, 8th edition, 2008. ISBN 978-0-13-502645-8.

2 Yutaka Urino, Takanori Shimizu, Makoto Okano, Nobuaki Hatori, Masashige Ishizaka, Tsuyoshi Yamamoto, Takeshi Baba, Takeshi Akagawa, Suguru Akiyama, Tatsuya Usuki, Daisuke Okamoto, Makoto Miura, Masataka Noguchi, Junichi Fujikata, Daisuke Shimura, Hideaki Okayama, Tai Tsuchizawa, Toshifumi Watanabe, Koji Yamada, Seiichi Itabashi, Emiko Saito, Takahiro Nakamura, and Yasuhiko Arakawa. First demonstration of high density optical interconnects integrated with lasers, optical modulators, and photodetectors on single silicon substrate. *Optics Express*, 19(26):B159–B165, 2011. https://doi.org/10.1364/OE.19.00B159.

3 James D Meindl. Beyond Moore's Law: The interconnect era. *Computing in Science Engineering*, 5(1):20–24, 2003. https://doi.org/10.1109/MCISE.2003.1166548.

4 Mikhail Haurylau, Guoqing Chen, Hui Chen, Jidong Zhang, NicholasA. Nelson, DavidH. Albonesi, EbyG. Friedman, and PhilippeM. Fauchet. On-chip optical interconnect roadmap: Challenges and critical directions. *IEEE Journal of Selected Topics in Quantum Electronics*, 12(6):1699–1705, 2006. https://doi.org/10.1109/JSTQE.2006.880615.

5 Gerd Keiser. Optical fiber communications. In *Wiley encyclopedia of telecommunications*. John Wiley & Sons, Inc., 2003. ISBN 978-0-471-21928-6. https://doi.org/10.1002/0471219282.eot158.

6 René-Jean Essiambre, Gerhard Kramer, Peter J Winzer, Gerard J Foschini, and Bernhard Goebel. Capacity limits of optical fiber networks. *Journal of Lightwave Technology*, 28(4):662–701, 2010. https://www.osapublishing.org/abstract.cfm?uri=jlt-28-4-662.

7 AndyEu-Jin Lim, Junfeng Song, Qing Fang, Chao Li, Xiaoguang Tu, Ning Duan, KokKiong Chen, Roger Poh-Cher Tern, and Tsung-Yang Liow. Review of silicon photonics foundry efforts. *IEEE Journal of Selected Topics in Quantum Electronics*, 20(4):405–416, 2014. https://doi.org/10.1109/JSTQE.2013.2293274.

8 Jurgen Michel and Lionel Kimerling. Electronics and photonics: Convergence on a silicon platform. In *33rd European Conference and Exhibition of Optical Communication*, pages 1–2, September 2007. https://doi.org/10.1049/ic:20070261.

9 Dong J Shin, Kwan S Cho, H-C Ji, Beom S Lee, Sung G Kim, Jin K Bok, Sang H Choi, YH Shin, JH Kim, SY Lee, K. Y. Cho, B. J. Kuh, J. H. Shin, J. S. Lim, J. M. Kim, H. M. Choi, K. H. Ha, Y. D. Park, and C. H. Chung. Integration of silicon photonics into DRAM process. In *2013 Optical Fiber Communication Conference and Exposition and the National Fiber Optic Engineers Conference (OFC/NFOEC)*, pages 1–3, March 2013.

10 Gunther Roelkens, Utsav D. Dave, Alban Gassenq, Nannicha Hattasan, Chen Hu, Bart Kuyken, Francois Leo, Aditya Malik, Muhammad Muneeb, Eva Ryckeboer, Dorian Sanchez, Sarah Uvin, Ruijun Wang, Zeger Hens, Roel Baets, Yosuke Shimura, Federica Gencarelli, Benjamin Vincent, Roger Loo, Joris Van Campenhout, Laurent Cerutti, Jean-Baptiste Rodriguez, Eric Tournié, Xia Chen, Milos Nedeljkovic, Goran Mashanovich, Li Shen, Noel Healy, Anna C. Peacock, Xiaoping Liu, Richard Osgood, and William M. J. Green. Silicon-based photonic integration beyond the telecommunication wavelength range. *IEEE Journal of Selected Topics in Quantum Electronics*, 20(4):394–404, 2014. https://doi.org/10.1109/JSTQE.2013.2294460.

11 Rajeev K Dokania and Alyssa B Apsel. Analysis of challenges for on-chip optical interconnects. In *Proceedings of the 19th ACM Great Lakes Symposium on VLSI*, GLSVLSI '09, pages 275–280, New York, NY, USA, 2009. ACM. ISBN 978-1-60558-522-2. https://doi.org/10.1145/1531542.1531607

12 Yutaka Urino, Tsuyoshi Horikawa, Takahiro Nakamura, and Yasuhiko Arakawa. High density optical interconnects integrated with lasers, optical modulators and photodetectors on a single silicon chip. In *2013 Optical Fiber Communication Conference and Exposition and the National Fiber Optic Engineers Conference (OFC/NFOEC)*, pages 1–3, March 2013. https://doi.org/10.1364/OFC.2013.OM2J.6.

13 Petr Markov, Jason G Valentine, and Sharon M Weiss. Fiber-to-chip coupler designed using an optical transformation. *Optics Express*, 20(13):14705–14713, 2012. https://doi.org/10.1364/OE.20.014705

14 Gunther Roelkens, Pieter Dumon, Wim Bogaerts, Dries Van Thourhout, and Roel Baets. Efficient fiber to SOI photonic wire coupler fabricated using standard CMOS technology. In *2005 IEEE LEOS Annual Meeting Conference Proceedings*, pages 214–215, October 2005. doi: 10.1109/LEOS.2005.1547946.

15 Hoyeol Cho, Pawan Kapur, and Krishna C Saraswat. Power comparison between high-speed electrical and optical interconnects for interchip communication. *Journal of Lightwave Technology*, 22(9):2021, 2004.

16 Qun Jane Gu. THz interconnect: The last centimeter communication. *IEEE Communications Magazine*, 53(4):206–215, 2015. https://doi.org/10.1109/MCOM.2015.7081096.

17 JB Pendry, L Martín-Moreno, and FJ Garcia-Vidal. Mimicking surface plasmons with structured surfaces. *Science (New York, N.Y.)*, 305(5685):847–848, 2004. https://doi.org/10.1126/science.1098999.

18 Francisco J Garcia-Vidal, Luis Martín-Moreno, and John B Pendry. Surfaces with holes in them: New plasmonic metamaterials. *Journal of Optics A: Pure and Applied Optics*, 7(2):S97, 2005. https://doi.org/10.1088/1464-4258/7/2/013.

19 Xiaopeng Shen, Tie Jun Cui, Diego Martin-Cano, and Francisco J Garcia-Vidal. Conformal surface plasmons propagating on ultrathin and flexible films. *Proceedings of the National Academy of Sciences*, 110(1):40–45, 2013. https://doi.org/10.1073/pnas.1210417110.

20 Xiaopeng Shen and Tie Jun Cui. Planar plasmonic metamaterial on a thin film with nearly zero thickness. *Applied Physics Letters*, 102(21):211909, 2013. https://doi.org/10.1063/1.4808350.

21 Hui Feng Ma, Xiaopeng Shen, Qiang Cheng, Wei Xiang Jiang, and Tie Jun Cui. Broadband and high-efficiency conversion from guided waves to spoof surface plasmon polaritons. *Laser & Photonics Reviews*, 8(1):146–151, 2014. https://doi.org/10.1002/lpor.201300118.

22 Yuan Liang, Hao Yu, Junfeng Zhao, Wei Yang, and Yuangang Wang. An energy efficient and low cross-talk CMOS sub-THz I/O with surface-wave modulator and interconnect. In *2015 IEEE/ACM International Symposium on Low Power Electronics and Design (ISLPED)*, pages 110–115, July 2015a. https://doi.org/10.1109/ISLPED.2015.7273499.

23 Yuan Liang, Hao Yu, Hao Chi Zhang, Chang Yang, and Tie Jun Cui. On-chip sub-terahertz surface plasmon polariton transmission lines in CMOS. *Scientific Reports*, 5:14853, 2015b. https://doi.org/10.1038/srep14853.

24 Yuan Liang, Hao Yu, Jincai Wen, Anak Agung Alit Apriyana, Nan Li, Yu Luo, and Lingling Sun. On-chip sub-terahertz surface plasmon polariton transmission lines with mode converter in CMOS. *Scientific Reports*, 6:30063, 2016. https://doi.org/10.1038/srep30063.

25 David AB Miller. Attojoule optoelectronics for low-energy information processing and communications. *Journal of Lightwave Technology*, 35(3):346–396, 2017. https://doi.org/10.1109/JLT.2017.2647779.

26 David AB Miller. Optical interconnects to electronic chips. *Applied Optics*, 49(25):F59–F70, 2010. https://doi.org/10.1364/AO.49.000F59.

27 Mikhail Erementchouk, Soumitra Roy Joy, and Pinaki Mazumder. Electrodynamics of spoof plasmons in periodically corrugated waveguides. *Proceedings of the Royal Society A*, 472(2195):20160616, 2016. https://doi.org/10.1098/rspa.2016.0616.

28 Anastasia Rusina, Maxim Durach, and Mark I Stockman. Theory of spoof plasmons in real metals. *Applied Physics A*, 100(2):375–378, 2010. https://doi.org/10.1007/s00339-010-5866-y.

29 Kyungjun Song and Pinaki Mazumder. Active terahertz spoof surface plasmon polariton switch comprising the perfect conductor metamaterial. *IEEE Transactions on Electron Devices*, 56(11):2792–2799, 2009. https://doi.org/10.1109/TED.2009.2030838.

30 Shun-Lien Chuang. A coupled mode formulation by reciprocity and a variational principle. *Journal of Lightwave Technology*, 5(1):5–15, 1987. https://doi.org/10.1109/JLT.1987.1075409.

31 Wei-Ping Huang. Coupled-mode theory for optical waveguides: An overview. *JOSA A*, 11(3):963–983, 1994. https://doi.org/10.1364/JOSAA.11.000963.

32 David M Pozar. *Microwave engineering*. Hoboken, NJ, Wiley, 4th edition, 2011. ISBN 978-0-470-63155-3.

33 Josh A Conway, Subal Sahni, and Thomas Szkopek. Plasmonic interconnects versus conventional interconnects: A comparison of latency, crosstalk and energy costs. *Optics Express*, 15(8):4474–4484, 2007. https://doi.org/10.1364/OE.15.004474.

5

Augmented Bandwidth by Spoof Plasmonics

5.1 Introduction

In the preceding chapters, we delved into the multifaceted realm of performance metrics for spoof plasmon interconnects, encompassing factors such as bandwidth density, propagation length, noise characteristics, and the pursuit of an optimal design strategy to extract maximum bandwidth from a spoof plasmonic data bus. One of the remarkable capabilities of spoof plasmonic waveguide is to suppress crosstalk at high frequencies—and the frequency of high confinement is at the disposal of the designer. This attribute is in stark contrast with conventional metallic waveguides, composed of a signal path and a return path (i.e., the ground), where the crosstalk increases as frequency increases. Consider an isolated "ideal" channel, where all the factors limiting signal integrity, such as ohmic loss, radiation, dispersion, impedance mismatch, and thermal noise are absent. Such a hypothetical channel would have an infinite bandwidth. However, when several of such ideal waveguides are brought into close proximity to each other to form a data bus, the bandwidth of the bus becomes constrained solely by the degree of crosstalk. Within the confines of this chapter, our objective is to dissect and scrutinize the concept of bandwidth limitation imposed by crosstalk in both electrical and spoof plasmonic data buses. Furthermore, we aim to explore the intriguing notion of enhancing the bandwidth of an electrical data bus through the incorporation of spoof plasmonic modes.

5.2 Background Studies: Severity of Crosstalk

To illustrate how severely the crosstalk alone can affect the signal integrity in a dense array of parallel electrical channels, we simulated two parallel electrical wires with ideal conductivity, supported by a ground plane at the bottom of the

Artificial Plasmonics for VLSI Interconnects: Bridging the Gap between Electronics and Optics, First Edition. Soumitra R. Joy and Pinaki Mazumder.

Companion Website: https://www.wiley.com/go/artificialplasmonics

substrate, with the following geometry: length of wire is 3 inches, the ratio of wire width to interwire spacing is 0.6, relative permittivity of substrate is 3.38, and substrate thickness is 20 mil. The assumption of ideal conductivity of the metal results in a noise-free, infinite bandwidth of an isolated channel. We will shortly see that, even under such perfect conditions, the ability of signal transfer of an interconnect becomes limited by the crosstalk. Figure 5.1 shows the degree of crosstalk in a two-channel electrical data bus. It reveals that, while signal integrity can remain unperturbed at the low-frequency regime (≤ 10 GHz), it begins to aggravate at high frequencies, as other undesired signals, especially the far-end crosstalk begin to compete with or even can exceed the power level of the transmitted signal. Thus, even in the absence of any material loss, and even with the assumption of infinite bandwidth of a standalone channel, the high-frequency regime becomes the most troublesome spectrum for data transfer with signal integrity. The loss of signal integrity originates from the geometric arrangement of the channels in a densely packed parallel data bus. In the following sections, we shall discuss some state-of-the-art techniques for addressing the issue of crosstalk in VLSI chips. Then we shall discuss their limitations of the standard techniques, and demonstrate how we can engineer the geometric morphology of the wire to enable spoof plasmonic mode propagation in order to suppress crosstalk without running into the problem associated with current methods of crosstalk reduction. We shall also experimentally verify the proposed technique.

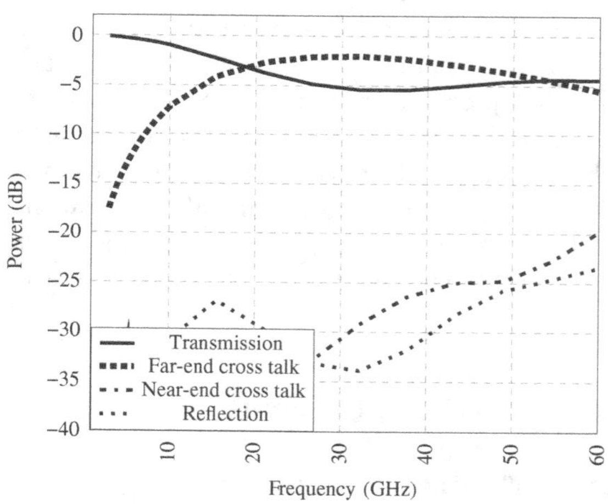

Figure 5.1 Near-end and far-end crosstalk in parallel electrical data bus. At frequencies over 10 GHz, the signal intended to be received at the destination-end of the channel can become indistinguishable from the signal emanating from the crosstalk.

5.3 Conventional Strategies for Crosstalk Reduction

Crosstalk originates from electromagnetic coupling between two channels. Typically, the coupling is characterized by mutual inductance and mutual capacitance. The parasitic coupling increases with higher frequencies and also with denser channel arrangement. Wire parasitic reduction is a significant challenge in high-density device and packaging design due to three main factors: stringent power requirements, difficulties in heat management, and limited area footprint [1–3]. We can classify different techniques of crosstalk minimization at different levels of system abstraction. At the network level, crosstalks are reduced by spatial dilation or wavelength dilation techniques. The practitioners maintain sufficient space between the traces that carry high-speed signals. A rule of thumb is to interpose between adjacent traces a space at least three times their width. In spatial dilation technique, a 2×2 optical switch with crosstalk ϵ is replaced by a dilated switch, with some redundant/unused ports. This dilation would reduce the crosstalk to ϵ^2. A drawback of dilation is that it cannot be achieved without a significant increase in the number of switches. Usually, the number of switches is doubled. Another approach to reduce switch crosstalk in a WDM network is to use wavelength dilation in the switches. This is particularly useful if a single switch is to handle multiple wavelengths. To reduce the interchannel cross-talk, we can use two switches instead of one. The first switch handles the odd-numbered channels, and the second the even-numbered channels. This effectively doubles the channel spacing as far as cross-talk is concerned. However, the cost is that twice as many switches are required. cross-talk reduction techniques at the circuit level typically involve differential signaling, virtual grounding etc. Differential signaling involves two signal lines with the same amplitude but opposite polarity that form a single high-speed signal. Since the receiver obtains the signal as the difference between the voltages of the two traces, and with the assumption that the electromagnetic noise equally affects both traces, the signal is supposed to maintain high integrity even in the presence of significant external noise. However, this technique may not work if there is an imbalance in the two traces (stemming from temperature gradient, or process variations). Virtual grounding technique is also discussed among current literature as a means to address crosstalk in circuits. Two-dimensional resistive sensor arrays that utilize shared row and column connections to simplify the interconnect complexity suffer from the crosstalk problem among their elements introduced due to the interconnection overloading. Saxena et al. [4] proposed mitigating crossalk under such circumstances by putting all of the row nodes at virtually equal potential using virtual ground of high-gain operational amplifiers (opamps) in negative feedback. The circuit solves the crosstalk problem to a large extent and provides faster scanning. The drawback of the technique is that it requires a large number of opamps, and is therefore energy-hungry. Therefore, traditional solutions like space dilation, virtual grounding, and compensation circuit design often prove impractical due to high idle-state power consumption.

Key techniques for reducing parasitic at the device and package levels include incorporating new low-k dielectric materials [5], new organization of interconnect networks [6], and new processing of substrate [7]. With the continuous increase in integration density of IC devices, there is an increase in resistance and capacitance, and obvious increase in the parasitic effect, which results in increased power consumption, signal lag, and crosstalk. The calculation formula of signal lag time depends on the dielectric constant ϵ of the interlayer dielectric material (ILD), as follows:

$$\tau = RC = 2\rho\epsilon_r\epsilon_0\left(\frac{4L^2}{P^2} + \frac{L^2}{T^2}\right) \tag{5.1}$$

where τ is the signal delay, R is the resistance, C is the capacitance, ρ is the specific resistance of the conductor, L is the length of the conductor, T is the thickness of the conductor, and P is the distance between two neighboring conducting wires.

According to Eq. (5.1), the reduction of the dielectric constant of the ILD material would result in lower signal delay. Thus, the study and application of new materials possessing low-k ($k \leq 3.0$) and ultra-low-k ($k \leq 2.0$) values have gained considerable interest. Polyimide (PI) is a high-performance engineering polymeric material. Although PI contains polar group such as $C = O$, the carbonyl groups in the conjugated imide ring are symmetrical, so the mobility of free electron is greatly restricted, and they endow excellent electrical insulation property to PI. However, the k of common aromatic PI is generally about 3.4, it cannot meet the requirement of microelectronics industry for the continuous decrease in k of dielectric materials. According to Clausius–Mossotti equation, there are two ways to reduce dielectric constant k: one, by reducing molecular polarizability or two, by decreasing the number of polarized molecules per unit volume. The other possible option is to introduce air-hole/porosity. Unfortunately, these efforts either offer limited reduction in the dielectric constant-k value, or they come with a price of increased phonon scattering, reduced thermal conductivity [8], and mechanical stability [9, 10], see Figure 5.2.

To address these issues, we, the authors, propose a novel approach to achieving effective ultralow-k ($k \leq 1$) values in the devices. Instead of altering the substrate itself, the method focuses on periodically patterning the metallic striplines, with the potential reduction of electromagnetic coupling among wires, effectively lowering the dielectric constant while using standard materials. The reduction of the effective dielectric constant eventually results in subdued crosstalk, low radiation loss, high bandwidth, low latency, and high signal integrity.

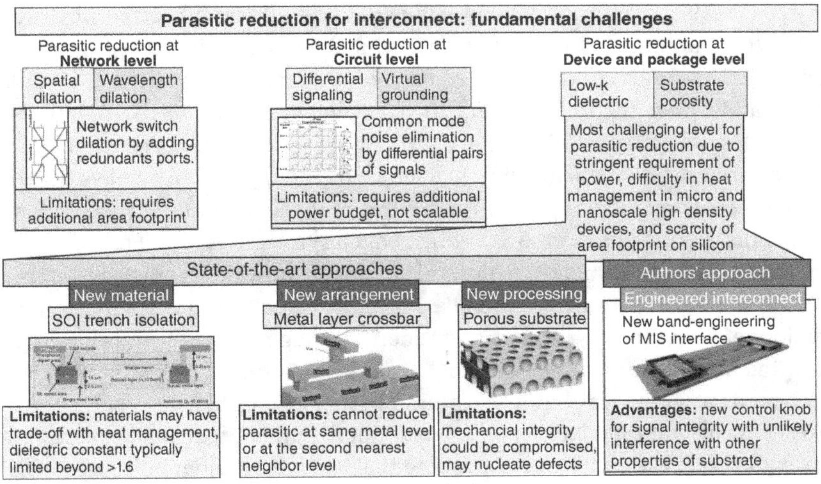

Figure 5.2 A summary of different techniques for interconnect parasitic reduction at device and packaging level. State-of-the-art parasitic reduction techniques primarily involve altering the substrate, which has met with significant challenges for trade-off between electrical, mechanical, and thermal integrity. The authors of this book propose a new pathway to reduce wire parasitic by engineering the metallic interconnect without any chemical or structural change to the Si substrate.

5.4 Authors' Contribution: Dealing with Crosstalk in Data Bus

As we are moving toward higher-speed and higher-spectrum technologies, crosstalk noise in VLSI chips is growing as a serious issue. Besides the techniques of crosstalk reduction discussed in the previous section, some other ideas include new fabrics or new architectures that can minimize crosstalk, such as the crossbar architecture at different metallic layers on-chip. Another technique includes a new encoding mechanism for raw data, so that the signal can withstand crosstalk corruption.

We, the authors, laid out a new path of research direction in addressing the issue of crosstalk. We investigated the possibility of crosstalk suppression at the device level. We showed that by changing the mechanism of electromagnetic waveguiding, the device can achieve an inherent resilience to crosstalk.

5.4.1 Advantages of Proposed Interconnect

Standard interconnect structures typically consist of plain metallic striplines on a substrate. The transverse electric mode (TEM) in regular electrical interconnects is subject to severe cross-coupling among neighboring interconnects at high frequencies. This unintended phenomenon is characterized by increased parasitic capacitance C_p, leading to compromised bandwidth, represented as $1/R(C_s + C_p)$, where C_s and C_p are the self and parasitic capacitances, respectively, and excessive power loss, expressed as $P \propto (C_s + C_p)V^2$. By artificially engineering interconnects, we can mitigate the parasitic capacitance by modifying the band structure in phase space using various geometric features of the waveguide, such as indentation depths, periodicities, groove widths, edge asymmetries, etc. For instance, indentation depth of the metal can control channel bandwidth, while edge asymmetry helps to reduce parasitic crosstalk. The engineered waveguide's hyperbolic dispersion relation can prevent undesired energy leakage (crosstalk and charging/discharging loss) into the side-channels. Figure 5.3 summarizes the mathematical models we, the authors, have employed to describe these engineered features. Our primary model is suggestive of improvement in signal propagation by the engineered features. This has the potential to enable a coherent engineering methodology for optimizing the control of signal routing through tailored electromagnetic band structures.

The first part of this chapter portrays a detailed theoretical model for the estimation of crosstalk in a standard electrical data bus (comprised of microstrip lines over a ground plane) as well as spoof plasmonic data bus (comprised of a single metallic conductor without any ground plane), and analyzes the maximum bandwidth that can be utilized in a crosstalk-mediated environment. The second part proposes a novel concept of bandwidth augmentation of electrical data bus by showing that the channels of the standard bus, if periodically patterned, can convert a fraction of the signal energy into spoof plasmonic TM mode, which are immune to crosstalk. This can reduce the bit-error rate by at least one order and can also increase the speed of baseband data transfer by a factor of 3.

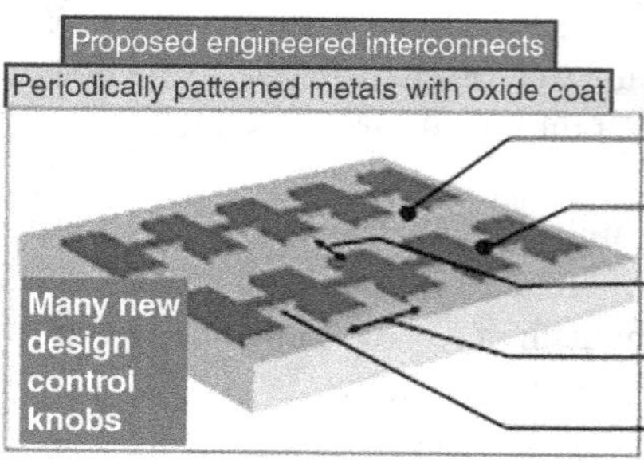

	Engineered features	What does it control?	How is it modeled?
1	Indentation depth	Channel bandwidth	2D-Bloch-wave theory
2	Groove width	Signal's group velocity	Poisson's electrostatic
3	Edge-asymmetry	Signal's leakage	Rigorous couple wave theory
4	Periodicity	Modal confinement	Fourier diffraction

Figure 5.3 In contrast to standard interconnect network consisting of plain metals and substrate, the authors propose a design of periodically patterned metal, which offers new control knobs for superior control over electromagnetic wave steering, and can reduce wire parasitic.

5.4.2 Crosstalk Limited Bandwidth in Lossless Electrical Bus

For a pair of interconnects with impedance-matched terminals, whose modal transmission is described by a propagation vector β and coupling factor K_c, the far-end crosstalk signal $a_2(x)$ can be described as the following:

$$a_{cx}(L) = ie^{i\beta L}\sin(K_c L)a_{in}(0) \tag{5.2}$$

where $a_{in}(0)$ is the injected signal at the input-end ($x = 0$) of the aggressor wire. The received signal transmitted through the aggressor wire $a_{tr}(L)$ is the following:

$$a_{tr}(L) = e^{i\beta L}\cos(K_c L)a_{in}(0) \tag{5.3}$$

In a pair of lossless (i.e., resistance of the wire $R = 0$, and substrate conductance $G = 0$) electrical interconnect, if we assume that the only coupling occurs through the mutual capacitance C_m, then the scalar propagating wave $V_i = V_{i0}e^{-\gamma x}$ abides by the following relation,

$$H_{el}\,\underline{V} = \gamma^2\,\underline{V} \tag{5.4}$$

where $\underline{V}$ is the two-elements column vector containing the propagating voltage V_1 and V_2, and the operator H_{el} is a 2×2 matrix:

$$H_{el} = \begin{bmatrix} U & Q \\ Q & U \end{bmatrix} \tag{5.5}$$

where $U = -\omega^2 LC$ and $Q = -\omega^2 LC_m$. The propagation constant γ of the electrical interconnect-pair can be found from the solution of the following equation: $det(H_{el} - \gamma^2 I) = 0$. It can be shown that the propagation constant of the interconnect-pair is

$$\gamma_{\pm} = i\omega\sqrt{L(C \pm C_m)} \approx i\omega\sqrt{LC}(1 \pm \frac{C_m}{2C}) = \gamma_0 \pm \Delta\gamma \tag{5.6}$$

where the approximation is made on the assumption that the mutual capacitance C_m is much weaker than self-capacitance, C.

Thus, for a pair of lossless electrical interconnects, Eq. (5.2), with the replacement of $\beta = \beta_{el} = \omega\sqrt{LC}$ and $K_c = K_{el} = \frac{1}{2}\beta_{el}\frac{C_m}{C}$ describes the crosstalk signal.

In order to distinguish the true bit from the erroneous bit emerging from crosstalk, the power of the signal transmitted to the desired receiver P_{tr} should be greater than the power P_{cx} unintentionally delivered to the receiver due to cross-coupling between adjacent wires. So, the bandwidth in a pair of lossless electrical interconnects can be defined by the spectral width for which the ratio $\frac{P_{cx}}{P_{tr}}$ is less than unity. For frequency components for which $\frac{P_{cx}}{P_{tr}} \geq 1$, we will have no way to isolate true signal from the false one.

The ratio of power components $\frac{P_{cx}}{P_{tr}}$ in a lossless electrical interconnect pair can be related to the signal amplitudes of Eqs. (5.2) and (5.3) as follows:

$$\frac{P_{cx}}{P_{tr}} = |\frac{a_{cx}}{a_{tr}}|^2 = \tan^2(K_c L) = \tan^2\left(\omega\sqrt{LC}\frac{C_m}{2C}L\right) \tag{5.7}$$

The above equation tells us that, with the increase of frequency, the ratio of power components $\frac{P_{cx}}{P_{tr}}$ diverges. This is why an electrical data bus can only support data transfer at the lower end of the spectrum. Figure 5.4 shows the various components of power flow in a typical electrical interconnect pair, resulting from the theoretical analysis. The length of each interconnect is taken as 120 mm, the width as 20 mm, substrate's dielectric constant is taken as 4, and the substrate

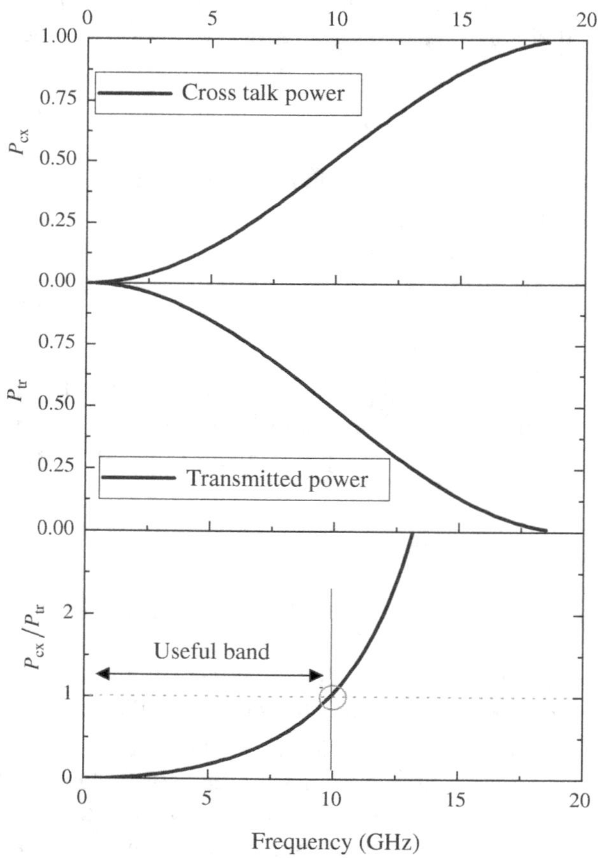

Figure 5.4 Various components of power flow in a 120 mm long electrical interconnect-pair, whose self capacitance is ~ 33 times of mutual capacitance.

thickness is taken as 10 mm, the ratio of mutual capacitance-to-self capacitance of the wire is taken as 0.03. The crosstalk power P_{cx} increases with frequencies, and the transmitted power P_{tr} reduces. The profile of power ratio $\frac{P_{cx}}{P_{tr}}$ can act as a figure of merit for estimating the available bandwidth in parallel electrical interconnects. As long as $\frac{P_{cx}}{P_{tr}}$ remains below unity, we will be able to distinguish between true signal and crosstalk-engendered erroneous signal, given the system is noiseless and lossless. The bottom plot of Figure 5.4 shows the spectral profile of power ratio $\frac{P_{cx}}{P_{tr}}$ and indicates the useful bandwidth for information transfer. Other practical considerations, such as noise and non-ideal signal detectors reduce the available bandwidth further, however.

5.4.3 Crosstalk Limited Bandwidth in Lossless SSPP Bus

The dominant tangential field components guided by the lateral edge of the SSPP waveguide are the following:

$$E_z = \hat{z} A_0 e^{-\kappa(z+s/2)} e^{i\beta x} \tag{5.8}$$

$$H_y = \hat{y} i \frac{1}{\omega\mu} \frac{\partial}{\partial x} E_z = \frac{\beta}{\omega\mu} E_z \tag{5.9}$$

where z is the lateral direction on the plane of substrate, x is the longitudinal direction/ direction of propagation, and s is the spacing between neighbor SSPP channels. The normalization condition of total power propagation is

$$\frac{1}{2} \int_{-s/2}^{\infty} E_z H_y^* dz = 1 \tag{5.10}$$

which yields the following expression of field magnitude,

$$|A_0|^2 = 4\kappa \frac{\omega\mu}{\beta} \tag{5.11}$$

Then, the coupling constant for SSPP wave can be found from the overlap integral:

$$\begin{aligned} K'_{sp} &= n^2 \epsilon_0 \frac{\omega}{4} \int_{-s/2}^{s/2} |A_0|^2 e^{-2\kappa s/2} dz \\ &= \kappa s e^{-\kappa s} \frac{k^2}{\beta} \end{aligned} \tag{5.12}$$

where $k = \omega\sqrt{\mu\epsilon_0\epsilon_r}$ is the free space wavenumber in the homogeneous medium. Equation (5.12) actually overestimates the coupling strength of the SSPP mode in the thin waveguide, as it is derived assuming all the power is guided by the edge of the SSPP interconnect, whereas, in reality, the mode exists around the entire cross-section of the waveguide. The fraction of the SSPP field guided by the edge is, however, the most dominant contributor to crosstalk due to the proximity to the neighboring waveguide. We can introduce a form factor *FF* in Eq. (5.12) to rewrite it as $K_{sp} = FF \times K_{sp}'$ in order to account for the fraction of the SSPP mode that does not have significant contribution to the crosstalk. We expect the form factor to be related to the geometry of the SSPP waveguide in the following manner:

$$FF \sim \frac{t_{\text{eff}}}{t_{\text{eff}} + h} \tag{5.13}$$

where h is the groove length and is the limiting dimension of the width of SSPP waveguide, and t_{eff} is the effective thickness of the thin SSPP waveguide.

We need to be mindful of the fact that Eq. (5.2) accounts for only one mechanism of crosstalk: coupling between the nearest neighbor channels. There might be another mechanism of crosstalk: due to the finite lateral extent of the propagating mode, a fraction of the power of the wave arriving at the receiving end of the intended channel is captured by the receiver of the neighbor channel. This second mechanism is especially important to take into account for SSPP wave at low frequencies, since SSPP mode increasingly extends out from the edge of the waveguide as we move toward lower frequencies. Thus, with the consideration of the modal extent of the wave at the receiving end, and the finite electromagnetic cross-section area of the receiver, we can write the following equations for transmitted power P_{tr} and crosstalk power P_{cx} received at the end of a parallel SSPP waveguide pair.

$$\begin{aligned} P_{\text{tr}} &= \frac{1}{2} Z_0 |A_0|^2 e^{-\kappa s} \int_{-s/2}^{0} \cos^2(K_{\text{sp}} L) e^{-2\kappa z} dz \\ &= \cos^2(K_{\text{sp}} L)(1 - e^{-\kappa s}) \end{aligned} \tag{5.14}$$

$$\begin{aligned} P_{\text{cx}} &= \sum (P_{\text{vm}}, P_{\text{im}}) \\ &= \frac{1}{2} Z_0 |A_0|^2 e^{-\kappa s} \int_{0}^{s/2} (\cos^2(K_{\text{sp}} L) e^{-2\kappa z} + \sin^2(K_{\text{sp}} L) e^{2\kappa z}) dz \\ &= \left[\cos^2(K_{\text{sp}} L)(e^{-\kappa s} - e^{-2\kappa s})\right] + \left[\sin^2(K_{\text{sp}} L)(1 - e^{-\kappa s})\right] \end{aligned} \tag{5.15}$$

where Z_0 is the impedance $\frac{E_z}{H_y} = \frac{\beta}{\omega\mu}$, and injected power at a given channel $P_{\text{inj}} = 1$. The electromagnetic cross-width of the receiver is assumed to be half of the spacing between the two parallel SSPP channels. The first term of the expression of crosstalk power at a given frequency P_{cx} in Eq. (5.15) indicates the fraction of power of the intended signal that is received by the receiver of the neighbor channel. Since this term corresponds to the power contained in the vestige of the exponential tail of the propagating mode that is off-aligned to the intended receiver, we will call it P_{vm}, where the subscript vm stands for vestigial mode.

On the contrary, the second term in Eq. (5.15) corresponds to the power of the guided mode in the neighbor channel that is electromagnetically induced from the mode propagating through the intended channel, let us call the associated power as P_{im}, where the subscript im stands for induced mode. The total crosstalk power P_{cx} is thus made up of the contribution of P_{vm} and P_{im}. The cross-width profile of the propagating mode in the intended channel is described by the form of $e^{-\kappa(z+\frac{s}{2})}$, spanning in the spatial range of $-\frac{s}{2} < z < \infty$, and that of the unintentionally coupled mode in the neighbor channel is described by the form of $e^{\kappa(z-\frac{s}{2})}$, where the coupled mode spans within the range of $-\infty < z < \frac{s}{2}$.

Figure 5.5 shows the result of the theoretical analysis of transmitted power and crosstalk power in a pair of SSPP waveguides. The specifications of the corresponding SSPP channel-pair are the following: groove length $h = 4$ mm, groove width $a = 2$ mm, period $d = 6$ mm, spacing between the waveguide pair $s = 35$ mm, length of each channel $L = 120$ mm. The lateral size of the receiver at each channel is taken to be half of the waveguide spacing ($s/2$). The channel-pair is considered to be freely hanging in the air. The injected power at one channel is

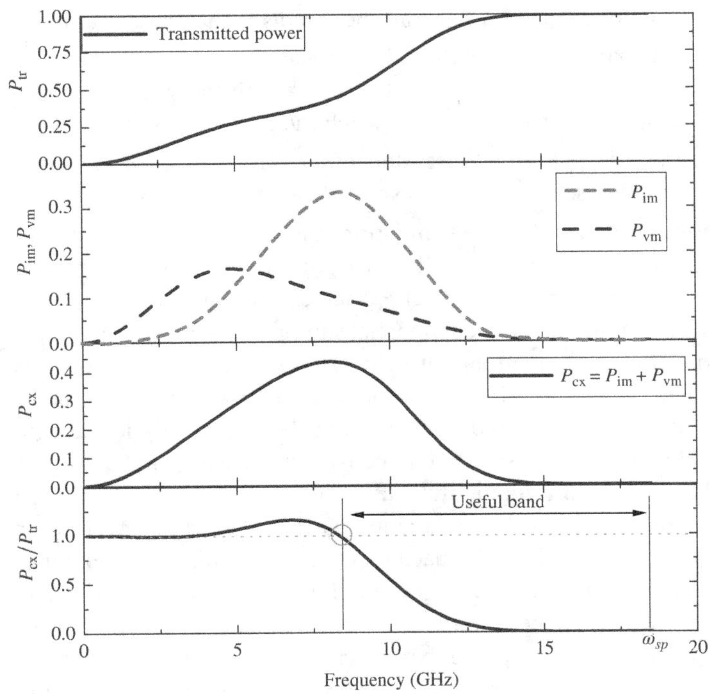

Figure 5.5 Various components of power flow in a 120 mm long lossless SSPP waveguide-pair, whose spoof plasmon resonance is at ~ 18 GHz.

taken as unity, i.e., $P_{inj} = 1$. The form factor *FF*, which is the ratio of power guided by the edge of SSPP to that guided by the entire cross-section of SSPP is taken as 1/10. As seen from Figure 5.5, the transmitted power P_{tr} captured by the receiver at the intended channel initially remains very low due to very poor confinement of field at lower frequencies, but gradually increases toward higher frequencies, as the field confinement near spoof plasmon resonance frequency strengthens. On the other hand, the crosstalk power P_{cx} consists of two components: induced power P_{im} at the victim waveguide, and the vestigial power P_{vm} contained in the lateral extent of SSPP mode that is captured by the receiver of the neighbor (victim) channel. The induced power P_{im} is small at low frequencies, since the electromagnetic coupling weakens around DC frequency. However, electromagnetic coupling in SSPP waveguide also diminishes at frequencies close to the resonant frequency ω_{sp}. In between the DC frequency and spoof plasmon resonant frequency, the magnitude of induced power can vary in different ways, depending on the form factor *FF*, and on the length *L* of the waveguides. The vestigial power P_{vm} is small close to DC frequencies, since the weak confinement of the mode leads to a too-wide cross-section of the guided mode to be captured efficiently by the receiver of finite size. P_{vm} also reduces at frequencies close to ω_{sp} for obvious reason. The most important figure among all these plots is the power ratio: $\frac{P_{cx}}{P_{tr}}$. If the ratio is greater than unity, we will have no means to distinguish between true signal and false signal, even if the detector is ideal and the system is noiseless. Therefore, the part of the spectrum over which the power ratio $\frac{P_{cx}}{P_{tr}} < 1$ is the useful band for information transfer. The bottom of Figure 5.5 shows the profile of $\frac{P_{cx}}{P_{tr}}$ and also indicates the useful band in our specific case of SSPP waveguide pair.

Figure 5.6 presents the 3D full-wave simulation results of various components of power propagating through a pair of SSPP waveguides. The trend of the transmitted power P_{tr} and that of crosstalk power P_{cx} agrees with our theoretical analysis. It can be noted that the standard commercial simulation tools, such as HFSS, provide conventional input and output ports that can only deal with TEM mode. Thus, we require a conversion mechanism at each end of the channel to study the behavior of the SSPP mode, and the converters have to be of large physical dimension for reasonable electrical-to-SSPP mode conversion efficiency. This "bulky converter" precludes us from bringing the SSPP wires too close to each other in order to study the impact of the SSPP mode coupling. Thus, in this particular setup of simulation, due to a relatively wide spacing between the adjacent SSPP wires (spacing = 33 mm), the crosstalk power essentially derives from the power of the vestigial mode. We also superimposed the transmission power through the same the SSPP wire with its neighbor being physically removed. As seen in Figure 5.6, the transmission power remains pretty much similar regardless of

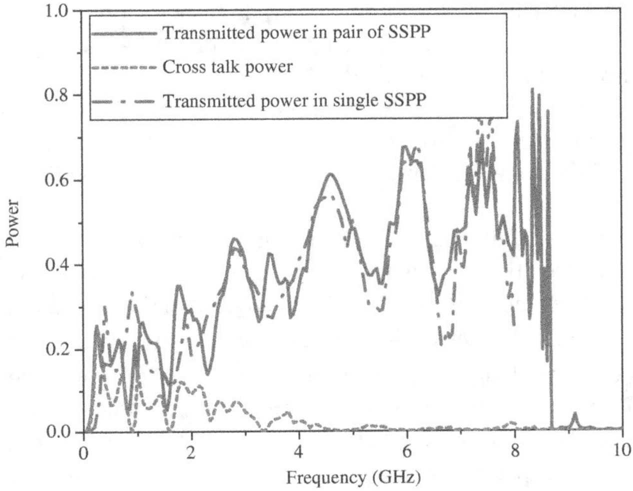

Figure 5.6 Simulation result of transmitted power and crosstalk power in a pair of SSPP waveguides of specific geometry, as shown in Figure 5.7.

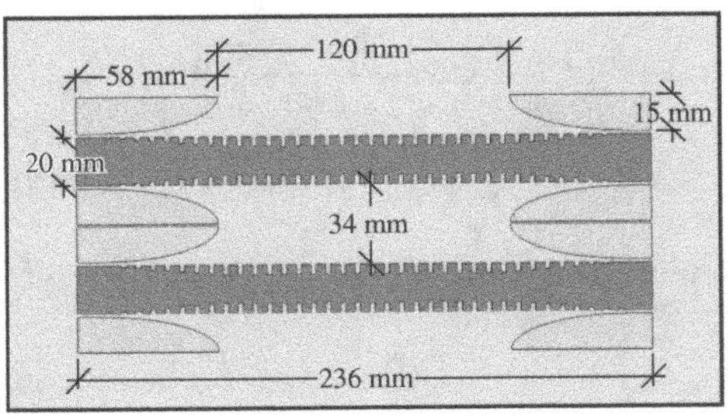

Figure 5.7 Geometry of pair of SSPP waveguides simulated to estimate the transmission power and crosstalk power.

whether the SSPP wire is accompanied by a neighbor channel or not in its proximity. This observation corroborates our analysis that the coupling strength between the SSPP wires at such a distance is too weak to make a significant impact.

5.5 Hybrid-SSPP Mode: Theory and Property Analysis

In Chapter 4, we informally introduced the concept of Hybrid spoof plasmonic mode, which can be supported by a patterned metallic stripline over a ground plane with substrate in between. We name it "Hybrid-SSPP" mode due to its mixed constituents of quasi-TEM mode and surface-wave TM mode (SSPP mode). Hybrid-SSPP mode is a promising means of data transfer, since it is capable of transmitting baseband signal without the need for any high-frequency carrier source, modulator, and detector. In current chapter, we shall provide a theory for describing the behavior of such hybrid-SSPP mode.

We will first show that such a transmission line supporting hybrid-SSPP mode can be constructed by attaching many capacitors C_p in parallel with the distributed inductive elements (L) of the transmission line. Figure 5.8 shows the modified transmission line model.

Let L stand for inductance per unit length of the transmission line, and C for per unit length capacitance between the metal wires and the ground plane. Therefore, the inductance of a small piece of transmission line of length Δz is $L\Delta z$, and the self-capacitance of the same is $C\Delta z$. And $\frac{C_p}{\Delta z}$ is the capacitance in parallel with each piece of transmission line of length Δz. For simplification, we shall ignore the resistance of the metallic wire, and will only consider the L, C, C_p parameters. Let the current propagating through an inductive element $L\Delta z$ is I_1, and the current propagating through parallel capacitor $\frac{C_p}{\Delta z}$ is I_2. Then, using Kirchhoff's voltage and current law, we get,

$$v(z,t) = v(z + \Delta z, t) + L\Delta z\frac{di_1(z,t)}{dt} \tag{5.16}$$

$$i_1(z + \Delta, t) + i_2(z + \Delta z, t) = i_1(z,t) + i_2(z,t) - C\Delta z\frac{dv(z + \Delta z, t)}{dt} \tag{5.17}$$

$$L\Delta z\frac{di_1(z,t)}{dt} = \frac{\Delta z}{C_p}\int i_2(z,t)dt \tag{5.18}$$

for $\Delta z \to 0$, we can rewrite the above-mentioned equations in the following way:

$$-\frac{dv(z,t)}{dz} = L\frac{di_1(z,t)}{dt} \tag{5.19}$$

$$\frac{di_1(z,t)}{dz} + \frac{di_2(z,t)}{dz} = -C\frac{dv}{dt} \tag{5.20}$$

$$L\frac{di_1(z,t)}{dt} = \frac{1}{C_p}\int i_2(z,t)dt \tag{5.21}$$

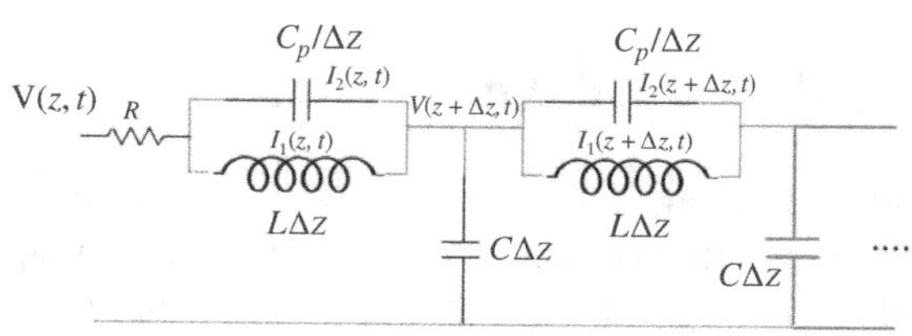

Figure 5.8 Modified transmission line model with distributed elements to represent hybrid-SSPP mode propagation.

If we assume the a priori solution of this system to be of the following form:

$$v = V_0 e^{ikz} e^{i\omega t} \tag{5.22}$$

$$i_1 = I_{1,0} e^{ikz} e^{i\omega t} \tag{5.23}$$

$$i_2 = I_{2,0} e^{ikz} e^{i\omega t} \tag{5.24}$$

Then, we can simplify the above equations as follows:

$$-ikV = i\omega L I_1 \tag{5.25}$$

$$ik(I_1 + I_2) = -i\omega CV \tag{5.26}$$

$$i\omega L I_1 = \frac{1}{i\omega C_p} I_2 \tag{5.27}$$

Using the above equations, we can write

$$i\omega CV = ik\frac{k}{\omega L} V(1 - \omega^2 L C_p) \tag{5.28}$$

which leads to the following relation between propagation wave vector k with the frequency ω

$$k = \frac{\omega\sqrt{LC}}{\sqrt{1 - \omega^2 L C_p}} \tag{5.29}$$

It is evident that k will be real as long as $\omega^2 L C_p < 1$. Also, the value of k is approximately $\omega\sqrt{LC}$ as long as $\omega << \frac{1}{LC_p}$. As the frequency approaches the resonance frequency of the SSPP channel, the value of k will become larger than $\omega\sqrt{LC}$. Beyond the SSPP resonance frequency, k will be imaginary, indicating bandstop zone for the SSPP waveguide.

By comparing the propagation constant of a regular *RLCG* transmission line $k = \omega\sqrt{LC}$, with Eq. (5.29) for a transmission line where a capacitor $\frac{C_p}{\Delta z}$ is in parallel with the inductor $L\Delta z$, we can say that the latter one can be modeled as a regular *RLCG* transmission line by replacing the C with an effective capacitance C_{eff} such that

$$C_{\text{eff}} = \frac{C}{1 - \omega^2 L C_p} \tag{5.30}$$

Notice that if ω approaches a value of $\frac{1}{\sqrt{LC_p}}$, the C_{eff} can take a value much larger than the capacitance C between the wire and the ground. Thus, we can increase the effective value of self-capacitance arbitrarily at any chosen design frequency. This behavior at frequencies close to the cut-off frequency $\omega = \frac{1}{\sqrt{LC_p}}$ is analogous to the behavior of spoof plasmonic mode.

In practice, we cannot attach capacitors in parallel to infinitesimally small pieces of the wire. However, if we create periodic grooves on the edges of otherwise plain metal striplines, the teeth of the grooved structures would form a capacitance between them, and result in the property that can be described by Eq. (5.29). The only modification would be the following: instead of the propagation constant k approaching infinity, the maximum value would be limited by $\frac{\pi}{d}$, where d is the periodicity of the grooved structure, i.e., $k_{\max} = \frac{\pi}{d}$. Notice that k approaches ∞ if $d \rightarrow 0$.

We can also notice that, the added parallel capacitance C_p would create a displacement current along the direction of signal propagation, in addition to the displacement current created by the self-capacitance C between the signal wire and the ground plane. Thus, there should be two different modes of signal propagation: the (quasi-)TEM mode where the dominant electric field is transverse to the propagation axis, and the TM mode where the dominant electric field is parallel to the propagation axis.

From the theoretical model, we can also calculate the ratio between the quasi-TEM mode extending between the signal wire and ground plane and the spoof plasmonic TM mode grazing along the patterned edge of the striplines. For TM mode, there is an electric field component E_x along the propagation axis x, which extends within the teeth of a groove (the teeth of a given groove act as parallel plate capacitor C_p). The voltage V_{C_p} dropped across the parallel capacitor C_p versus the voltage V_C dropped across the self-capacitor C will give the ratio of TM mode vs TEM mode, as follows:

The characteristic impedance Z_0 of such structure supporting hybrid-SSPP mode can be written as:

$$Z_{0,\text{hyb}-\text{SSPP}} = \sqrt{\frac{L}{C_{\text{eff}}}} \tag{5.31}$$

$$\frac{\text{TMmode(SSPP)}}{\text{TEMmode(stripline)}} = \frac{V_{C_p}}{V_C} = \frac{|Ve^{ikd} - V|}{|V|} \tag{5.32}$$

For $k = k_{\max} = \frac{\pi}{d}$, the ratio reaches the maximum value of 2.

$$\frac{\text{TM} - \text{mode(SSPP)}}{\text{TEM} - \text{mode(stripline)}}\Big|_{\max} = \frac{|Ve^{i\pi} - V|}{|V|} = 2 \tag{5.33}$$

We validate this result by independent numerical simulation in commercial software HFSS, as presented in Figure 5.9. It is evident that the ratio of E_x/E_z and V_x/V_z reaches the value of 2 near the transmission cut-off frequency of 4 GHz.

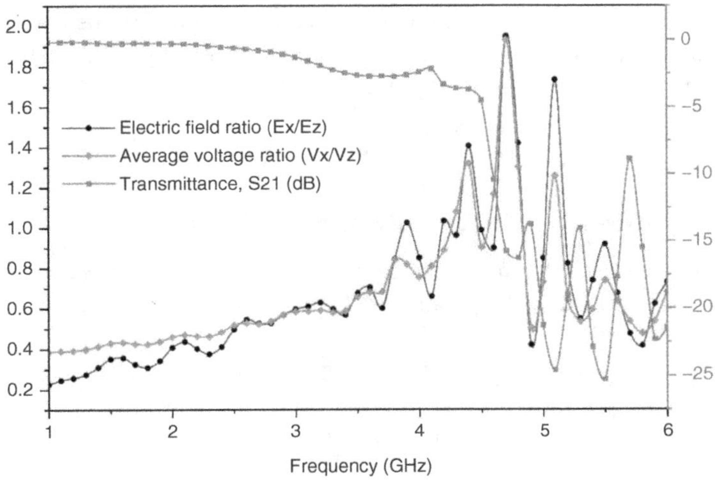

Figure 5.9 The ratio of TM-mode vs TEM-mode in hybrid-SSPP structure. The longitudinal E_x field corresponding to TM mode is calculated by taking the average E_x field extending within the teeth of a groove, while the transverse E_z field is calculated by taking the average E_z extending between patterned metal wire and the ground plane, as indicated in Figure 5.10.

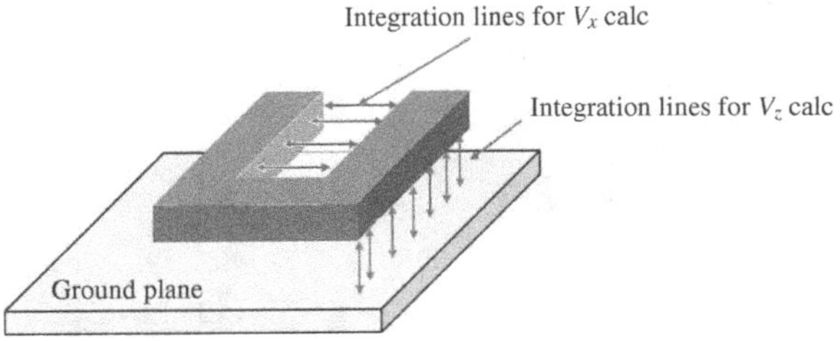

Figure 5.10 Unit cell of a hybrid-SSPP structure, and the integration lines taken to calculate longitudinal voltage V_x and transverse voltage V_z on the cell.

Since the maximum possible value of propagation constant k is limited to π/d due to the finite periodicity of the structure, the band-edge frequency $\omega_{\max}$ corresponding to $k = \frac{\pi}{d}$ of the dispersion diagram of the first Brillouin zone for the unit cell of the structure will be slightly detuned from the resonant frequency $\omega_r = \frac{1}{\sqrt{LC_p}}$, i.e., $\omega_{\max} = \omega_r - \delta$. From our developed theory above, we can also characterize the detuning parameter delta by equating Eq. (5.29) with $k_{\max} = \frac{\pi}{d}$ and replacing ω with $\omega_{\max}$ that corresponds to $k_{\max}$.

$$\omega_{\max} = \frac{\pi}{\sqrt{d^2LC + \pi^2LC_p}} \tag{5.34}$$

It is evident from Eq. (5.34) that there are two ways the detuning can be made vanishingly small: either by making the period d very small, or by making the resonance frequency $1/\sqrt{LC_p}$ very low, such that the following condition satisfies: $d << \frac{\pi\sqrt{C_p}}{\sqrt{C}}$.

5.5.1 Coupling Between Two Channels of Hybrid-SSPP Mode

Let us consider two parallel channels in close proximity. The coupling between the channels is characterized by the parameter C_m, where C_m stands for mutual capacitance per unit length.

The general expression for a transmission line coupled with the neighbor lines is

$$\left(C - \frac{k_z^2}{\omega^2 L} + 2C_m\right) V_i - C_m V_{i-1} - C_m V_{i+1} = 0 \tag{5.35}$$

If there are N number of transmission lines in the parallel bus system, then we can write the above equation in matrix form as

$$\mathbf{A}\underline{V} = 0 \tag{5.36}$$

where $\mathbf{A}$ is an $N \times N$ matrix

$$A = \begin{bmatrix} C_0 + C_m & -C_m & 0 & 0 & \cdots & & \\ -C_m & C_0 + 2C_m & -C_m & 0 & \cdots & & \\ 0 & -C_m & C_0 + 2C_m & -C_m & 0 & \cdots & \\ \vdots & \ddots & & & & & \\ \cdots & 0 & 0 & 0 & -C_m & C_0 + C_m & \end{bmatrix}$$

where $C_0 = C - \frac{k^2}{\omega^2 L}$ and

$$\underline{V} = \begin{bmatrix} V_0 \\ V_1 \\ V_2 \\ \vdots \\ V_{N-1} \end{bmatrix}$$

For nontrivial solution, the determinant of $\mathbf{A}$ must be zero.

For a two-channel system, the matrix **A** of Eq. (5.36) is the following:

$$A = \begin{bmatrix} C_0 + C_m & -C_m \\ -C_m & C_0 + C_m \end{bmatrix}$$

where $C_0 = C - \frac{k_z^2}{\omega^2 L}$. In order to have a vanishing determinant, the propagation constant k_z should follow the relation,

$$k^2 = \omega^2 L(C' \pm C_m) \tag{5.37}$$

where $C' = C + C_m$. These are the two eigenvalues of k_z. Using Taylor series expansion and assuming $C_m << C'$, we can write the two eigenvalues as $k_{z+} \approx \omega\sqrt{LC'}(1 + \frac{C_m}{2C'}) = k_{z0} + k_m$, and $k_{z-} \approx \omega\sqrt{LC'}(1 - \frac{C_m}{2C'}) = k_{z0} - k_m$. Here

$$k_{z0} = \omega\sqrt{LC'} \tag{5.38}$$

$$k_m = \omega\sqrt{LC'}\frac{C_m}{2C'} \tag{5.39}$$

Plugging these eigenvalues back into the Eq. (5.36), we can find the eigenvectors. The eigen-vector corresponding to $k = k_{z+}$ will be,

$$vv_+ = \begin{bmatrix} 1 \\ -1 \end{bmatrix}$$

Likewise, the eigenvectors corresponding to eigenvalue $k = k_{z-}$ will be

$$vv_- = \begin{bmatrix} 1 \\ 1 \end{bmatrix}$$

Then, following the method shown in Chuang [11] Chapter 8 p. 305, we can show that the general solution of the coupled two-channel transmission line is

$$\begin{bmatrix} Va(z) \\ Vb(z) \end{bmatrix} = \begin{bmatrix} 1 & 1 \\ -1 & 1 \end{bmatrix} \begin{bmatrix} e^{ik_{z+}z} & 0 \\ 0 & e^{ik_{z-}z} \end{bmatrix} \begin{bmatrix} 1 & 1 \\ -1 & 1 \end{bmatrix}^{-1} \begin{bmatrix} Va(z=0) \\ Vb(z=0) \end{bmatrix}$$

If we excite only one channel, i.e., if $Va(z = 0) = 1$, and $Vb(z = 0) = 0$, then

$$Va(z) = \frac{1}{2}e^{ik_{z0}z}\left[e^{ik_m z} + e^{-ik_m z}\right] = e^{ik_{z0}z}\cos(k_m z) \tag{5.40}$$

$$Vb(z) = -\frac{1}{2}e^{ik_{z0}z}\left[e^{ik_m z} - e^{-ik_m z}\right] = -ie^{ik_{z0}z}\sin(k_m z) \tag{5.41}$$

Thus, to see a full transfer of signal from one channel to other, the channel length $L_{\text{ch,side-steer}}$ should be the following

$$k_m L_{\text{ch,side-steer}} = \frac{\pi}{2} \Rightarrow L_{\text{ch,side-steer}} = \frac{\pi}{2\omega\sqrt{LC'}} \frac{2C'}{C_m} \tag{5.42}$$

At a length of $L_{\text{ch,side-steer}}$, a channel would completely transfer its signal to the neighbor channel through its lateral coupling. We can also find a critical length for crosstalk protection $L_{\text{cr,cx}}$ such that a waveguide with length $L < L_{\text{cr,cx}}$ will have its intended signal (by direct transmission) stronger than the unintended signal (by cross-transmission from the side-channel).

$$\tan^2(k_m L_{\text{ch}}) < 1 \Rightarrow L_{\text{ch}} < L_{\text{cr,cx}} = \frac{\pi}{4\omega\sqrt{LC'}} \frac{2C'}{C_m} \tag{5.43}$$

In the case of hybrid SSPP waveguide, $C' = C_{\text{eff}} + C_m$, where C_{eff} follows Eq. (5.30) and can become larger than the self-capacitance C at a designed frequency. The maximum value of C_{eff} can be determined by substituting ω with ω_{max} in the expression of Eq. (5.30) as

$$C_{\text{eff,max}} = \frac{C}{1 - \omega_{\text{max}}^2 L C_p} \tag{5.44}$$

where ω_{max} follows Eq. (5.34).

Thus, at a given operating frequency, if we replace the standard microstrip line waveguide with a hybrid-SSPP waveguide, we can make signal propagate a longer distance before it gets corrupted by crosstalk. The factor F by which signal propagation length could be increased with parallel bus of a hybrid-SSPP compared to a parallel bus of microstrip lines is the following:

$$F = \left.\frac{L_{\text{cr,cx-hybridSSPP}}}{L_{\text{cr,cx-microstrip}}}\right|_{\omega=\omega_{\text{max}}} = \sqrt{\frac{C_{\text{eff,max}} + C_m}{C + C_m}} = \sqrt{\frac{1 + (1 - \omega_{\text{max}}^2 L C_p) C_m / C}{(1 - \omega_{\text{max}}^2 L C_p)(1 + C_m / C)}} \tag{5.45}$$

As evident from the relation, we can make this factor F arbitrarily large at a design frequency by making $\omega_{\text{max}} \to \frac{1}{\sqrt{LC_p}}$ through proper choice of periodic pattern of the metallic waveguide. However, this formula is derived with the assumption of perfect conductivity of metal, and therefore may not remain valid at very high frequencies, where metallic loss becomes dominant.

5.5.2 Coupling Among a Large Number of Parallel Hybrid-SSPP Channels

For analysis of a large number of parallel wires, we can consider an infinite number of parallel lines consisting of identical wires extended along z axis, each separated from the neighbor by an amount of S along y axis with nearest-neighbor capacitive coupling C_m. Because this system is symmetric upon discrete lateral translation of S along the y axis, the supporting modes across the infinite set of wires would be such that the voltage at n–th wire would differ from the voltage at other wires by a phase factor $k_y Sn$ along the lateral direction, where k_y is the propagation constant along the lateral direction.

$$V_n = V_0 e^{ik_z z} e^{i\omega t} e^{ik_y Sn} \tag{5.46}$$

Then, we can rewrite Eq. (5.35) as

$$(C - \frac{k_z^2}{\omega^2 L} + 2C_m) - C_m \frac{V_{i-1}}{V_i} - C_m \frac{V_{i+1}}{V_i} = 0 \tag{5.47}$$

Using the relation between voltages at the parallel wires, i.e., Eq. (5.46), we can write $\frac{V_{i-1}}{V_i} = e^{-ik_y S}$ and $\frac{V_{i+1}}{V_i} = e^{+ik_y S}$. Therefore, the previous equation can be written as

$$C + C_m(2 - e^{-ik_y S} - e^{ik_y S}) = \frac{k_z^2}{\omega^2 L} \tag{5.48}$$

which leads to the following dispersion relation for infinite number of parallel lines with nearest neighbor capacitive coupling.

$$k_z^2 = \omega^2 LC \left[1 + \frac{2C_m}{C}(1 - \cos(k_y S))\right] \tag{5.49}$$

Using this dispersion relation, we can find out the group velocity of an envelope along the longitudinal (v_{gz} along z-axis) and lateral (v_{gy} along y-axis) direction.

$$v_{gz} = [\frac{\partial k_z}{\partial \omega}]^{-1} = \frac{1}{\sqrt{LC}} \frac{1}{\sqrt{[1 + \frac{2C_m}{C}(1 - \cos(k_y S))]}} \tag{5.50}$$

$$v_{gy} = [\frac{\partial \omega}{\partial k_y}] = -\frac{k_z}{k_y \sqrt{LC}} \frac{C_m}{C} \frac{(k_y S)\sin(k_y S)}{[1 + 2\frac{C_m}{C}(1 - \cos(k_y S)]^{3/2}} \tag{5.51}$$

Notice that v_{gy} is zero for $k_y S = \pm n\pi, n = 0, 1, 2, \ldots$. Therefore, in a data-transmission system with high frequency modulation, we could prevent any lateral propagation (i.e., crosstalk) if all the wires are fed with identical carrier signals with the same magnitude and phase at a given time (that corresponds to $k_y S = 0$), or the carrier signals at each of the wires are 180° out of phase with their nearest neighbor wires (that corresponds to $k_y S = \pi$). If the carriers at all channels are not identical, or if the carriers are not properly synchronized so that they fall short in satisfying $k_y S = \pm n\pi$ condition, then the modulating message signal will experience an undesired non-zero lateral propagation velocity v_{gy}, proportional to the ratio of mutual vs self-capacitance, $\frac{C_m}{C}$. In fact, maintaining such a rigorous condition of synchronized carriers at all channels is a difficult feat, since there could be many factors that cannot be fully controlled in a data-transmission system (such as process variation, temperature fluctuations affecting signal delay, etc.), especially one with a large number of wires. We will see later in the section how hybrid-SSPP mode can offer a solution by reducing the ratio $\frac{C_m}{C}$ in order to address the issue of crosstalk even when the carrier signals at all channels deviate from $k_yS = \pm n\pi$ condition.

5.5.3 Design Technique of Hybrid-SSPP for High Frequency Modulated Data Transmission

The parallel transmission lines have to be designed in such a way so that the input signal pulse reaches the receiving end of the intended channel earlier than the pulse traverses laterally toward the receiving end of the nearest neighbor channel. If L_{ch} is the length of the channel, then the time to reach the receiving end of the intended channel is

$$t = \frac{L_{\text{ch}}}{v_{\text{gz}}} \tag{5.52}$$

at this time t, the pulse signal propagates laterally a length of s toward the neighbor channel. Therefore, to keep crosstalk at a bare minimum level, the channel spacing S must have a lower bound.

$$S > s = v_{\text{gy}}t = v_{\text{gy}} \times \frac{L_{\text{ch}}}{v_{\text{gz}}} \tag{5.53}$$

Thus, the theoretical maximum channel density ρ_{max} that we can achieve in a given parallel data bus system is

$$\rho_{\text{max}} = \frac{1}{s} = \frac{v_{\text{gz}}}{v_{\text{gy}}}\frac{1}{L_{\text{ch}}} \tag{5.54}$$

It can be seen from Eq. (5.54) that the maximum channel density limited by the crosstalk is a function of channel length: a longer channel length would result in lower channel density, since crosstalk increases with channel length. Remind that, in case of hybrid-SSPP channels, the capacitance C can be replaced by C_{eff}, where C_{eff} follows Eq. (5.30). Now, C_{eff} impacts both longitudinal group velocity (v_{gz}) and lateral group velocity (v_{gy}) of signal. Increasing the value of C_{eff} at a given frequency would reduce the ratio of $\frac{C}{C_{\text{eff}}}$ and thus would reduce v_{gy} as per Eq. (5.51). However, an act of increasing C_{eff} would also reduce v_{gz} as per Eq. (5.50) (since, $v_{\text{gz}} \propto \frac{1}{\sqrt{C_{\text{eff}}}}$). Thus, in hybrid-SSPP channels, there would be an optimum choice of C_{eff} at a given operating frequency such that the ratio of $\frac{v_{\text{gz}}}{v_{\text{gy}}}$ would reach a maximum, resulting in the maximum possible wire density ρ_{max}.

For a given set of identical carrier signals at operating frequency ω fed into the parallel lines with their relative states described by lateral wave vector k_y. The theoretical limit of $\rho_{\max}$ in hybrid-SSPP channel can be found by sweeping the value of C in Eq. (5.49) of the longitudinal wave vector k_z, then using Eqs. (5.50) and (5.51) to find the ratio of $\frac{v_{gz}}{v_{gy}}$, and pick the value of $C = C_{\mathrm{opt}}$ that results in the maxima of $\frac{v_{gz}}{v_{gy}}$ and plugging in the value in Eq. (5.54). In order to know the right design frequency of SSPP resonance $\omega_r = \frac{1}{\sqrt{LC_p}}$, we need to equate $C_{\mathrm{opt}} = C_{\mathrm{eff}}$ in Eq. (5.30), and find the value of LC_p at the carrier frequency ω. Remind that, in Eq. (5.30), C is self-capacitance and is not a function of frequency, while C_{eff} is a function of frequency. The extracted LC_p value can then be used to determine the groove length h of the hybrid-SSPP waveguide using the following relation: $\omega_r = \frac{1}{\sqrt{LC_p}} = \frac{\pi}{2h}$. The process is straightforward, albeit a bit tedious.

However, the above process of determining the design parameters of hybrid-SSPP works well in the case of data transmission by high-frequency carrier modulation scheme. Hybrid-SSPP can also transmit baseband data without the need for a carrier frequency. In order to harvest the full potential of hybrid-SSPP waveguide for baseband data transmission, we need to abide by some other design criteria. The next section will describe the same.

5.6 Optimal Design Technique for Hybrid-SSPP Waveguide for Baseband Communication

Hybrid-SSPP waveguide proposed by the authors has a unique advantage over standard SSPP waveguide: it can transmit baseband data. That means the waveguide can be used for data transmission without the need for a high-frequency carrier signal and signal modulation and demodulation circuit, which will save energy consumption/bit, footprint on the substrate, and enable reduced battery consumption when the system is in low power/idle state mode. In regular microstrip lines, the crosstalk monotonically increases with higher frequencies, and therefore they are good only at low frequencies. In standard SSPP waveguide, the crosstalk reduces as we get closer to spoof plasmon resonant frequency, and thus they are better only at high frequencies (frequencies close to SSPP resonance). Hybrid-SSPP waveguide brings the benefit from both these two technologies.

In hybrid-SSPP waveguide, crosstalk remains low at both low frequencies and at high frequencies close to SSPP resonance. crosstalk for hybrid-SSPP could actually be worse at some frequencies in the middle. We can find the condition where the ratio of the transmitted power to the crosstalk power is minimalminimum.

This minimal value must be greater than 1 in order to have data transmission without crosstalk power overtaking transmitted power. The condition can be expressed as:

$$\cot^2(k_m L_{\text{ch}}) > 1 \tag{5.55}$$

$$=> \omega\sqrt{L}\frac{C_m}{2\sqrt{C'}}L_{\text{ch}} < \frac{\pi}{4} \tag{5.56}$$

For hybrid-SSPP, the function $(k_m L_{\text{ch}})$ hits a minimum at a frequency $\omega = \omega_{\text{valley}}$ that can be determined as follows:

$$\frac{d}{d\omega}\left[\omega L_{\text{ch}}\sqrt{L}\frac{C_m}{2\sqrt{C'}}\right]\Bigg|_{\omega=\omega_{\text{valley}}} = 0 \tag{5.57}$$

Then plugging in the value of $C' = C_{\text{eff}} + C_m$ in the above equation and using Eq. (5.30), we can find the value of ω_{valley}. For the case of low value of mutual coupling capacitance, that is $\frac{C_m}{C} << 1$, an approximate value of ω_{valley} can be found as follows:

$$\omega_{\text{valley}} \approx \frac{1}{\sqrt{2LC_p}} = \frac{1}{\sqrt{2}}\omega_r \tag{5.58}$$

The equation shows that, for a hybrid-SSPP waveguide, the crosstalk from aggressor channel would become strongest at a frequency of approximately $\frac{1}{\sqrt{2}}$ times that of the SSPP resonant frequency. Hence, the parameter ω_{valley} is a critical one in the design of hybrid-SSPP so that there remains a minimum margin between signal power and crosstalk power at $\omega = \omega_{\text{valley}}$.

For a given channel length L_{ch}, in order to transfer the maximum possible baseband data, the hybrid-SSPP channel's resonant frequency $\omega_r = \frac{1}{\sqrt{LC_p}}$ should be designed in such a way so as to abide by the following relation:

$$\omega_{\text{valley}}\frac{\sqrt{L}}{2\sqrt{C_{\text{eff,valley}} + C_m}}C_m L_{\text{ch}} < \frac{\pi}{4} \tag{5.59}$$

where $C_{\text{eff,valley}}$ is the effective self-capacitance of the hybrid-SSPP waveguide at the frequency of $\omega = \omega_{\text{valley}}$

$$C_{\text{eff,valley}} = \frac{C}{1 - \omega_{\text{valley}}^2 LC_p} \tag{5.60}$$

5.7 Experimental Characterizations of SSPP Data Bus

SSPP interconnect can accommodate two different kinds of modes: electro-SSPP mode (the hybrid-SSPP mode) and opto-SSPP mode. In hybrid-SSPP mode, an SSPP channel transfers signal with the assistance of a ground wire. In opto-SSPP mode, a standalone SSPP wire suffices to transport information. Provision of a ground wire is redundant in this mode of operation, as the signal sticks to the surface of the metallic wire due to the coupling with the resonant unit cells of metasurface.

We, the authors, have experimentally verified both of these modes in a system incorporating a parallel data bus of SSPP waveguide, refer to Figure 5.11.

5.7.1 Verification of Electronic-SSPP Mode

In order to get a proof-of-concept demonstration of the information capacity of spoof plasmon interconnect, we designed a basic SSPP channel with a momentum converter. In actual implementation, the carrier frequency of the channel is envisioned to be hundreds of gigahertz for ultrafast data transfer. However, for the convenience of measurement, we designed the channel to have a cut-off frequency of 10 GHz. The momentum converter, despite being redundant for carrying electrical signal through SSPP channel, will become a necessity when we conduct the experiment on the transport capacity of opto-SSPP signal through the same physical structure. The main section of the designed channels comprises unit cells of metamaterial with spoof plasmon resonance frequency at 10 GHz. The electromagnetic scattering properties of the fabricated parallel channels are tested with the channels terminated with standard 50 Ω input and output resistance. The results of the measurement are shown in Figure 5.12, which illustrates the transmission magnitude of a single channel. The transmission coefficient remains almost unity

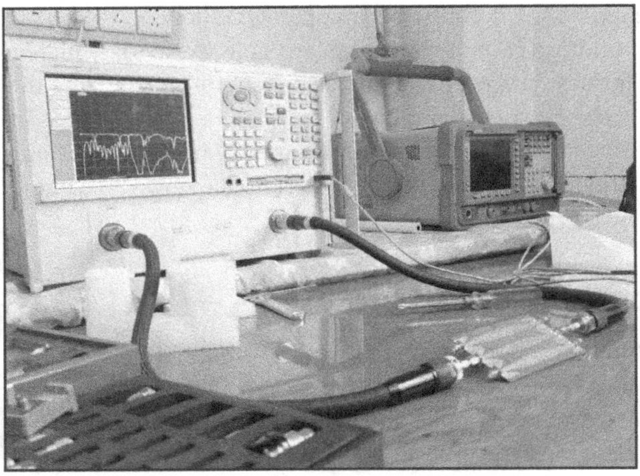

Figure 5.11 Experimental set-up for the verification of dual-mode accommodation in SSPP parallel data bus.

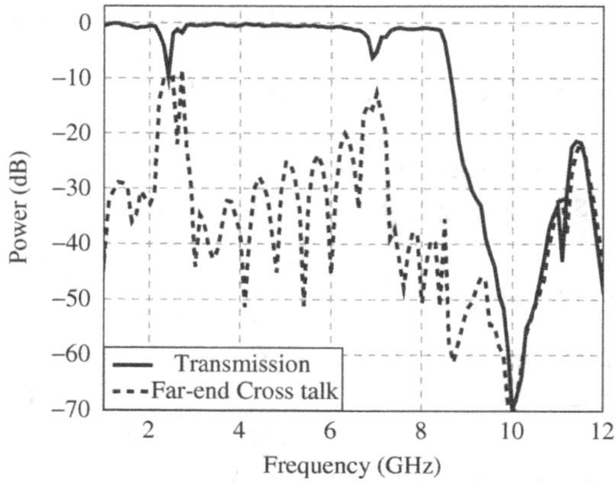

Figure 5.12 Experimental verification of the electro-SSPP mode in SSPP channel. The spectral region from 2-7 GHz corresponds to the electro-SSPP mode. Transmission and crosstalk are measured in a three-channel parallel data bus.

unless we reach the dispersion band-edge of SSPP metastructure, (in this specific design, at 8.5 GHz). The sharp dip of transmission (−70 dB attenuation) appears at 10 GHz, which is exactly where spoof plasmon resonance is designed. Hence, spoof plasmon resonance is a measure of the maximum frequency that can propagate through the metasurface channel.

To measure the far-end crosstalk, we provided signal to the middle channel, measured signal power from the receiving-end of one of the nearest neighbors of the middle channel. The remaining ports were kept open-circuited.

The crosstalk, for most of the part of the band remains below −20 dB, which is quite a promising result for designing dense data bus system for maximizing the bandwidth density. In the low-frequency prototype, the space between each pair of neighboring channels is kept twice the length of the groove. In the final design for 300 GHz operating frequency, we plan to aggressively scale down this guard space to increase the traffic density in the data bus.

5.7.2 Verification of Optical-SSPP Mode

In order to make point-to-point signal transmission via opto-SSPP mode, we designed an appropriate adapter at the transmitting end to convert the transverse electromagnetic mode into a surface-bound mode. The adapter comprises of a Vivaldi antenna, and the length of the grooves of the adapter has been gradually increased so as to match the TEM mode with SSPP mode over a broadband. On the receiving side, we employed the mirrored structure of the transmitting adapter. No ground plane is provided at the bottom of the substrate, unlike the previous experiment where we tested the transmissibility of electro-SSPP mode.

The dispersion relation of spoof plasmon metasurface suggests that, the field confinement is negligible for frequencies far away from resonance, and the confinement increases as we choose to operate close to resonance. Thus, the lower part of the dispersion diagram could not be utilized in a parallel data bus system for the sake of shielding the neighboring channels from cross-interference. We have shown by an analytical model that the upper 1/3 fraction of the fundamental band of SSPP interconnect can suppress crosstalk during data transfer [12]. For instance, an SSPP data bus designed from 300 GHz carrier frequency is expected to carry bandwidth of approximately 100 GHz spanning from 200 to 300 GHz in an optimal set-up.

In order to verify the nature of crosstalk suppression in our prototype of 10 GHz chip, we provided input signal with 50 Ω impedance to the centered waveguide of the three-channel system and receive signal from the output terminal of the two remaining side channels. Figure 5.13 illustrates the measurement result of transmission and far-end crosstalk in an SSPP data bus. It clearly shows that, as expected from theory, the low frequency regime is severely plagued by crosstalk to the point that the cross-coupled signal power is almost comparable to the desired transmitted signal level. The crosstalk becomes suppressed as we tend toward the band-edge (in our specific case, band-edge is ~ 10 GHz). A fraction of the transmission spectrum, spanning from almost 7 to 10 GHz in Figure 5.13 exists where the crosstalk remains at least −3 dB low from the signal level. This measurement result substantiates the authors' earlier prediction that, in the circumstances of crosstalk limited bandwidth, an SSPP channel in an optimally designed parallel

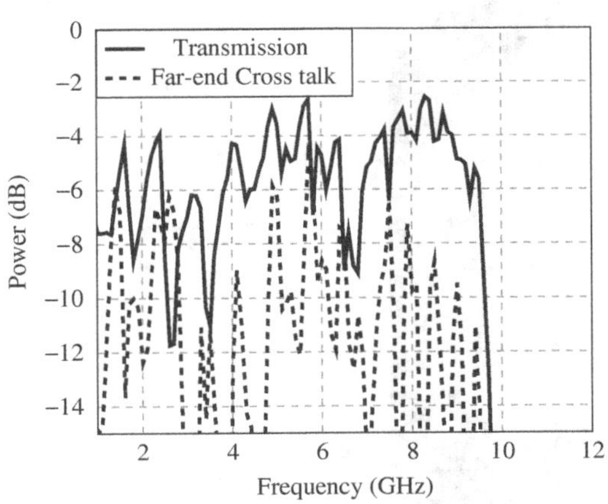

Figure 5.13 Experimental verification of the opto-SSPP mode in SSPP channel. Transmission and crosstalk are measured in a three-channel parallel data bus. The region from 7-10 GHz indicates the frequency spectrum we can utilize for data transfer with low crosstalk.

data bus can contribute to a maximum of 1/3 band for data transfer while maintaining signal integrity.

5.7.3 Microfabrication Process

The fabrication process of the spoof plasmon meta-structure that is designed to operate at 300 GHz carrier frequency involves the following sequential steps: (i) wet cleaning and plasma cleaning on quartz, (ii) spin coating of negative photoresist (PR), followed by soft bake, (iii) imprint by the standard photolithography with negative mask, (iv) chemical development, (v) oxygen (O_2) plasma collision for removal of the photoresist (PR) remnant, (vi) electron beam evaporation for 200 nm Chromium (Cr)+Aluminum (Al) layer deposition, as a 10 nm Cr layer for better cohesion of Al, and (vii) liftoff of PR to fabricate metal microstructure array on quartz.

Figure 5.14 shows a microscopic image of the SSPP data bus chip comprising three parallel channels with an operating frequency of 300 GHz.

5.8 Mechanism for Bandwidth Augmentation

In a densely arranged parallel data bus, the bottleneck of interconnect speed derives from the degree of crosstalk among the neighbor channels. Under such cases, patterning the metallic interconnect can add extra safeguard against

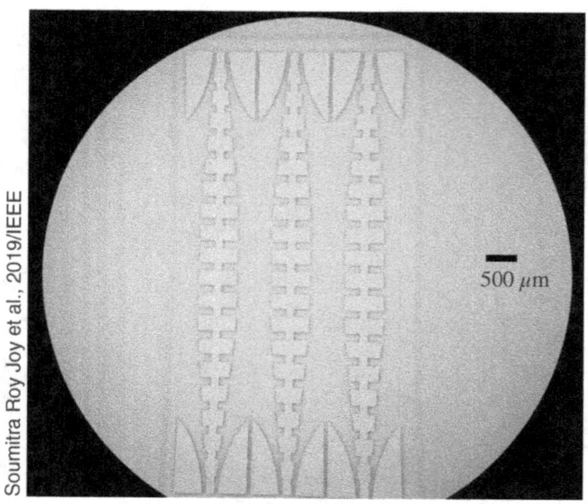

Figure 5.14 Optical micrograph of fabricated parallel SSPP channels. The cut-off frequency is designed at 300 GHz. *Source*: Joy et al. [12]/IEEE.

crosstalk at elevated frequencies, and therefore can augment bandwidth. To prove our point, two electrical interconnects are placed in parallel with the following structural specification: the length of each interconnect is 250 mm, wire width = 20 mm, spacing between wires = 12 mm, dielectric constant of the substrate's material = 4.0, substrate's thickness = 10 mm, and the bottom of the substrate is considered to be a large ground plane. The material of the substrate is considered lossless, and the metal of the interconnect is taken to be perfect conductors, so that we can see the impact of crosstalk alone in limiting the bandwidth of the channel. A pair of interconnects is a four-port system, whose transport properties can be described by 16 different S-parameters ($S_{i,j}$, where i, $j = 1, 2, 3, 4$). Because of the reciprocity of the system, and because the two channels are identical, it suffices to know only four S-parameters, namely S_{11}, S_{21}, S_{31}, and S_{41} to get the full description of the system.

In order to make a fair comparison between the performance of the electrical interconnect and the newly proposed electro-SSPP interconnect, we designed the electro-SSPP interconnect-pair in such a way so that the geometric footprint of the electro-SSPP channel-pair is same as that of electrical interconnect-pair; i.e., the two kinds of interconnect-pairs share same length of wire, same width of wire, same wire spacing, and same substrate with ground at the bottom. The only uniqueness of the electro-SSPP interconnect is the judicious pattern on it. The specifications of the metallic patterns that constitute SSPP waveguide are as follows: groove length $h = 4$ mm, groove width $a = 2$ mm, period $d = 6$ mm. In addition, the electro-SSPP interconnect pair includes a converter on each of the signal transmission and reception side, since the standard mode of waveguide input and output is TEM mode with 50 Ω resistance. The geometric specification of the converter came from a simple educated guess, rather than a mathematical optimization technique. Figure 5.15 shows the top view of the electrical and electro-SSPP interconnect-pair with corresponding geometric specifications that are considered in the simulation process.

Figure 5.16a shows the transmission band of a single, standalone regular microstrip line interconnect – with no other neighbor channel around, and that of a single hybrid-SSPP interconnect channel. A standalone microstrip line, especially with lossless dielectric and perfect electric conductor has a very large bandwidth (ideally infinite bandwidth), which is eventually limited by radiation loss. A hybrid-SSPP interconnect on the other hand, has its bandwidth limited by the spoof plasmon resonant frequency, and in Figure 5.16a in particular, the cut-off frequency happens to be near 7 GHz. Thus, it may appear counter-intuitive why hybrid-SSPP waveguide can bring any benefit over regular electrical interconnect.

The answer would become clear once we begin to consider the case of multiple wires in parallel. The presence of neighbor wires in the close vicinity would trigger crosstalk among them, which can significantly limit the transmission bandwidth.

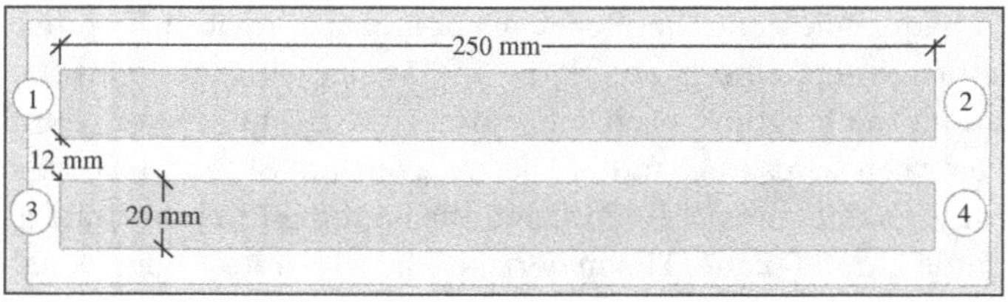

(a) Geometry of electrical interconnect pair

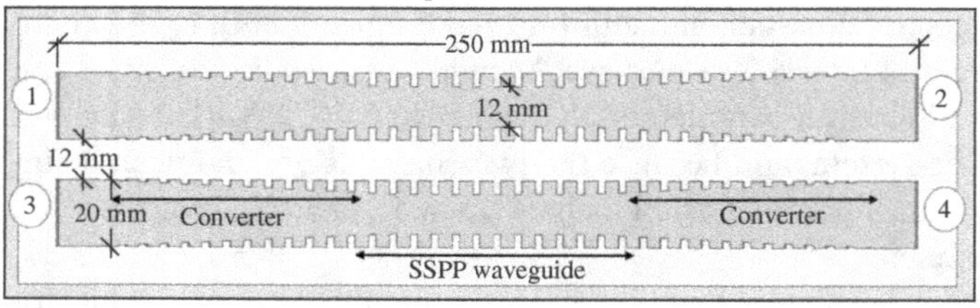

(b) Geometry of electro-SSPP interconnect pair

Figure 5.15 Geometry of electrical and electro-SSPP interconnect-pair. Each set of interconnect pair resides over 10 μm thick substrate, with a ground plane at the bottom. The relative permittivity of the substrate is 4. Each of the four ports in the interconnect pair is enumerated.

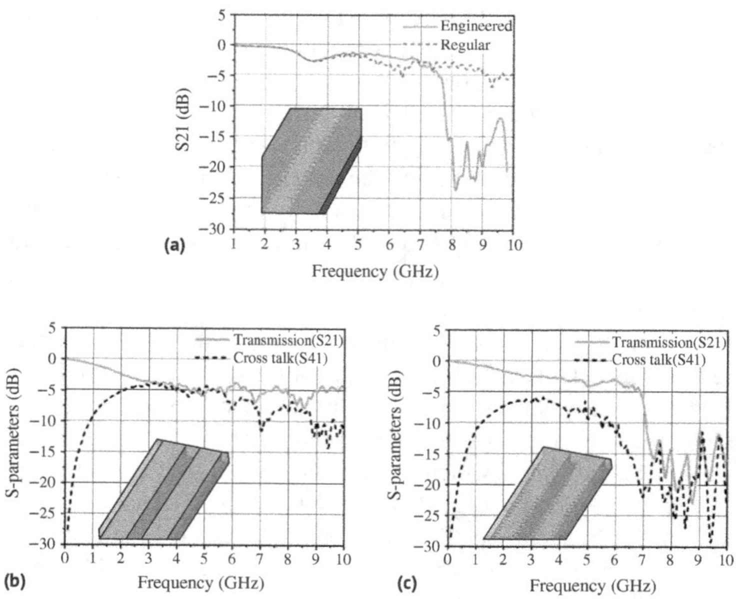

Figure 5.16 Comparison between performance of regular microstrip line interconnect and engineered interconnect: (a) Transmission of a standalone regular and a standalone engineered interconnect. (b) Transmission band and crosstalk property of a pair of regular interconnect. (c) Transmission band and crosstalk of a pair of engineered interconnect.

Figure 5.16b shows the scattering parameters for a pair of electrical interconnect, and Figure 5.16c shows that of engineered hybrid-SSPP interconnect, obtained by 3D full-wave simulation. It indicates that, for electrical interconnect-pair, the transmission coefficient (S_{21}) and the crosstalk coefficient (S_{41}) become equal at 3 GHz frequency. Therefore, transmission of signal whose bandwidth increasingly exceeds 3 Gbps would cause an exponential rise in bit-error-rate. On the contrary, the electro-SSPP interconnect-pair, which shares the same geometric footprint on the substrate as that of the electrical interconnect-pair, has a propagation characteristics almost identical to electrical ones below 1 GHz, but introduces a gap between the trace of S_{21} and S_{41} parameter profile beyond 3 GHz, and keeps those trace separated until we reach 7 GHz. Thus, we can expect to transmit baseband signal with bandwidth reaching as high as 7 Gbps without suffering much bit-error-rate.

In order to illustrate the impact of the gap between the trace of S_{21} and S_{41} ranging from 3 to 7 GHz engendered by the electro-SSPP interconnect pair, we transmit a single rectangular pulse, modulated with a 5 GHz carrier, through the electro-SSPP channel-pair, and observe the time-domain profile of the pulse. Figure 5.17 shows the profile of pulses for various pulse bandwidths transmitted through electro-SSPP channel-pair. The input pulse at channel-1 of the interconnect-pair is drawn with rectangular lines, and the solid curves represent the transmitted pulse through channel-1 when no input is provided through channel-2. The dotted lines represent the induced signal at channel-2 leaked from channel-1. There is a clear distinction of signal levels between the transmitted signal and the crosstalk induced erroneous signal in this case, which indicates that we can resolve between the intended bit and the erroneous bit by setting a proper threshold value.

On the contrary, if the same signal modulated by a 5 GHz carrier is dispatched through the conventional electrical interconnect-pair, then, due to severe crosstalk, the signal retrieved at the receiving end becomes impossible to resolve between the intended bit and the erroneous bit. This is illustrated in Figure 5.18 where we see that the transmission through channel 1, while channel 2 remains silent, gives rise to a true bit at the receiver of channel 1 (solid curves) and a false bit at the receiver of channel 2 (dotted line) which are of equal signal strength. Therefore, no parallel data transmission is possible in conventional electrical interconnect over the spectrum of $3-7$ GHz, whereas electro-SSPP channel-pair can enable us sending information over the $3-7$ GHz bandwidth.

Next, we would like to see how the baseband signal transmission can be benefit from the electro-SSPP interconnect pair. Improvement of baseband signal quality would be of particular interest, as it liberates us of employing high-frequency carrier sources, modulators, and demodulators. This additional transceiver circuitry for high-frequency modulation scheme not only consumes a considerable amount of energy, but also occupies extra footprints on the substrate. A baseband data

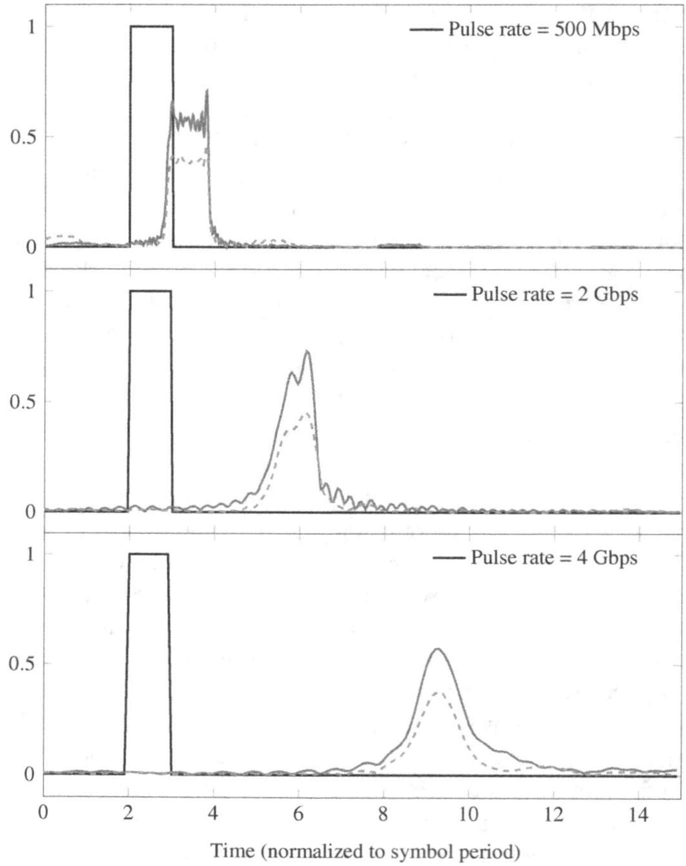

Figure 5.17 Pulse profile, before and after modulation, transmitted through channel 1 of the hybrid-SSPP interconnect-pair, where channel 2 is a victim of crosstalk from channel 1. Rectangular solid lines are input pulse at the input of channel 1, solid curves are output received at the end of channel 1, and the dotted lines are crosstalk induced erroneous pulses received at the output port of channel 2. The carrier frequency is chosen as 5 GHz.

transmission system can thus provide a solution for energy-proportional computing, which states that the energy consumed by the system should be proportional to the utilization of the system, i.e., when the system performs at a fraction of its peak capacity, the energy consumption should also scale down proportionally.

Figure 5.19a shows the output of a regular microstrip line channel in a data bus in response to a pulse of 0.33 ns pulse width. This corresponds to a 3 Gbps speed of pulse. It shows that, at this rate of pulse, the output response of a channel becomes almost indistinguishable from the false signal that cross-coupled to

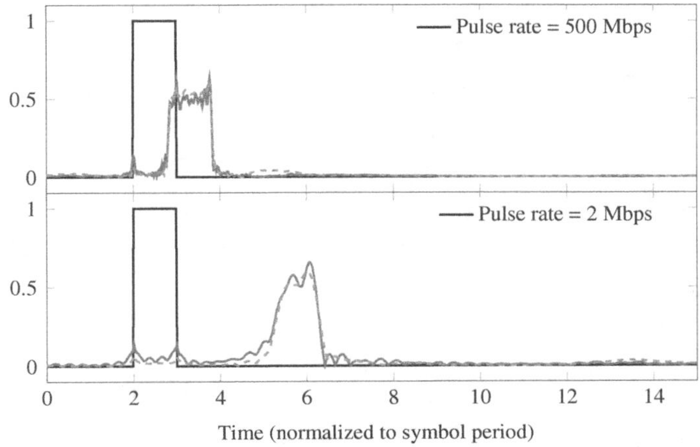

Figure 5.18 Pulse profile, before and after modulation, transmitted through channel 1 of the electrical interconnect-pair, where channel 2 is a victim of crosstalk from channel 1. Rectangular solid lines are input pulse at the input of channel 1, solid curves are output received at the end of channel 1, and the dotted lines are crosstalk induced erroneous pulses received at the output port of channel 2. Carrier frequency is chosen as 5 GHz.

the intended channel from a neighboring channel, and therefore the information cannot be recovered, until we use a repeater in the channel. The other thing to notice is the latency of pulse (the delay between the input and output pulse). For this particular case of regular wire with 25 cm channel length, the pulse latency appears to be 2 ns.

Let us now compare these performance metrics with a parallel data bus made of engineered hybrid-SSPP interconnects. Figure 5.19b shows the output from a channel in a parallel data bus made of hybrid-SSPP waveguides in response to a pulse of 0.33 ns width. For an input pulse of unity magnitude in a 25 cm length of data bus, the normalized output pulse reaches as high as 0.6, while the cross-coupled signal reaches an amplitude of 0.4. Thus, there remains at least a 50% safety margin between amplitude of the direct transmitted signal and the cross-coupled signal. Also note that the pulse latency is reduced down to less than 1 ns, a 50% improvement in latency compared to regular microstrip interconnect.

Figure 5.19c shows the response of data bus made of regular interconnects vs hybrid-SSPP interconnects. The data clearly shows the advantages of hybrid-SSPP interconnects: the output pulse strength is higher, and the output latency is lower compared to a regular interconnect. Note that the hybrid-SSPP attains this remarkable advantage over standard interconnect without consuming any additional power or occupying any additional footprint.

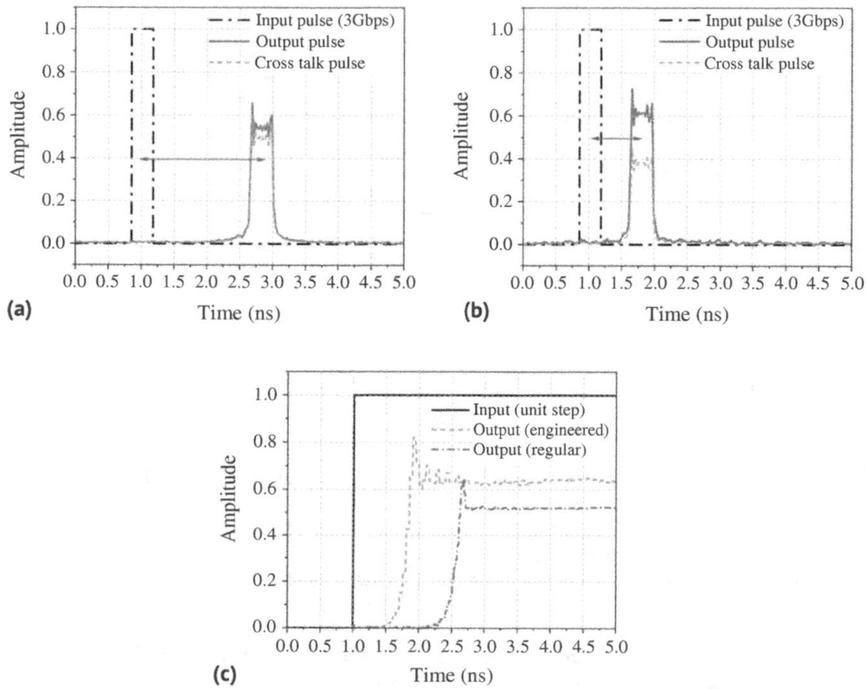

Figure 5.19 Comparison between the response of an engineered hybrid-SSPP and a regular microstrip line interconnect being in the proximity of similar neighboring wires for a binary, unmodulated pulse. (a) A pulse of 0.33 ns width (equivalent to 3 Gbps pulse speed) is applied to a particular channel in a parallel data bus consisting of regular microstrip line interconnects, and its response. (b) A pulse of 0.33 ns width applied to a channel in a data bus consisting of engineered hybrid-SSPP interconnects, and its response. (c) A unit impulse response of a parallel bus of regular interconnect vs engineered hybrid-SSPP interconnect.

Figures 5.20 and 5.21 show the time profile of random binary bitstream transmitted simultaneously through both channels 1 and 2 of the electro-SSPP interconnect pair. The geometry of the interconnect pair is kept the same as before, and we added Gaussian noise to the transmitted signal, where the signal-to-noise ratio (SNR) is taken as 20 dB. The baseband data is sent frame-by-frame, where a frame of data contains 10^6 binary symbols. The sampling rate is taken to be six times higher than the bit rate (for instance, when bit rate is 1 Gbps, sampling frequency is taken as 6 GHz) to circumvent any possible signal corruption engendered by aliasing in the digital domain.

Under such circumstances, the data received at the other end is corrupted due to three factors: uneven gain over the bandwidth of the pulse transmitted, the

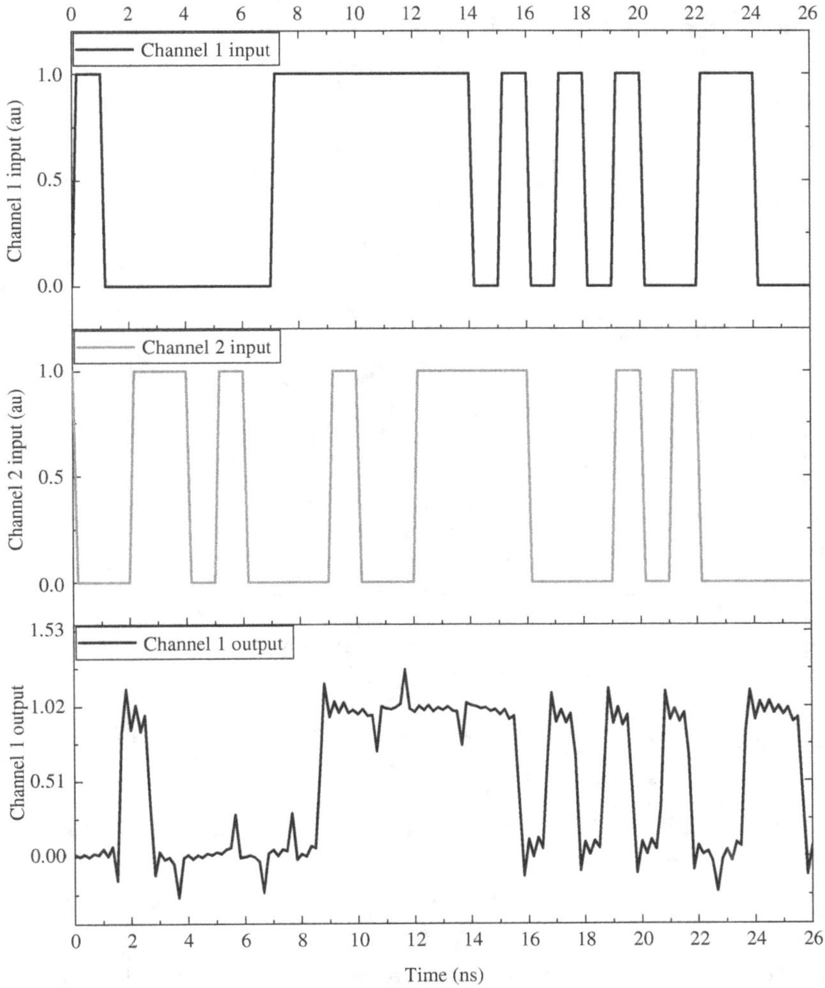

Figure 5.20 Time domain characterisitcs of binary data stream transmitted through channel 1 of the pair of electro-SSPP wires, and subjected to the crosstalk from channel 2. Speed of data through each channel is 1 Gbps.

crosstalk caused by the aggressor channel in the neighborhood, and the noise. In the analysis, we kept the noise level low enough (we chose SNR = 20 dB, i.e., noise power is 100 times weaker than the signal power) so that it does not play a vital role in signal corruption. Thus, the main contributing factors of bit errors come from crosstalk and uneven gain-frequency characteristics.

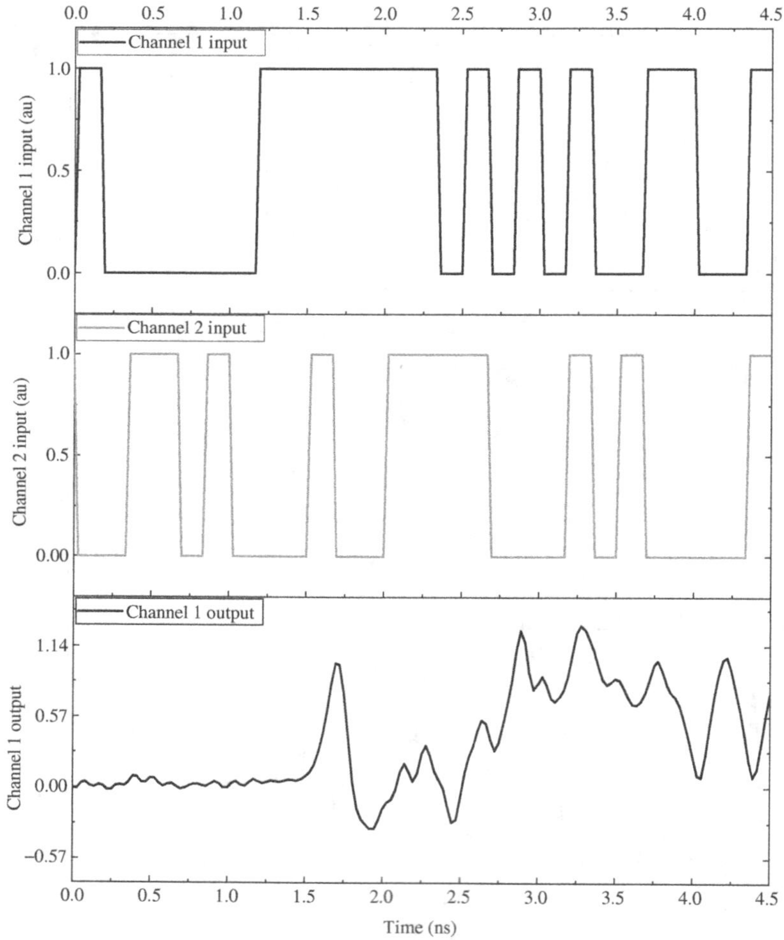

Figure 5.21 Time domain characterisitcs of binary data stream transmitted through channel 1 of the pair of electro-SSPP wires, and subjected to the crosstalk from channel 2. Speed of data through each channel is 6 Gbps.

We notice that, for 1 Gbps baseband data transmission (Figure 5.20), the output signal at channel 1 is quite recognizable and matches with the input bit pattern of channel 1 even visually. The level of crosstalk is also too low at that spectrum of 1 GHz and below to frustrate signal quality. However, when baseband data speed is raised to 6 Gbps, a significant degree of signal distortion appears at the output of channel 1. To quantify the level of signal corruption, we calculated the time delay between the input and output data frames of channel-1, and aligned the frame to calculate bit error rate (BER). This BER will be the yardstick for us to compare the performance of the electrical interconnect pair and electro-SSPP interconnect pair.

Then we provided a random bitstream to an electrical interconnect with the same noise level (i.e., SNR = 20 dB), varied the data rate of baseband signal, and computed the BER. The resultant BER for both electrical and electro-SSPP interconnect pair is plotted in Figure 5.22 which shows at least an order of reduction of BER due to the employment of electro-SSPP interconnect compared to that of electrical interconnect for signal bandwidths extending from 3 to 7 Gbps. Figure 5.22 also shows, by dotted lines, the BER when we remove the effect of crosstalk by providing data at only one channel of the interconnect pair at a time. The dramatic reduction of BER (approximately by a factor of 10^{-3}) due to elimination of crosstalk indicates that the most dominant mechanism of signal corruption in our case occurs through crosstalk. This figure is suggestive of the fact that the designer should take special consideration to the mechanism of crosstalk suppression in a densely packed data transmission bus.

Figure 5.23 shows the eye diagrams for data transmission through electrical and electro-SSPP interconnect pairs for various rate of data transfer. Notably, an eye diagram is a tool of measuring the signal integrity. The clearer the shape of the eye, and the wider the opening of the eye, the better the quality of the transmitted

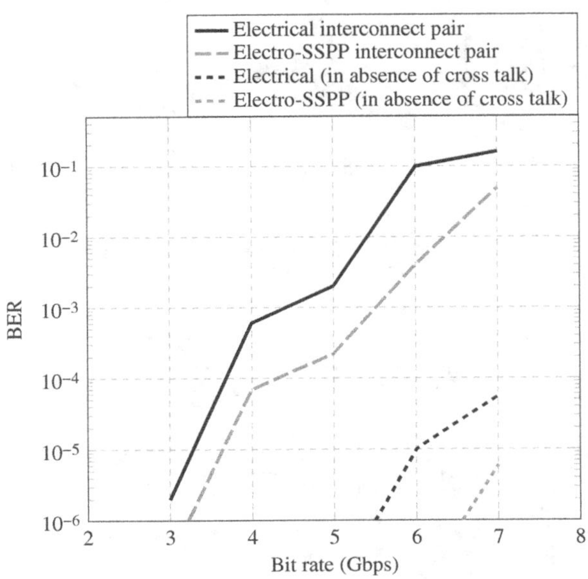

Figure 5.22 Comparison of bit-error-rate (BER) for two sets of interconnect pair: electrical interconnects, and electro-SSPP interconnect pair.

signal. It is evident that the eye of the signal is almost closed for regular interconnect at 3 Gbps speed, while that for hybrid-SSPP channels is wide open, indicating a high degree of signal integrity in engineered interconnect as opposed to regular microstrip line interconnect.

5.8.1 Experimental Validation of Bandwidth Enhancement

In order to validate the simulation results, two microstrip line pairs (regular and grooved) are designed, as shown in Figure 5.24. S-parameters are measured using a Copper Mountain VNA (100 kHz–8 GHz), with unused terminals terminated by 50 Ω resistors. Tapers optimized for 50 Ω matching are used at all ports. Grooved channels are designed face-to-face to test the worst possible case of crosstalk in hybrid-SSPP.

The dimensional parameters of the fabricated interconnects are shown in Table 5.1. The model is fabricated on FR4 substrate.

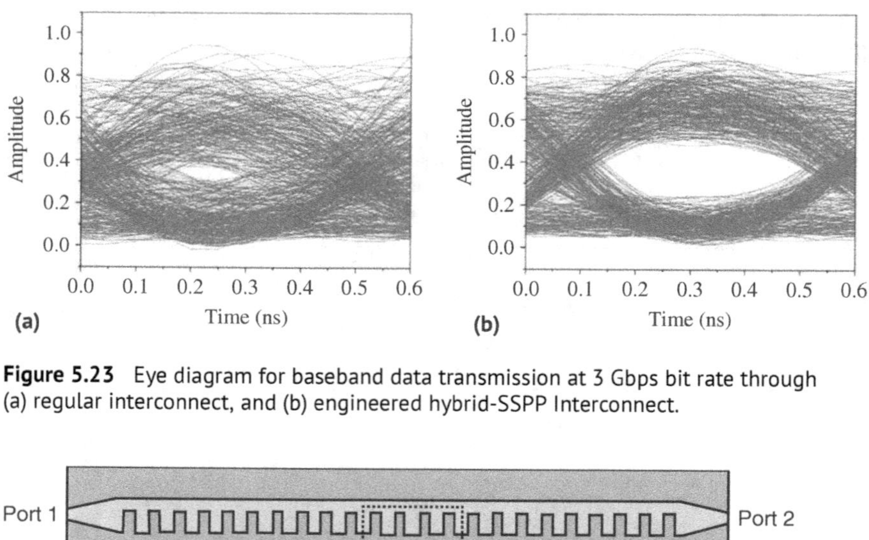

Figure 5.23 Eye diagram for baseband data transmission at 3 Gbps bit rate through (a) regular interconnect, and (b) engineered hybrid-SSPP Interconnect.

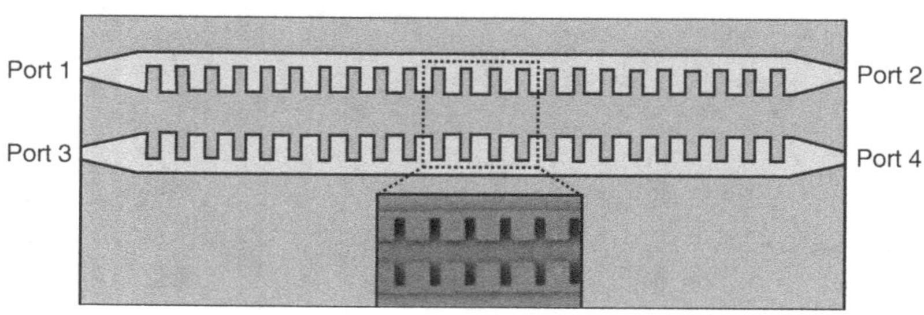

Figure 5.24 Experimental setup of dual channel engineered hybrid-SSPP interconnect.

Table 5.1 Description of gemetric features of Hybrid-SSPP interconnect and a regular microstriplines of equal footprint.

Parameter	Regular interconnect (mm)	Hybrid-SSPP interconnect (mm)
Channel spacing	4	4
Length	228	228
Width	5	5
Substrate thickness	1.2	1.2
Tapering length	19	19
Tapering-end terminal width	2	2
Groove length (*h*)	N/A	4
Groove width (*a*)	N/A	2
Cell period (*d*)	N/A	6

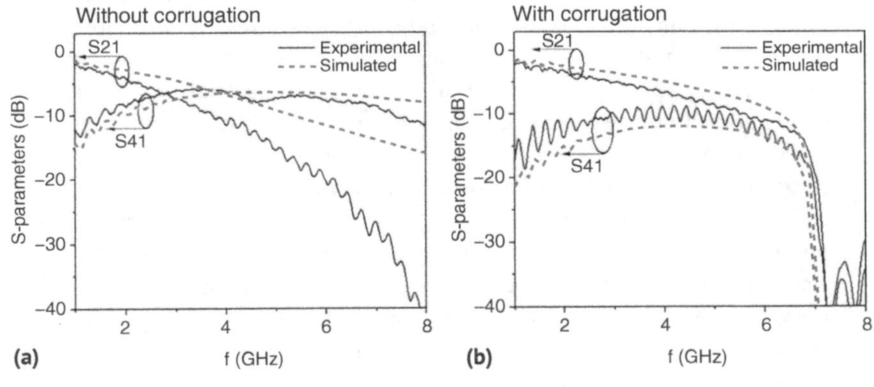

Figure 5.25 Experimental validation of bandwidth enhancement by SSPP. (a) Transmission band and cross-coupled signal within 8 GHz window for a pair of regular microstrip lines. (b) Transmission band and cross-coupled signal for a pair of hybrid-SSPP channels. The experimental data matches reasonably well with simulated data and shows a clear evidence of bandwidth increment by hybrid-SSPP waveguide.

Figure 5.25 illustrates the bandwidth improvement in engineered interconnects. As shown in Figure 5.25, engineered interconnects exhibit superior crosstalk suppression across a wideband. While regular interconnects show the maximum crosstalk of −5.87 dB at 3.55 GHz, engineered interconnects reduce it to −9.99 dB. This suppression enhances transmission, improving from −9.39 dB in regular interconnects to −5.80 dB in engineered ones.

5.9 How Spoof Plasmon Advances the Engineering of Interconnect

Key Features of an Effective Interconnect System In order to implement a physical interconnect with high energy efficiency and high bandwidth, the following five features should be made attainable by an interconnect technology.

1. **Low Material Loss**
   The material of the interconnect must exhibit minimal energy absorption across a wide frequency spectrum. In conventional metallic interconnects, such as microstrip lines, striplines, and substrate-integrated waveguides, material loss increases with frequency due to the skin effect, thereby reducing bandwidth. Optical interconnects, in contrast, can exhibit low absorption within specific spectral windows—for example, optical fibers demonstrate minimal loss near 1.55 and 1.3 μm wavelengths, enabling efficient long-haul data transmission.

   Spoof plasmonic interconnects, however, face challenges in reducing material loss. The resonant behavior of corrugated metals at quarter-wavelength dimensions can exacerbate energy absorption. Addressing this limitation necessitates significant research, potentially exploring the concept of implementing spoof plasmonic waveguide on advanced materials such as superconductors at cryogenic temperatures or topological insulators, which have shown promise in mitigating electron scattering.
2. **Low Electromagnetic Radiation and Interference**
   In high-density wiring systems, parallel wires can introduce parasitic coupling, leading to electromagnetic interference (EMI) and crosstalk noise. According to Faraday's law of electromagnetic induction, the strength of such noise increases with frequency, often limiting bandwidth in otherwise ideal channels. Both electrical and optical interconnects are susceptible to crosstalk in high-density configurations.

   Spoof plasmonic interconnects offer a potential solution by strongly confining signals to the metallic surface at high frequencies. Designers can mitigate crosstalk noise by optimizing the spoof plasmon resonance frequency, enhancing performance in dense wiring environments.
3. **Linear Energy Scaling with Utilization**
   Efficient interconnect systems should exhibit energy consumption that scales linearly with utilization across varying system power levels. High-frequency modulation techniques often achieve energy efficiency at peak capacity but suffer significant losses at lower utilization rates due to the fixed overhead energy consumption of oscillators, modulators, and detectors, along with

overhead footprint of the additional circuitry on the substrate. This inefficiency is particularly problematic, as servers frequently operate at 10 – 50% of their peak capacity.

Baseband data transmission systems offer a viable alternative, characterized by low idle-state power consumption and simpler transceiving circuitry. Our research demonstrates that hybridizing spoof plasmonic modes in corrugated metal wires, which is a mix of the TM mode of spoof plasmonics and TEM modes in microstrip lines, supports baseband data transfer, facilitating linear energy scaling with system utilization.

4. **High Conversion Efficiency Between Data Storage and Propagation Modes**
Electrons are the primary medium for data storage and processing due to their strong material interactions, but this property also results in significant scattering losses during transmission. Photons, with weaker material interactions, are better suited for data propagation but require energy-intensive conversion between electronic and photonic states. This inefficiency makes photonic interconnects less practical for short-range communication, such as inter-chip distances spanning a few centimeters.
Hybrid spoof plasmonic interconnects circumvent the need for such conversion mechanisms by directly utilizing electrons for signal propagation, preserving efficiency at short scales.
5. **Scalability of Wire Dimensions**
To support diverse connection levels, from intra-chip to inter-chip communication, interconnects must maintain consistent physical properties across varying wire dimensions. Current spoof plasmonic designs, reliant on quarter-wavelength corrugations, face scalability challenges. Fractal metal corrugations could provide a pathway to achieving wire scalability by compactly integrating quarter-wavelength resonators.
6. **Bandwidth Considerations**
Readers may note the absence of "high bandwidth" as an explicit feature in the above criteria. Bandwidth limitations typically arise from material losses and electromagnetic parasitics. Consequently, achieving the aforementioned five properties inherently ensures high bandwidth capacity.
Spoof plasmonic interconnects have demonstrated partial success in meeting critical interconnect requirements. However, significant advancements in materials and design techniques are necessary to address remaining challenges and realize their full potential. Future research must focus on integrating advanced materials and innovative design strategies to complete the puzzle of high-performance interconnect systems. Figure 5.26 shows a visual illustration of the key desired features of physical interconnect and how spoof plasmonic interconnect technology offers solutions to part of the problem, while the remaining challenge remains open to further research explorations.

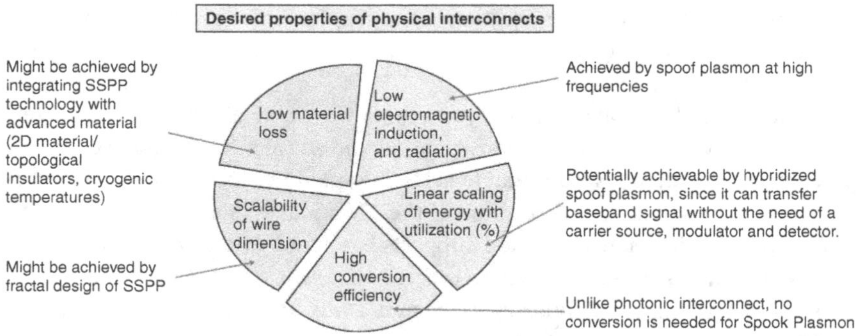

Figure 5.26 Desired properties of physical interconnects in a high-density wiring system to ensure high energy efficiency across all levels of system power and utilization. The diagram highlights potential improvements enabled by spoof plasmonic interconnects and areas requiring further research to address unresolved challenges.

5.10 Summary

In this chapter, we analyzed how much bandwidth we can avail from a lossless, noiseless parallel data bus for different kinds of interconnect technologies, namely electrical interconnect and spoof plasmonic interconnect. We showed that these two technologies are superior to each other at two different parts of the transmission spectrum; electrical interconnects are good at crosstalk suppression for low-frequency transmission, while spoof plasmon interconnects are good for high-frequency transmission. We also proposed the idea of bandwidth augmentation by merging the two different kinds of waveguiding technique: electronic and spoof-plasmonic on the same platform. The simulation shows that, even with suboptimal design of electronic to SSPP mode converter, the bandwidth of a lossless electrical data bus can be augmented by at least three times by judicious patterning on the metal wires, which accommodates spoof plasmon mode.

References

1 Tim Pfahler, Sascha Breun, Lukas Engel, Christian Geissler, Jan Schür, and Martin Vossiek. Broadband low-loss fan-in chip-to-package interconnect enabling system-in-package applications beyond 220 GHz. *IEEE Transactions on Components, Packaging and Manufacturing Technology*, 14(5):880–890, 2024.

2 Mohamed S Abouelyazid, Sherif Hammouda, and Yehea Ismail. Fast and accurate machine learning compact models for interconnect parasitic capacitances considering systematic process variations. *IEEE Access*, 10:7533–7553, 2022.

3 Lisheng Wang, Wenbo Wang, Raymond JE Hueting, Gert Rietveld, and Jan Abraham Ferreira. Review of topside interconnections for wide bandgap power semiconductor packaging. *IEEE Transactions on Power Electronics*, 38(1):472–490, 2023.

4 Raghvendra Sahai Saxena, RK Bhan, Navneet Kaur Saini, and RJ Muralidharan. Virtual ground technique for crosstalk suppression in networked resistive sensors. *IEEE Sensors Journal*, 11(2):432–433, 2010.

5 Willi Volksen, Robert D Miller, and Geraud Dubois. Low dielectric constant materials. *Chemical Reviews*, 110(1):56–110, 2010.

6 Sang K Kim, Christianto C Liu, Lei Xue, and Sandip Tiwari. Crosstalk reduction in mixed-signal 3-D integrated circuits with interdevice layer ground planes. *IEEE Transactions on Electron Devices*, 52(7):1459–1467, 2005.

7 Lei Guo, Xiangdong Qin, and Francisco Zaera. Chemical treatment of low-k dielectric surfaces for patterning of thin solid films in microelectronic applications. *ACS Applied Materials & Interfaces*, 8(9):6293–6300, 2016.

8 Md Tarekul Alam, Raghu A Pulavarthy, Jeffery Bielefeld, Sean W King, and Md Aman Haque. Thermal conductivity measurement of low-k dielectric films: Effect of porosity and density. *Journal of Electronic Materials*, 43(3):746–754, 2014.

9 Kris Vanstreels, Chen Wu, and MR Baklanov. Mechanical stability of porous low-k dielectrics. *ECS Journal of Solid State Science and Technology*, 4(1):N3058–N3064, 2014.

10 Shou-Yi Chang, Hui-Lin Chang, Yung-Cheng Lu, Syun-Ming Jang, Su-Jien Lin, and Mong-Song Liang. Mechanical property analyses of porous low-dielectric-constant films for stability evaluation of multilevel-interconnect structures. *Thin Solid Films*, 460(1-2):167–174, 2004.

11 Shun Lien Chuang. *Physics of photonic devices*. John Wiley & Sons, 2012.

12 SR Joy, M Erementchouk, H Yu, and P Mazumder. Spoof plasmon interconnects—communications beyond RC limit. *IEEE Transactions on Communications*, 67(1): 599–610, 2019. https://doi.org/10.1109/TCOMM.2018.2874242.

6

Signal Modulation by Spoof Plasmonics

6.1 Introduction

In order to transmit digital data over an SSPP channel through a high-frequency modulation mechanism, we need to find a way of modulating the transmission properties of the SSPP channel (see Figure 6.1). In microwave wireless communications, we usually switch the carrier source by the data generator. However, in the case of SSPP channel-assisted data transfer, we target to send information through a densely packed parallel bus, where each channel of the bus may share the same carrier frequency at a time. Therefore, for the SSPP data bus, we have to modulate the transmission properties of the channels rather than switching the oscillator so that each channel of the data bus can be independently controlled. This chapter discusses some potential mechanisms of modulating the transmission coefficient of the SSPP channel.

In this chapter, our goal will be finding a way so that we can significantly alter the transmission properties of the SSPP channel by introducing minimal change in its physical state. Our novel technique of spoof plasmon mode modulation will rely on the existence of an electromagnetic bandgap (BG) in the spoof plasmon channel.

6.2 Background Studies: Design of Modulator

A standard practice of designing high-speed modulators is to couple the waveguide with a resonator, and then change the resonant frequency ω_r by altering the electromagnetic property of the resonator. This technique capitalizes on the fact that

Artificial Plasmonics for VLSI Interconnects: Bridging the Gap between Electronics and Optics, First Edition. Soumitra R. Joy and Pinaki Mazumder.
 Companion Website: https://www.wiley.com/go/artificialplasmonics

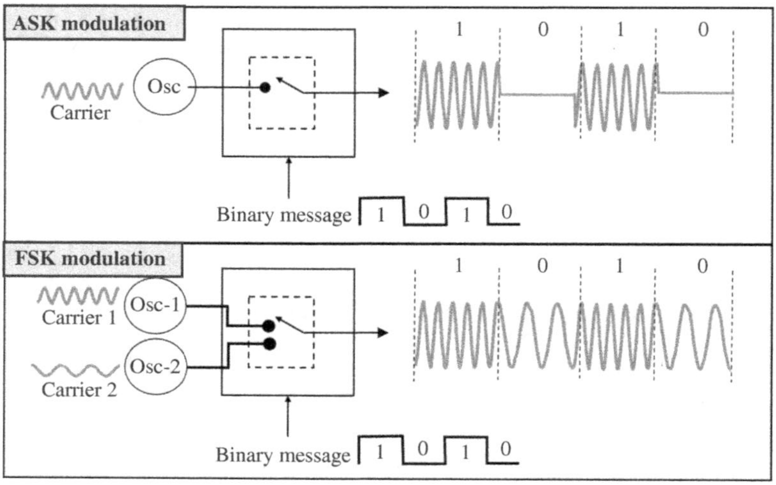

Figure 6.1 A generic block diagram of a digital signal modulation scheme: ASK and frequency shift keying (FSK).

the transmission of a resonator-coupled waveguide at a given frequency ω is ultra-sensitive to the change of ω_r. Fortunately, a spoof plasmonic waveguide is intrinsically made of quarter-wavelength resonators, giving us an opportunity to design modulator by changing the resonance frequency of the in-built resonators. However, an SSPP waveguide consists of periodic resonators, and we definitely cannot afford to control a large number of resonators—as this would result in issues with synchronization and large footprint of the modulator. Thus, the question remains as to how many of those resonators' properties we have got to change in order to make a modulator with a high extinction (ON/OFF) ratio.

It is a commonly held belief that an effective medium description of a metastructure breaks down as the number of functional elementary cells becomes too small [1]. In contrast to this argument, we show here that spoof surface plasmon polariton (SSPP) structures, consisting of narrow grooves with sub-wavelength width on a plain metal surface (see Figure 6.2a), can preserve the same "effective medium" description even when the size of the structure is reduced down to minimal number of unit cells. This facilitates building a very short and simple one-dimensional (1D) SSPP grating structure, which is able to manifest excellent mode confinement quality, i.e., high Q-factor ($\sim Q_{\max} \approx 10^5$) and high Purcell factor. The structure is simple in the sense that it does not require a high-index or negative-index medium. Additionally, the spoof plasmon states defined by the dispersion relation medium are well-preserved in finite structures, as periodicity plays a very nominal role in determining their propagation character, which is in strong contrast to crystal structures [2] in general.

One of the standard techniques of controlling signal transmission in waveguides is the incorporation of pre-engineered defects in plasmonic [3, 4] and photonic structures [5]. High speed modulation, coherent electron–photon interactions [6], ultra-small filters, low-threshold lasers [7, 8], photonic chips, multiplex biosensing [9] and quantum information processing [10]; all these sophisticated applications almost invariably seek for a mode-confining structure with both high quality factor, Q, and small modal volume, V.

For example, a point defect embedded in a large array of photonic crystals can yield a high-Q resonant mode inside the photonic BG. However, many applications require devices with stringent conditions of small sizes. In those cases, the wavelength-scale photonic crystal-based device falls markedly short of, especially at low frequencies. Metamaterials, with their ability to offer deep subwavelength confinement [11, 12], stand out to be better candidates for small devices. Efforts for making a small cavity size of good quality have so far resorted to embedding the defect in a large array of unit cells, the number of cells reaching an order of 10^2 or even higher [13–15]. The reasons are apparent. In both photonic crystals and metamaterials, the resonant mode lies inside the BG of a host structure formed as a result of interference between multiple scattered waves. Therefore, one has to leave a long chain of unit cells around the defect in order to reproduce the effect of periodic structure. Reports on appearance of BG in disordered metamaterials also resort to a large ($\approx 10^2$) number of unit cells to host a defect [16]. High-quality mode proposed or realized by far, thus, inexorably bringing large structures.

6.3 Authors' Contribution in the Field of Controlling Spoof Plasmon

Our observations of the properties of an SSPP waveguide reveal that the transmission and dispersion characteristics of these waveguides obviate the need for a long structure. In fact, a single unit cell suffices to conserve the main electromagnetic scattering properties of a periodic structure. This can enable to realize a high-Q resonant tunneling of spoof plasmon mode in a very short host medium. We illustrate this observation by considering a chain of three unit cells of the SSPP waveguide, with the left and right grooves having groove height h_h, and the groove at the middle having height h_d. We show that, for $h_d \neq h_h$, a resonant tunneling mode inside the SSPP BG is formed. Thus, we can reduce the entire structure of "a defect in host" system to its minimal possible length. In the subsequent sections, we shall refer to this minimal-length "host-defect-host" hetero-junction system as a (1–1–1) structure. The first part of this chapter discusses how the incorporation of a pre-engineered defect structure in the spoof plasmon waveguide can provide means to control the transmission properties of the channel. The second part of this chapter discusses how the dynamic alternation of the waveguide's transmission is possible to realize by incorporating transistors in each of the defect cells and develops a theoretical model for the dispersion behavior of the transistor-incorporated spoof plasmon waveguide. Based on our developed model, we analyzed various crucial figures of merit of this newly proposed mechanism of the modulation of spoof plasmon signals, such as modulation speed and extinction ratio (ER).

6.4 Transmission Spectra of Homogeneous and Heterogeneous Structures

The schematic of the structure under study is presented in Figure 6.2a. Periodic structures were a subject of intensive research [17–20], and their basic properties are well established. The SSPP dispersion diagram in a periodic structure is presented in Figure 6.2b. The gaps in the spectrum indicate the frequency regions where finite SSPP structures demonstrate subsided transmission, as is illustrated by spectra shown in Figure 6.3a: the transmission coefficient of the waveguide stays close to unity in a wide frequency range with slight modulation by Fabry–Perot ripples and shows strong reflection in the BG zone.

A thorough treatment of the optical response of a good conductor's surface with periodic narrow grooves shows that such structures are characterized by an effective Drude model with a spoof plasma frequency:

$$\omega_p = \frac{\pi c}{2h\sqrt{\epsilon}} \tag{6.1}$$

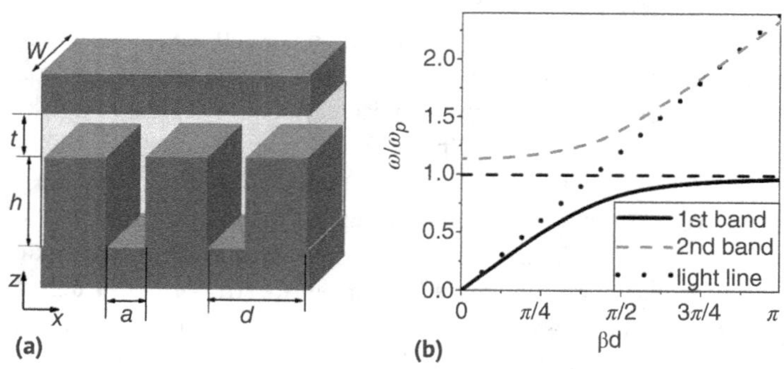

Figure 6.2 (a) The geometry of a homogeneous SSPP structure. The dark regions of the schematic are metal, and light regions are dielectric. (b) The dispersion diagram of a periodic SSPP waveguide with $h/d \sim 1$, $t/d \sim 0.3$, $a/d \sim 0.1$, and $w \sim \lambda$. The frequency is normalized by the spoof plasma frequency $\omega_p = \pi c/2h$. The bands are formed due to anti-crossing between light-line and ω_p. For $2h \gg d$, the anti-crossing happens far away from the Brillouin zone boundary.

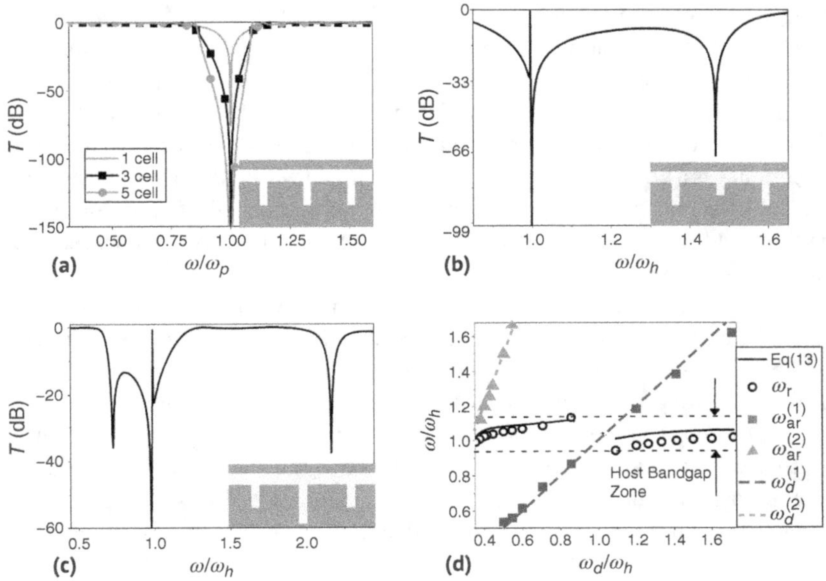

Figure 6.3 (a) Transmission spectra of homogeneous SSPP waveguides with different numbers of unit cells. Inset shows the corresponding SSPP structure. (b, c) Transmission spectra of heterogeneous structures with the defect cells characterized by $h_d/h_h = 0.75$ and $h_d/h_h = 1.25$, respectively. Anti-resonances (transmission dips) at $\omega/\omega_d \neq 1$ are induced by zeroth and higher order spoof plasmon resonance modes of defect cells. In addition, a transmission resonance emerges inside the BG. The insets show corresponding heterogeneous geometry. (d) Trajectories of transmission resonance (ω_r) and anti-resonances ($\omega_{ar}^{(1)}$ and $\omega_{ar}^{(2)}$) in heterogeneous SSPP structures as functions of the spoof plasma frequency of the defect cell. The solid black curves are loci of transmission resonances obtained via Eq. (6.13). Note that the locus of transmission resonances ω_r demonstrate discontinuity around $\omega_d/\omega_h \sim 1$, as at this condition, the SSPP structure becomes homogeneous and contains no resonance inside the BG. The dots of ω_r and $\omega_{ar}^{(n)}$ show the loci of the resonance and the nth antiresonance, respectively, obtained from the full-wave simulation.

where h is the height of the grooves, c is the speed of light in a vacuum, and ϵ is the dielectric function of the medium filling the groove. It should be emphasized that this frequency does not derive from material properties, but from the geometry of the structure. Generally, taking into account the higher resonances, the nth spoof plasma frequency can be written as:

$$\omega_p^{(n)} = (2n-1)\omega_p \tag{6.2}$$

with $n = 1, 2, \ldots$.

Figure 6.3a shows that the major characteristics of transmission spectra of a long periodic SSPP waveguide are successfully retained in a finite waveguide even when its size is reduced to a single unit cell. This result implies the immunity of SSPP waveguide's dispersion to finite-size effects. Manifestation of this transmission dip or anti-resonance of a cell with a single groove is utilized to propose bandstop filter in other works as well [21], albeit in a different context not related to the spoof plasmon concept. In fact, the attenuated transmission around cavity resonant frequency is a characteristic of a laterally coupled waveguide-cavity system in general [22]. Here, however, we emphasize that a single cell of spoof plasmon waveguide yields not only the "bandgap effect" but also the "dispersion effect" of a periodic waveguide.

To demonstrate the above-mentioned phenomenon, we study the transmission behavior of a composite (1–1–1) structure made of three unit cells of SSPP waveguide, all but the middle unit cell having groove height h_h. We shall call the middle cell a "defect" with groove height $h_d \neq h_h$ embedded in a host medium. The term "defect" in this context means an intentional breaking of translational symmetry of an otherwise regular structure. As we will show, the transmission properties of the composite medium can be successfully explained by considering each of the single host and defect cells as an effective individual SSPP medium, characterized by ω_h and ω_d spoof plasma frequencies, respectively. To put it differently, the effective medium description of SSPP metastructure is valid for a single cell as well. The apparently simple concept has far greater significance than being of a mere theoretical interest. This helps in designing, as we shall show, a very short spoof plasmon structure with giant Q-factor and Purcell factor, which is remarkable from an application point of view.

The transmission spectrum of the composite structure is a result of the interplay between characteristic frequencies describing states of the field in periodic structures made of either host or defective unit cells. A straightforward impact of series addition of a new functional cell to a background regular SSPP waveguide is added spectral filtering (Figure 6.3b,c). Due to short attenuation length, as is demonstrated in Figure 6.4b, each cell acts as a band-rejection notch filter, which suppresses transmission around its own spoof plasma frequency ω_p.

A nontrivial case of transmission resonance as well may arise for a specific configuration of the composite SSPP system: a structure with a defect cell inserted in the middle of a regular host waveguide. In this case, the narrow transmission resonance (see Figure 6.3b,c) corresponding to a resonant tunneling would emerge inside the host SSPP BG. The characteristic field distribution of this special resonant tunneling mode is shown in Figure 6.4b. Figure 6.3d depicts the trajectories of the transmission resonances ω_r and anti-resonances ω_{ar} with variation of ω_d. In the next section, we show how the numerically obtained trajectory of ω_r and ω_{ar} in a (1–1–1) composite structure can be explained by describing the propagating state in each single cell as spoof plasmon mode. In this regard, it should be emphasized

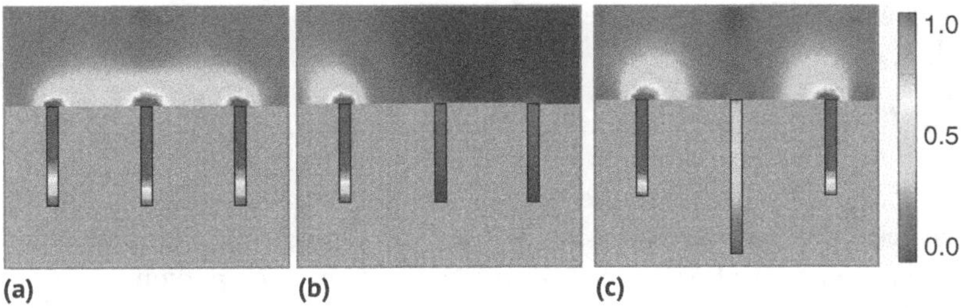

Figure 6.4 The field distribution inside a three-cell structure with incoming wave from the left side. (a) Propagating SSPP mode in a homogeneous structure at a frequency corresponding to passband. (b) Non-propagating SSPP at a frequency inside the stopband. (c) Resonant tunneling SSPP mode in a heterostructure incorporating a defect cell in the middle.

that, as Figure 6.4 demonstrates, all modes, propagating, non-propagating and resonant tunneling, are concentrated close to the surface of the structure. the appearance of such kind of resonant tunneling mode of spoof plasmon with high Q-factor supported by SSPP in short structures is the principal message of the present work.

6.5 SSPP Scattering in Heterogeneous Structures

Despite significant efforts invested into theoretical studies of SSPP, a full yet tractable description of the propagation of spoof plasmons in finite structures is yet to be developed. Basic transport properties of such structures, however, can be understood within a simplified picture based on the assumption that the propagation within a single cell can be effectively characterized by a Bloch wavenumber and imposing continuity on average conditions at the interfaces between different cells.

In what follows, we will denote host and defect cells with subscripts "h" and "d," respectively. Retaining a single mode, the electric and magnetic fields can be presented as:

$$H_y = e^{iq(\omega)z}\{f_+ e^{i\beta(\omega)x} + f_- e^{-i\beta(\omega)x}\} \tag{6.3}$$

$$E_z = -\frac{\beta(\omega)}{\omega\epsilon} e^{iq(\omega)z}\{f_+ e^{i\beta(\omega)x} - f_- e^{-i\beta(\omega)x}\} \tag{6.4}$$

where the coordinate axes are chosen as shown in Figure 6.2a, f_+ and f_- are amplitudes of the waves propagating forward and backward, respectively, $q(\omega) = \sqrt{(\omega/c)^2 - \beta^2(\omega)}$ is the transverse wavenumber and $\beta(\omega)$ is found for the host and defect cells from the respective dispersion equations $D_{h,d}(\omega, \beta) = 0$, where Xu et al. [23]

$$D(\omega, \beta) = \frac{q}{\omega/c} \tan(qt) + \frac{a}{d} \text{sinc}^2(a\beta/2) \tan(\omega h/c) \tag{6.5}$$

The possibility to characterize the state of the electromagnetic field within a single cell with the help of dispersion Eq. (6.5) is suggested by the following consideration. In a typical periodically perturbed system, say a photonic crystal, shown in Figure 6a the corresponding band diagram can be obtained by replicating the dispersion diagram of the background at each reciprocal lattice site. The replicated diagrams intersect at the boundary of the Brillouin zone, and the cross-resonance between them leads to opening a gap between the bands. Figure 6.5c shows how gap formation occurs in a typical photonic crystal. Periodicity thus plays the most crucial role for BG formation in photonic crystals. Far away from the boundary of the Brillouin zone, the band modification due to periodicity is weak. BG in spoof plasmon, shown in Figure 6b however, is born of a different type of mode coupling. The coupling here occurs between

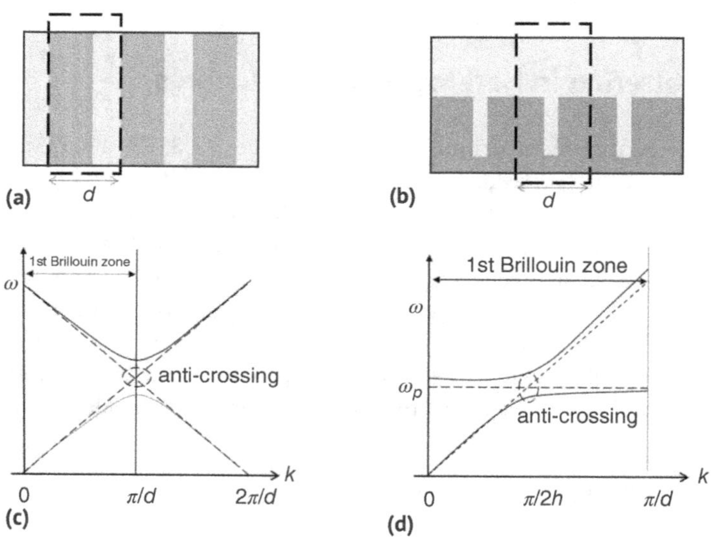

Figure 6.5 (a, b) 1D photonic crystal and SSPP structure, respectively. The dashed box indicates the unit cells and periodicity *d*. (c, d) Band diagrams of a photonic crystal and the SSPP periodic waveguide, respectively. Anti-crossing (dashed circles) between bands of the perturbed background (dashed lines) leads to the formation of new bands (solid lines). For the SSPP, the anti-crossing location ($k = \pi/2h$) must occur far away from the boundary of the Brillouin zone in order to have well-developed SSPP. As a result, SSPP modes are immune to broken periodicity and withstand the finite-size effect.

states inside the groove and that outside the groove. The mechanism of SSPP BG creation due to crossover between the light-line in dielectric outside of the groove and the resonant mode inside the groove at frequency ω_p is shown in Figure 6.5d. We notice that well-defined SSPP are formed when $h > d/2$ [24], which holds when the intersection between light line and the resonant mode is far from the boundary of the Brillouin zone. Thus, in a well-defined SSPP supporting structure, periodicity becomes a nominal concern for band formation, and one can regard Eq. (6.5) as describing an effective medium even for a very short SSPP structure with a minimal number of cells.

Assuming that the interface between the cells is situated at $x = 0$, the continuity of average conditions for the tangential component of the electric field at the interface has the form:

$$\int_0^t E_{z(d)}|_{x=0} dz = \int_0^t E_{z(h)}|_{x=0} dz \tag{6.6}$$

while the continuity of the total energy flux across the interface leads to:

$$\int_0^t E_{z(d)} H^*_{y(d)} dz = \int_0^t E_{z(h)} H^*_{y(h)} dz \tag{6.7}$$

Enforcing these conditions, we obtain the transfer matrix for the interface between host and defect cells:

$$\begin{pmatrix} f_{h+} \\ f_{h-} \end{pmatrix} = \frac{1}{2} \begin{pmatrix} t_1 & t_2 \\ t_2 & t_1 \end{pmatrix} \begin{pmatrix} f_{d+} \\ f_{d-} \end{pmatrix} = T_{hd} \begin{pmatrix} f_{d+} \\ f_{d-} \end{pmatrix}$$

Here, $t_i = A - (-1)^i \beta_d / A\beta_h$ and, for small arm thickness t, one has $A = \frac{q_d}{q_h} \frac{e^{iq_h t} - 1}{e^{iq_d t} - 1} \approx 1$. Here, q_d and q_h are transverse wave vectors of the defect and host cells, respectively.

An SSPP structure with a defect cell in the middle is described by the transfer matrix:

$$T = T_{eh} T_h T_{hd} T_d T_{dh} T_h T_{he} \tag{6.8}$$

where $T_{dh} = T^{-1}_{hd}$, $T_{h,d} = \text{diag}\left(e^{i\theta_{h,d}}, e^{-i\theta_{h,d}}\right)$, with $\theta_{h,d} = \beta_{h,d} d$, are the transfer matrices between the host and defect cells, and $T_{eh} = T^{-1}_{he}$ are the transfer matrices between the SSPP structure and environment, which we will take in the form of an interface transfer matrix:

$$T_{eh} = \frac{1}{2} \begin{pmatrix} s_1 & s_2 \\ s_2 & s_1 \end{pmatrix} \tag{6.9}$$

with some parameters $s_{1,2}$, the specific form of which is of low importance for the following.

Since we are interested in the frequency range inside the BG of the host structure, where $\beta = i\kappa_h$, we present $\theta_h = iNd\kappa_h$, where N is the number of host cells on either side of the defect cell; and obtain from Eq. (6.8) the transmission coefficient:

$$t = \frac{16e^{i\theta_d}s_1 s_2 t_1 t_2}{e^{2\kappa_h N}\mathcal{D}^{(+)} + e^{-2\kappa_h N}\mathcal{D}^{(-)} + \mathcal{D}^{(0)}} \tag{6.10}$$

where

$$\begin{aligned} \mathcal{D}^{(\pm)} &= (s_1 \pm s_2)^2 \left[(t_1 \pm t_2)^2 - e^{2i\theta_d}(t_1 \mp t_2)^2\right] \\ \mathcal{D}^{(0)} &= 2(1 - e^{2i\theta_d})\left(s_1^2 - s_2^2\right)\left(t_1^2 - t_2^2\right) \end{aligned} \tag{6.11}$$

The resonant tunneling occurs at frequency, where the exponential attenuation of the transmission with the length of the host cancels. This yields:

$$\mathcal{D}^{(+)}(\omega) = D_+(\omega)D_-(\omega) = 0 \tag{6.12}$$

where $D_\pm(\omega) = t_1 e^{-i\theta_d/2} \pm t_2 e^{i\theta_d/2}$. Using the obtained transfer matrices, it can be shown that $D_+(\omega) = 0$ and $D_-(\omega) = 0$ define frequencies of states confined to the defect cell embedded into an infinite host structure, even and odd with respect to the center of the defect cell, respectively. It's not difficult to see, however, that $D_+(\omega) = 0$ does not have solutions inside the first Brillouin zone. Thus, the resonant transmission is supported only by odd local states, or when:

$$\frac{\beta_d(\omega)}{\kappa_h(\omega)} = \cot\left(\frac{d\beta_d(\omega)}{2}\right) \tag{6.13}$$

Basic features of the transmission resonance can be discussed directly on the base of Eq. (6.13) taking into account that inside the bandgap above host spoof plasma frequency ω_h, (i.e., $\omega_h < \omega < \omega_h^{(H)}$), we have:

$$\kappa_h = \frac{\omega}{c}\sqrt{\frac{\omega_h^{(H)} - \omega}{\omega - \omega_h}} \tag{6.14}$$

Here, $\omega_h^{(H)}$ is the upper frequency edge of the SSPP BG, which is found to be:

$$\omega_h^{(H)} = \omega_h\left(1 + \frac{4ah}{\pi^2 dt}\right) \tag{6.15}$$

As the height of the groove in the defect cell increases, the transmission resonance ω_r emerges near $\omega_h^{(H)}$ and moves deeper into the host BG ($\omega_h^{(H)} - \omega_r \sim \sqrt{h_d - h_h}$). Defect cells with shorter grooves can be of particular interest, however. It yields resonant states with an evanescent transverse field profile in defect cells. The surface-bound resonant mode for shorter defect groove may be exploited to design a high-Q cavity with no physical enclosure around defect cell. The comparison of the analytical estimation following from Eq. (6.13) and the numerical results is provided in Figure 6.3d. The very well agreement with the numerically obtained data validates the concept that spoof plasmon mode can survive even in a short meta-structure consisting of a minimal number of cells. It should be noted that when the resonance frequency ω_r approaches ω_h, the point $\beta_d(\omega_r)$ gets close to π/d, where the characterization of the state of the electromagnetic field inside the defect cell in terms of dispersion Eq. (6.5) is no longer valid, and Eq. (6.13) breaks down.

We conclude this section with a note about the defect-induced anti-resonances in transmission spectra, for instance, at $\omega/\omega_h \approx 1.45$ in Figure 6.3b and at $\omega/\omega_h \approx 0.7$ and 2.2 in Figure 6.3c. They, as well, can be studied using Eq. (6.10). It's not difficult to see that they emerge at frequencies inside the stopband of the defect cell. As a result, with the variation h_d, they follow defect cell spoof plasma frequencies, $\omega_d^{(n)}$.

6.6 Q-factor and Enhanced Radiation Rate

We consider two figures of merit for the resonant mode inside the SSPP BG: the Q-factor and the Purcell factor. In a defect incorporated finite waveguide, for a given frequency inside the BG, the fraction of power scattered out through the structure can be written as:

$$|t|^2 \sim e^{-4\kappa_h d N} \tag{6.16}$$

As the Q-factor is inversely proportional to the power radiated out of the structure, taking into account Eq. (6.14) for κ_h, we can estimate for frequencies not too close to ω_p

$$\log Q \sim N\sqrt{\frac{\omega_h^{(H)} - \omega_r}{\omega_r - \omega_h}} \tag{6.17}$$

As has been discussed above, with gradually changing the height of the groove in the defect cell, the resonance frequency approaches ω_h, which results in significant increase of κ_h, yielding very high-quality modes localized on the defect cell.

Figure 6.6a shows how the quality factor changes with the variation of the defect's spoof plasma frequency. The fact that electromagnetic transmission essentially stops at ω_h results in an exponential increase of the Q-factor. Thus, a very high-quality resonating mode is possible to realize in the SSPP host medium consisting of only a single groove on each side of the defect cell. Although the divergence in Eq. (6.17) is an artifact of the approximation retaining only a single mode inside the grooves, the COMSOL MULTIPHYSICS full-wave simulation justifies very high Q-factor in the (1–1–1) structure, which hits a maximum of

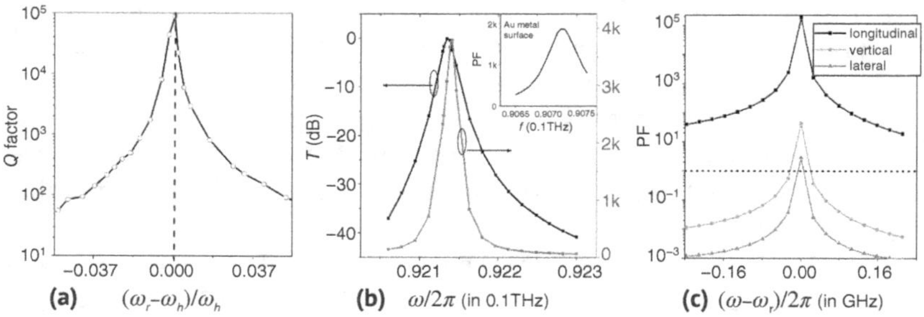

Figure 6.6 (a) The Q-factor as a function of the difference between resonant mode and host's spoof plasma frequency. Physical height of each grooves = 0.8 mm, period = 1 mm, arm = 300 μm, defect's optical height = 2.5 mm. (b) The Purcell factor, F_p, in the limit of perfect electric conductor in the (1−1−1) heterostructure is plotted together with the transmission spectrum. Inset shows the Purcell factor around the resonant frequency for a structure of the same geometry but with Au metal surface. (c) The Purcell factor in (1−1−1) SSPP heterostructure with the above mentioned geometry for source with three different dipole orientations, positioned near the groove opening in the defect cell. The region above the dotted line corresponds to $F_p > 1$.

value of 10^5 when resonant mode frequency coincides with the host spoof plasma frequency ω_h.

6.7 Dynamic Switching of SSPP Transmission Property

While the previous sections discussed how we can get resonant mode in SSPP waveguide with high quality factor, we are yet to discuss how this property can be leveraged to make modulators, i.e., how to dynamically alter the effective height of the groove.

There are two possible options for dynamic modification of effective groove height:

- by loading the interior of the groove with electro-optic material, such as lithium niobate ($LiNbO_3$) that shows Kerr effect, and applying voltage to get the electro-optic effect;
- by placing transistors across each of the grooves of spoof plasmonic waveguide. Switching the transistors ON or OFF can drastically alter the effective resonating length of grooves.

While the first option is commonplace in the design of optical modulators, the second one is still unexplored. In addition to that, the second approach does not rely on finding the right material with the desired electro-optic property, and hence can considerably simplify the design strategy. In the subsequent sections, we will discuss the spoof plasmon modulation strategy based on the second approach, reported by Nayem et al. [25].

6.7.1 SSPP Dispersion in the Presence of Modulator: Theoretical Framework

A schematic illustration of the proposed structure for signal modulation of the SSPP interconnect is shown in Figure 6.7a. The structure consists of periodic rectangular grooves of length h_0 with variable resistors connected in between the teeth at position $|y| = h_m$, where $h_m < h_0$ (Figure 6.7b). The variable resistor can be implemented by the gate-controlled channel resistance of a field-effect transistor (FET). The basic mechanism of the proposed SSPP modulator is as follows. The conductivity state of the electronic modulator would alter the cutoff frequency of the structure and thus would modulate the transmission of a carrier frequency. The cutoff frequency of a static SSPP structure of groove length h_0 is approximately $f_c \approx \frac{c}{4h_0}$ where c is the free- space velocity of electromagnetic wave (a more accurate estimation would be given in Section 2.2). During OFF-state of the electronic

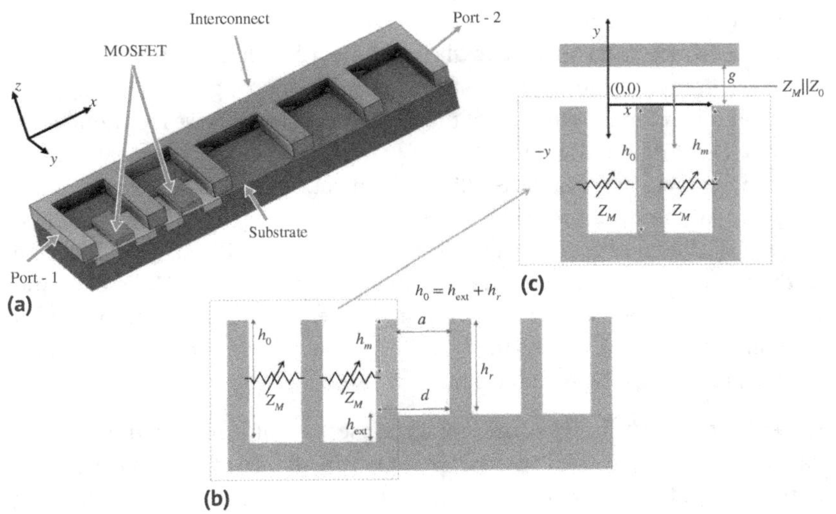

Figure 6.7 Schematic illustration of the SSPP interconnect with modulator: (a) three-dimensional (3D) schematic of the proposed SSPP interconnect with adjustable resistor across the teeth, envisioned to be implemented by a gate-controlled FET; (b) two-dimensional (2D) view showing groove length (h), period (d), groove width (a) and modulating resistors (Z_M) located at position h_m of the anomalous grooves; the dashed box represents the region that enables signal modulation; (c) extended 2D view of the dashed region shown in (b), with a metal bar placed adjacent to the interconnect at $y = g$.

modulator (i.e., when gate voltage is less than the threshold voltage of FET), it provides no conductive path across the teeth, and therefore does not impact the electromagnetic state of the system. On the contrary, during ON-state, it creates a very low impedance path between the teeth at the position $|y| = h_m$, thereby effectively reducing the optical length of the groove to h_m. This in-turn increases the cutoff frequency of the structure upto $\frac{c}{4h_m}$. Thus, a carrier frequency chosen between $\frac{c}{4h_0}$ and $\frac{c}{4h_m}$ can alternate between high transmission and high backscattering if the controllable resistor is switched between a low and a high impedance value. In a real system, the SSPP modulator is supposed to feed a static SSPP interconnect of groove length h_r. It can be shown that, in order to match the mode between the transmission state of the SSPP modulator and the static SSPP interconnect, the best position of placing the controllable resistor inside the SSPP modulator is $h_m = h_r$. Thus, a natural design choice for the controllable resistor-implanted SSPP modulator would be having its groove length h_0 longer than the groove length h_r of the static SSPP interconnect to which the modulating signal is fed.

In our subsequent analysis, the width of all grooves is considered as a, and the groove periodicity as d (Figure 6.7b). It may be noted that SSPP interconnects are not limited by RC-delay as they can transmit signals without metallic ground. Hence, in accordance with previous studies [26–29], no metallic ground is considered in this work. It has been shown that only a handful of grooves suffices to produce the transmission properties of a chain of periodic grooves in the SSPP structure [29]. Therefore, for the rest of the paper, unless otherwise mentioned, we designed the SSPP modulator consisting of two grooves, slightly elongated from the rest of the grooves in the static SSPP interconnect. The geometry of the modulating segment of the structure is shown in Figure 6.7c. To generalize our theoretical framework, the modulator is considered to have a metallic stripline situated next to the grooves in the same plane and at a distance of $y = g$. While the parameter g gives us another knob for controlling the electromagnetic properties of the interconnect, it will be shown that the SSPP waveguide can propagate signals without the presence of the metallic bar.

a. **Impedance Without Modulator:** The insertion of a modulating resistor having impedance Z_M will change the input impedance of the groove at position $y = -h_m$. For a TEM mode, the equivalent input impedance (Z_{in}) looking through the anomalous groove is given by:

$$Z_{\text{in}} = Z_M \parallel Z_0 \tag{6.18}$$

Here, Z_0 is the impedance of the groove looking at the position $y = -h_m$ in the absence of modulating resistance. Z_0 is defined as:

$$Z_0 = \left.\frac{E_{g,x}}{H_{g,z}}\right|_{Z_M \to \infty} \tag{6.19}$$

where $E_{g,x}$ and $H_{g,z}$ are spatial components of electric and magnetic fields inside the groove. The spatial distribution of $E_{g,x}$ has the following form:

$$E_{g,x}\big|_{Z_M=\infty} = \sum_m G_m \cos\left(\frac{m\pi}{a}(x - a/2)\right) \sin\left(Q_m(y + h_0)\right) \tag{6.20}$$

For a perfect metal, the tangential field component $E_{g,y}$ inside the groove vanishes at the metal interface at $x = \pm\frac{a}{2}$. Consequently, the first mode of nonzero field component $E_{g,y}$ would correspond to higher-order modes having mode numbers $m \geq 1$ and frequencies higher than $\frac{\pi}{a}$. Considering a narrow groove approximation [24], which suggests that the operating frequency for information transfer through the SSPP channel is lower than $\frac{\pi}{a}$, we can set $E_{g,y} = 0$ in our frequency regime of interest.

The magnetic field component inside the grooves in the absence of modulating resistance is given by:

$$H_{g,z}\big|_{Z_M=\infty} = \sum_m \frac{k^2a^2}{k^2a^2 - (m\pi)^2} \frac{iQ_mG_m}{\omega\mu} \cos\left(\frac{m\pi}{a}(x - a/2)\right) \cos\left(Q_m(y + h_0)\right) \tag{6.21}$$

where G_m is the amplitude of mth Fourier component of the field and $Q_m^2 = k_0^2 - \left(\frac{m\pi}{a}\right)^2$ is the wave vector corresponding to mode of order m. By setting $y = -h_m$ in Eqs. (6.20) and (6.21), the impedance Z_0 can be expressed in terms of the geometry of the structure as:

$$Z_0 = \frac{E_{g,x}}{H_{g,z}}\bigg|_{y=-h_m, Z_M\to\infty} = -i\frac{\omega\mu}{Q_0} \tan\left(Q_0\left(h_0 - h_m\right)\right) \tag{6.22}$$

b. Impedance with Modulator: For impedance Z_M inserted at the position $y = -h_m$, Eqs. (6.20) and (6.21) can be rewritten by decomposing the fields into upward ($+y$ direction) and downward ($-y$ direction) propagating components having amplitudes G_{m+} and G_{m-}, respectively, as follows:

$$E_{g,x} = \sum_m \left(G_{m+}e^{iQ_my} + G_{m-}e^{-iQ_my}\right) \cos\left(\frac{m\pi}{a}\left(x - \frac{a}{2}\right)\right) \tag{6.23}$$

$$H_{g,z} = -\frac{i}{\omega\mu}\frac{\partial E_{g,x}}{\partial y} = \sum_m \frac{1}{Z_c}\left(G_{m+}e^{iQ_my} - G_{m-}e^{-iQ_my}\right) \cos\left(\frac{m\pi}{a}\left(x - \frac{a}{2}\right)\right) \tag{6.24}$$

Assuming the fundamental mode to be the dominant mode, we obtain the expression of the equivalent impedance to be:

$$Z_{\text{in}} = \frac{E_{g,x}}{H_{g,z}}\bigg|_{y=-h_m} = \frac{G_{0+}e^{iQ_0h_m} + G_{0-}e^{-iQ_0h_m}}{Q_0/(\omega\mu) \times \left(G_{0+}e^{iQ_0h_M} - G_{0-}e^{-iQ_0h_m}\right)} \tag{6.25}$$

Considering $\tilde{Z}_{\text{in}} = \frac{QZ_{\text{in}}}{\omega\mu}$ to be the normalized input impedance, the terms r and θ can be defined as follows:

$$r = \frac{\tilde{Z}_{\text{in}} + 1}{\tilde{Z}_{\text{in}} - 1} \tag{6.26}$$

$$\theta = -2Q_0 h_m \tag{6.27}$$

According to these definitions, r depends on the magnitude of Z_M, and θ depends on the location of inserting Z_M inside the groove. Utilizing r and θ in Eq. (6.25), it can be shown that:

$$G_{0+} = G_{0-} r e^{i\theta} \tag{6.28}$$

c. Relating Field Components Inside and Outside the Groove: Outside the groove, i.e., in the region $0 \leq y \leq g$, assuming no variation of the field components along z direction (i.e., $k_z = 0$), the electric field component can be written as follows:

$$E_{a,x} = \sum_n (A_{n+} e^{ik_y y} + A_{n-} e^{-ik_y y}) e^{i\beta_n x} \tag{6.29}$$

Here, $\beta_n = \beta_0 + \frac{2n\pi}{d}$, where β_n and β_0 are the propagation constants for the nth and zeroorder modes respectively. Based on the assumption of no variation of the field along z-direction (i.e., $\frac{\partial E_z}{\partial z} = 0$), we can apply Gauss's law in a source-free region ($\nabla.\bar{\mathbf{E}} = 0$) to get:

$$E_{a,y} = -\int \frac{\partial E_{a,x}}{\partial x} dy \tag{6.30}$$

Taking into account the time dependence, the magnetic field component can be written as:

$$H_{a,z} = \frac{i}{\omega\mu}\left(\frac{\partial E_y}{\partial x} - \frac{\partial E_x}{\partial y}\right) = \sum \frac{k^2}{\omega\mu k_y}(A_{n+} e^{ik_y y} - A_{n-} e^{-ik_y y}) e^{i\beta_n x} \tag{6.31}$$

From the boundary condition that the tangential electric field at the metal wall vanishes, we obtain the following relation from Eq. (6.29), between upward and downward propagating electric field components outside the groove.

$$A_{n+} = -A_{n-} e^{-2ik_y g} \tag{6.32}$$

Again, from the continuity condition of the tangential electric field E_x at the SSPP-air interface ($y = 0$), we get the following from Eqs. (6.23) and (6.29):

$$\sum_m (G_{m+} + G_{m-}) \cos\left(\frac{m\pi}{a}\left(x - \frac{a}{2}\right)\right) = \sum_n (A_{n+} + A_{n-})\, e^{i\beta_n x} \tag{6.33}$$

Multiplying Eq. (6.33) with $e^{-i\beta_{n'}x}$, and integrating both sides of the equation over the SSPP period d, we get the following:

$$\sum_m (G_{m+} + G_{m-}) \int_{-a/2}^{a/2} e^{-i\beta_{n'}x} \cos\left(\frac{m\pi}{a}\left(x - \frac{a}{2}\right)\right) dx = (A_{n'+} + A_{n'-})\, d \tag{6.34}$$

where on the left-hand side, we leveraged the fact that the tangential electric field component $E_{x,a}$ is zero at the metallic interface of SSPP ($a/2 \leq |x| \leq d/2$). Here, on the right-hand side of the equation, we applied the following mathematical identity:

$$\int_{-d/2}^{d/2} e^{\frac{i2n\pi}{d}x} dx = \begin{cases} d & \text{for n} = 0 \\ 0 & \text{for } n \neq 0 \end{cases} \tag{6.35}$$

To define coupling between mth mode of the field inside the groove and nth mode of the field outside, we introduce the shorthand notation $\mathcal{K}_{mn} = \int_{-a/2}^{a/2} e^{-i\beta_n x} \cos\left(\frac{m\pi}{a}\left(x - \frac{a}{2}\right)\right) dx$, where n' has been replaced by n. Then Eq. (6.34) reduces to the following form:

$$\sum_m (G_{m+} + G_{m-})\mathcal{K}_{mn} = (A_{n+} + A_{n-})d \tag{6.36}$$

Considering only the dominant fundamental modes, the relation can be simplified as:

$$(G_{0+} + G_{0-})\frac{a}{d}\text{sinc}\left(\frac{a\beta_0}{2}\right) = (A_{0+} + A_{0-}) \tag{6.37}$$

Similarly, by utilizing the continuity condition of the tangential magnetic field at the SSPP-air interface at $y = 0$, Eqs. (6.24) and (6.31) can be equated and integrated over the limit $-d/2 \leq x \leq d/2$ to arrive at the following relation:

$$\frac{Q_0}{\omega\mu}(G_{0+} - G_{0-}) = \frac{\omega\epsilon}{k_y}(A_{0+} - A_{0-})\text{sinc}\left(\frac{a\beta}{2}\right) \tag{6.38}$$

d. Dispersion Relation with Modulator: From Eqs. (6.37) and (6.38), we obtain the following relation between amplitudes of Fourier components of the field inside and outside the groove:

$$\frac{a}{d}\text{sinc}^2\left(\frac{a\beta}{2}\right)\frac{G_{0+} + G_{0-}}{G_{0+} - G_{0-}} = \frac{k_y}{k}\frac{A_{0+} + A_{0-}}{A_{0+} - A_{0-}} \tag{6.39}$$

Using Eqs. (6.11) and (6.15) and applying the mathematical identity that $\cot(x) = i\frac{e^{ix}+e^{-ix}}{e^{ix}-e^{-ix}}$, we obtain the dispersion relation of an SSPP waveguide having a lump impedance Z_M:

$$\left(\frac{k_y}{k_0}\right)^2 = 1 - \frac{(\beta' + i\beta'')^2}{k_0^2} = \left[i\frac{a}{d}\ \mathrm{sinc}^2\left(\frac{a\beta}{2}\right)\frac{1+re^{i\theta}}{1-re^{i\theta}}\ \cot(k_y g)\right]^2 \quad (6.40)$$

Here the longitudinal propagation constant β is explicitly shown to have a real (β') and imaginary (β'') component. As most works on SSPP interconnects focus on the passband, usually the imaginary part of β is not considered. In contrast, here we focus on the imaginary component of β and show that β'' can be altered by changing the value of Z_M such that the information transmitted through the SSPP channel is modulated.

Finally, if the metallic bar on the top of the interconnect is removed, k_y for a non-radiating SSPP mode becomes imaginary such that $k_y = i\kappa$, where κ is the attenuation constant along $+y$ axis away from the SSPP-air interface. Under this condition, the dispersion relation of the SSPP waveguide is obtained by evaluating Eq. (6.40) at the limit of $g \to \infty$:

$$i\kappa = \frac{\omega}{c}\frac{a}{d}\mathrm{sinc}^2\left(\frac{a\beta}{2}\right)\frac{1+re^{i\theta}}{1-re^{i\theta}} \quad (6.41)$$

If the grooves are terminated with a short-circuit (i.e., $Z_{\mathrm{in}} = 0$) at the point $y = -h_m$, then $r = -1$, and $\theta = -2Q_0 h_m$. Under this condition, the dispersion relation reduces to the following well-known form of a regular SSPP interconnect having groove width a and groove periodicity d:

$$\kappa = \frac{\omega}{c}\frac{a}{d}\mathrm{sinc}^2\left(\frac{a\beta}{2}\right)\tan\left(Q_0 h_m\right) \quad (6.42)$$

6.7.2 Prediction of Dispersion-limited Modulation Speed

In the previous section, we discussed how the dispersion relation of the SSPP waveguide is drastically altered by the introduction of a switching impedance Z_M. In this section, we will show how the SSPP signal modulation speed can be estimated from the modulated SSPP interconnects transmission characteristics. For this, we consider a carrier signal of unity amplitude with frequency ω_c, which coincides with the upper edge (ω_u) of the SSPP channel's BG. This BG is centered around the resonant frequency (ω_r) of the SSPP waveguide. The carrier signal is modulated by the mechanism of abrupt impedance change over a range of $Z_M = [Z_{\mathrm{OFF}}, Z_{\mathrm{ON}}]$, where Z_{OFF} and Z_{ON} are open and short circuit impedance, respectively. The carrier signal can pass through the waveguide if $Z_M = Z_{\mathrm{ON}}$ and is blocked when . If the modulation frequency is ω_m, for a square-wave message signal $\Pi(t)$, the modulated signal $S_m(t)$ becomes:

$$S_m(t) \approx \sin(\omega_c t) \times [\Pi_T(t)(S_{21,\text{ON}}(\omega) - \Pi_T(t-\tau)S_{21,\text{OFF}}(\omega)) \tag{6.43}$$

Here, $\tau = \frac{\pi}{\omega_m}$ and $S_{21,\text{ON}}$ and $S_{21,\text{OFF}}$ are the transmission coefficients of the waveguide for $Z_M = Z_{\text{ON}}$ and $Z_M = Z_{\text{OFF}}$, respectively. Considering the first Fourier component of $\Pi(t)$, we obtain:

$$S_m(t) \approx \frac{1}{2}\sin(\omega_c t)(S_{21,\text{ON}}(\omega_c) + S_{21,\text{OFF}}(\omega_c)) + c_1\cos((\omega_c - \omega_m)t) \tag{6.44}$$

where

$$c_1 \propto \frac{1}{2}\left[S_{21,\text{ON}}\left(\omega_c - \frac{\pi}{2\tau}\right) - S_{21,\text{OFF}}\left(\omega_c - \frac{\pi}{2\tau}\right)\right] \tag{6.45}$$

It can be seen that the modulated signal essentially becomes a monotonic wave of frequency ω_c if the Fourier component c_1 vanishes. Therefore, c_1 essentially corresponds to the modulation depth at the critical limit of modulation frequency. According to Eq. (6.45), c_1 would drastically drop once the modulation frequency ω_m becomes large enough to make $S_{21,\text{OFF}} = S_{21,\text{ON}}$ at the frequency $\omega = \omega_c - \omega_m$. This limiting frequency corresponds to the lower edge of the SSPP BG, which we denote as ω_e. Therefore, the theoretical maximum of modulation frequency in the SSPP waveguide will be approximately equal to the bandgap, $\text{BG} = \omega_u - \omega_e$ of the SSPP interconnect when $Z_M = Z_{\text{OFF}}$.

In order to estimate this BG, we set $Z_{\text{off}} \to \infty$ and obtain the following dispersion relation from the general expression of Eq. (6.40):

$$k_y = -\frac{\omega}{c}\frac{a}{d}\text{sinc}^2\left(\frac{a\beta}{2}\right)\cot(k_y g)\tan(Q_0 h) \tag{6.46}$$

The BG around the resonant frequency $\omega_r = \frac{\pi c}{2h}$ has two regions, namely $\omega_e < \omega < \omega_r$ and $\omega > \omega_r > \omega_u$. The lower band-edge frequency ω_e is given by:

$$\omega_e = \left(\frac{\pi}{2} - \delta_L\right)\frac{c}{h} \tag{6.47}$$

where δ_L is the lower detuning with respect to ω_r. Similarly, the upper band-edge frequency can be expressed as $\omega_u = \omega_r + \delta_u\frac{c}{h}$, where δ_u is the upper detuning parameter. From these definitions of ω_u and ω_e, we obtain

$$\text{BG} = (\delta_u + \delta_L)\frac{c}{2\pi h} \tag{6.48}$$

If δ_L is small, then $k_y^2(\omega_e) \approx \left(\frac{\pi}{2h}^2 - \beta^2(\omega_e)\right) = \left(\frac{\pi}{2h}^2 - \left(\frac{\pi}{d}\right)^2\right)$, and $\tan(Qh) \approx \frac{1}{\delta_L}$. Thus, the lower detuning parameter becomes:

$$\delta_L \approx \frac{\pi}{2}\frac{a}{d}\frac{\coth(\kappa_e g)}{\kappa_e h} \tag{6.49}$$

where $k_y = i\kappa$ and $\kappa_e = \kappa(\omega = \omega_e) = \sqrt{\left(\frac{\pi}{d}\right)^2 - \left(\frac{\pi}{2h}\right)^2}$. If the detuning δ_u is also small, we obtain the following expression by setting $\beta = 0$ $\left(\text{i.e., } k_y = \frac{\omega}{c}\right)$ in Eq. (6.46).

$$\delta_u \approx \frac{a}{d}\cot\left(\frac{g}{h}\right) \tag{6.50}$$

This relation is obtained within the approximation $\tan(Q_0 h) = \tan\left(\frac{\pi}{2} + \delta_u\right) \approx -\frac{1}{\delta_u}$. By replacing δ_L and δ_u of Eq. (6.48), with Eqs. (6.49) and (6.50), the BG—i.e., the maximum modulation frequency—can be expressed in terms of geometrical parameters of the SSPP interconnect. In subsequent sections, we shall validate this prediction by numerical analysis and finite element method-based simulations.

6.7.3 Analysis of Energy Efficiency

In order to analyze the energy efficiency of the modulation scheme, we need to account for the various power loss mechanisms in the system. Power loss can occur through the following pathways: Joule heating loss in the controllable resistor (i.e., channel of FET), carrier signal absorption loss in the patterned metal of the SSPP waveguide, insertion loss due to modal mismatch between the TEM wave feeder and the SSPP modulator, radiation loss due to waveguide bending, and interference loss due to electromagnetic coupling with neighboring channels. The radiation and interference loss analysis is not within the scope of the current work, as we focus on analysis of an isolated SSPP modulator in its most basic form. The insertion loss has been reduced to have nominal effect by designing an appropriate tapered adapter, as shown in previous works [28]. Thus, the most salient form of power loss in our considered structure is the I^2R loss in modulator's resistance ($P_{\text{loss},M}$) and the metallic absorption in SSPP waveguide ($P_{\text{loss},wg}$). Then the energy efficiency η can be calculated as:

$$\eta = (1 - P_{\text{loss},wg} - P_{\text{loss},M}) \times 100\% \tag{6.51}$$

In order to calculate the loss in SSPP wave propagation due to metallic absorption, let us assume we are applying a rectangular pulse of width τ in the time domain on the control terminal of the electronic modulator. Let us also assume that the electronic modulator's resistance is very low ($Z_M << Z_c$) during the ON-state of the pulse. Here, $Z_c = \frac{\omega\mu}{Q}$ is the characteristic impedance of the medium. For a real metal, its resistive loss can be accounted for by its skin depth l_s. It is shown in Rusina et al. [30] that the wave vector Q inside the SSPP groove can be expressed as a complex quantity for a lossy metal as,

$$Q = Q_R + iQ_I \approx \frac{\omega}{c}\left(1 + i\frac{l_s}{2a}\right) \tag{6.52}$$

For the low impedance condition of the electronic modulator during the ON-state (i.e., $Z_M << Z_c$), we can assume $r \approx -1$. Then, by replacing the term (Q_0) in Eq. (6.42) with the complex value of Q, we can solve for a complex value of β. Following Joy et al. [26], the attenuation constant, represented by $Im[\beta]$ along the direction of SSPP signal propagation can be related to Q as,

$$Im[\beta] = \left(\frac{\omega}{c}\right) \times Im\left[\sqrt{1 + \left(\frac{a}{d}\right)^2 \tan^2((Q_R + iQ_I)h_m)}\right] \tag{6.53}$$

Hence, if the carrier signal is modulated by a rectangular pulse of the message signal where the pulse-width of the message is τ in time domain, then the normalized power loss due to the propagation of double-sideband modulated signal with carrier through lossy SSPP waveguide of length L_{wg} would be,

$$\begin{aligned} P_{\text{loss.wg}} \approx P_{\text{loss,OFF}} + \Big[\frac{1}{4}(1 - \exp(-2Im[\beta_{c0}]L_{\text{wg}})) \\ + \frac{1}{\pi^2}(1 - \exp(-2Im[\beta_{c-}]L_{\text{wg}})) \\ + \frac{1}{\pi^2}(1 - \exp(-2Im[\beta_{c+}]L_{\text{wg}}))\Big] \end{aligned} \tag{6.54}$$

where $Im[\beta_{c0}], Im[\beta_{c+}], Im[\beta_{c-}]$ denotes an imagninary value of the complex β evaluated at $\omega = \omega_c$ (carrier frequency), $\omega = \omega_c + \frac{\pi}{\tau}$ (upper sideband), and $\omega = \omega_c - \frac{\pi}{\tau}$ (lower sideband), respectively. The $P_{\text{loss,OFF}}$ denotes the power loss during OFF-state of electronic modulator, which corresponds to no transmission of carrier. Let us assume that the ASK modulation with random bits of ones and zeros of equal pulse-width is implemented. Then, depending on whether the metal-bar at $y = g$ position in Figure 6.7c is present or not, during the OFF state of electronic modulator, half of the input power would be either reflected back ($P_{\text{loss,OFF}} = 0$), or lost through radiation ($P_{\text{loss,OFF}} = 0.5$).

6.7.4 Numerical Analysis

It is well known that the cutoff frequency (ω_c) of the fundamental mode of an SSPP waveguide is inversely proportional to its groove length. Quantitatively, $\omega_c = \frac{\pi c}{2\epsilon_r h}$, where c is the free-space velocity of the electromagnetic wave and ϵ_r is the dielectric constant of the material in between the grooves. The longitudinal attenuation constant (β''), which is the imaginary part of the longitudinal propagation constant β, is supposed to be high around ω_c. This is confirmed by Figure 6.8a, which shows the change of attenuation constant for groove lengths of 5 and 20 mm. Here it has been assumed that the variable resistor connected between the grooves has an impedance of $Z_M = 5$ kΩ, which effectively serves as a high impedance for the interconnect and therefore does not impact the field interior to the groove.

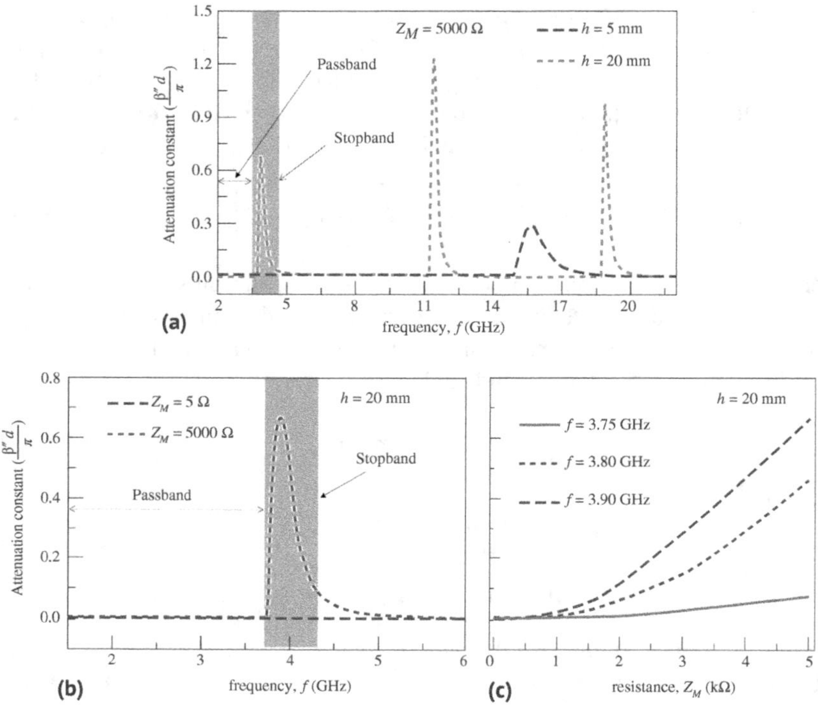

Figure 6.8 Frequency dependence of attenuation constant. (a) Normalized attenuation constant (β'') for two interconnects having different groove lengths (h) and a fixed modulating resistance $Z_M = 5000\ \Omega$. Normalized attenuation constant (β'') of an interconnect having h = 20 mm; (b) as a function of frequency for modulating resistances, $Z_M = 5\ \Omega$ and $Z_M = 5000\ \Omega$; and (c) as a function of Z_M, for three carrier frequencies within the stopband.

As observed in Figure 6.8a, the stopband of SSPP waveguide appears at every odd integer multiple of the frequency $\frac{c}{4h}$. For $h = 20$ mm, the first three stopbands occur at, 3×3.75, and 5×3.75. The first stopband of the interconnect coincides with the cutoff frequency of 3.75 GHz corresponding to $h = 20$ mm, implying that in principle, a carrier signal of 3.75 GHz can be modulated if the length of the SSPP groove can be dynamically changed from $h = 20$ mm. However, because the groove length cannot be physically changed in real-time, we aim to mimic the same effect by electrically changing the impedance Z_M.

To numerically calculate the effect of changing Z_M, the input impedance Z_{in} is calculated based on Eq. (6.18), and the result has been utilized to evaluate r and θ from Eqs. (6.26) and (6.27). Next, using Eq. (6.41) and based on the relation between wave vector components of the SSPP mode $\left(\beta^2 - \kappa^2 = Q^2 = \left(\frac{\omega}{c}\right)^2\right)$, we obtain the change of attenuation constant for two different values of modulating impedance Z_M in the SSPP channel of $h = 20$ mm. According to Figure 6.8b, when $Z_M = 5\ \text{k}\Omega$, the attenuation constant is high around $\omega_c = 3.75$ GHz and hence a stopband exists around this frequency. However, when Z_M is reduced to 5 Ω, the cutoff frequency shifts to a higher frequency, effectively making the frequency region below 16 GHz a passband. This suggests that for a fixed groove length, it is possible to attain the function of modulation by changing the impedance of the variable resistor. Hence, for a given carrier frequency within the stopband, the act of switching the modulating impedance Z_M would encode the message signal into the amplitude of the carrier.

6.7.5 Trade-off Between Modulation Speed and Energy-efficiency

If we can drastically switch the value of Z_M such that $Z_{M,\text{ON}} << Z_c$ and $Z_{M,\text{OFF}} >> Z_c \tan\left(\frac{\omega}{c}(h_0 - h_m)\right)$, then both the ON- and OFF-states of the SSPP modulator can be described with Eq. (6.42). However, for OFF-state, the variable h_m in Eq. (6.42) has to be replaced by h_0. Then the additional bandgap Δ_{BG} induced by the switching of Z_M from ON- to OFF-states would be:

$$\Delta_{\text{BG}}(\text{Hz}) = \frac{c}{4}\left(\frac{1}{h_m} - \frac{1}{h_0}\right) + \frac{c}{2\pi}\left(\frac{a}{h_0\sqrt{4h_0^2 - d^2}} - \frac{a}{h_m\sqrt{4h_m^2 - d^2}}\right) \tag{6.55}$$

As evident from Eqs. (6.54) and (6.55), the maximum possible modulation speed would be inversely proportional to groove length h, while the power loss increases with higher frequencies and therefore with shorter groove length. Thus, modulation speed and energy efficiency of an SSPP modulator hold a direct trade-off relation. In Figure 6.9a, based on our theory, we show how the act of scaling of the groove length h_0 modifies the modulation speed and how it takes a toll on the energy efficiency of the structure. While scaling the groove length h_0, we also scaled the other feature sizes of the SSPP waveguide, such as groove width a, groove period d, and modulator position hm in a way so that the following geometric ratio always holds: $a/h_0 = 0.2$, $h_m/h_0 = 0.8$, $d/h_0 = 1$. For all cases, we considered the total length of the SSPP waveguide to remain as $L_{wg} = 100$ mm. For the calculation of this trade-off relation, the carrier frequency for modulation is chosen as half of the summation of the band-edge frequencies ($f_{e,\text{ON}}, f_{e,\text{OFF}}$) of the SSPP waveguide (i.e., the frequency of surface-bound mode at longitudinal phase of 180°) corresponding to OFF- and ON-state of electronic modulator, respectively. For the double-side band modulation scheme considered here, the switching frequency of the controllable resistor is taken to be $f_m = f_{e,\text{ON}} - f_c = f_c - f_{e,\text{OFF}}$. It can be seen from Figure 6.9a that a double-sideband modulation scheme can yield ~ 10 Gbps data speed with $\geq 80\%$ energy-efficiency. By increasing the ratio of a/h_0

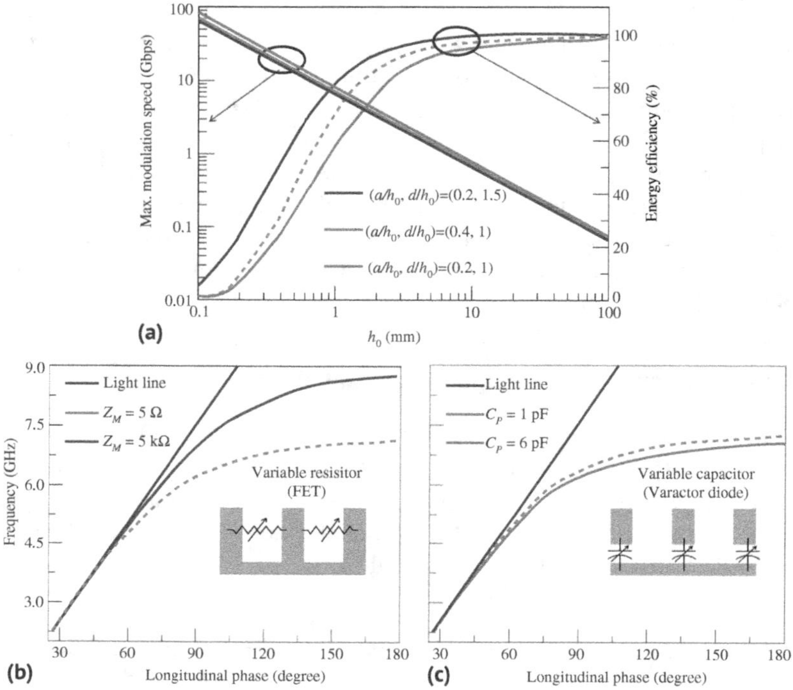

Figure 6.9 Trade-off between modulation speed and energy efficiency in controllable-resistor-based SSPP modulator and modulation of SSPP dispersion using different modulators. (a) The modulation speed in double-sideband (DSB) scheme increases with an inverse proportion to the scaling of groove length h, while the other geometric features of the SSPP waveguide are scaled accordingly. The energy efficiency is calculated for the corresponding SSPP structure having a length of 100 mm. (b) Modulation of geometric dispersion relation in the SSPP waveguide with the switching of a variable resistor between 5 Ω and 5 KΩ (c) Modulation of dispersion relation in the SSPP waveguide with the switching of a varactor between 1 and 6 pF.

from 0.2 to 0.4, we can increase the energy efficiency by 10% for 1 mm long groove length (corresponding to 75 GHz cutoff frequency) with nominal change in modulation speed. We may also increase energy efficiency by 20% by increasing the ratio d/h_0 from 1 to 1.5, albeit at the expense of reduced surface confinement.

6.7.6 Extension of the Theory for a Broad Class of SSPP Modulators

Our theoretical framework offers a unified approach to understanding the properties of various SSPP modulators, encompassing both previously reported behaviors from existing literature and the novel behavior introduced by our proposed modulator. Thus, our theoretical machinery provides a comprehensive explanation of modulator behavior. In order to prove our statement, we will show how our theory can be readily extended to produce the dispersion properties of a varactor diode-controlled SSPP modulator, previously reported in He et al. [31]. The extension of the theory would require to modify the input impedance of the groove, represented by Eq. (6.22) as follows:

$$Z_{\text{in}} = \left.\frac{E_{g,x}}{H_{g,z}}\right|_{y=-h_m} = -iZ_c \tan\left(Q_0\left(h_0 - h_m\right)\right) + \frac{i}{\omega C_M} \tag{6.56}$$

where C_M stands for the variable capacitance of the varactor. Note that the sign of the imaginary part of the additional impedance would depend on the assumption of the form of time-dependence of the electromagnetic wave. Since our theory is developed with the assumption of $e^{-i\omega t}$ time-dependence, any additional impedance in series with the natural input impedance of the groove would be $-i\omega L$ and $+\frac{i}{\omega C}$ for inductive and capacitive modulators, respectively. With that simple modification, if we apply the equation of dispersion relation of Eq. (6.41), then for an increasing value of C_M, we can see that the transmission band SSPP waveguide would decrease.

In Figure 6.9b,c, we plotted the dispersion diagram of an SSPP waveguide for different electrical impedance states of controllable resistors and varactors, respectively. For both the diagrams, the sample SSPP waveguide has the following geometry: groove length $h = 10$ mm, groove width $a = 2$ mm, period $d = 10$ mm. The modulating component (either transistor or varactor) is placed at the position of $h_m = 8$ mm inside into the groove. The controllable resistor is connected as a bridge between two neighboring teeth of the grooves, while the varactor is connected by splitting each tooth of the groove at $h_m = 8$ mm position and connecting the split parts of the tooth with the varactor.

In case of the resistor-controlled modulation scheme, the switching of the electronic modulator from 5000 Ω (OFF-state) to 5 Ω (ON-state) increases the cutoff frequency from 6.75 to ~8.625 GHz, a ~28% change in transmission band. Sweeping the value of resistance in this range can be practically achieved by state-of-the-art transistors with submicron gate length and a hundred microns of gate width [32, 33]. It can be readily seen from Eq. (6.55) that the controllable resistor-induced change in transmission band can be enhanced even further by placing the electronic modulator further away from the short-circuited terminal of the groove (i.e., by choosing smaller h_m in the design), albeit at the price of reduced surface confinement of the signal. In case of varactor-controlled modulation scheme, our theory shows how the transmission band decreases with the increase of the value of its capacitance (Figure 6.9c), which agrees with the findings of varactor-based SSPP modulation reported in He et al. [31].

6.8 Experimental Considerations and Signal Modulation

6.8.1 Design Considerations

Our vision for the SSPP modulator involves its integration into chip-scale communication systems, necessitating the selection of a modulating device with low on-resistance and minimal switching latency. Advancements in transistor scaling over the past years have yielded metal-oxide-semiconductor field-effect transistors (MOSFETs) and high electron mobility transistors (HEMTs) with exceptional switching speeds and reduced parasitic capacitances. Very recently, a record bandwidth of 400 GHz and a record low ON-state impedance of 500 Ωμm have been reported with GaN MOSHEMTs [33]. By capitalizing on the switching speed of such state-of-the-art transistors, it is anticipated that the choice of FET as the modulating component for SSPP waveguides will be highly advantageous. A schematic illustration of the envisioned scheme (shown in Figure 6.7a) suggests that, during experimental realization, the source and drain regions of the FET can posit right below the neighboring teeth of the groove to make good contact with the metallic part of the SSPP waveguide. One may choose the n-channel enhancement mode FET to capitalize on the relatively higher mobility of electrons compared to holes for designing a high-speed modulator. The following factors should play an important role in the design of the FET-based SSPP modulator:

- **FET ON-resistance:** In order to increase the modulation speed, the ON-resistance of the FET has to be small. This can be achieved by integrating a FET of small channel length (L_{ch}) and large width (W_g) into the interconnect. To reduce the contact resistance between the SSPP metal and FET source-drain regions, the SSPP metal width needs to be large. Quantitatively, the ON-state resistance of the FET can be written as:

$$R_{\text{on}} = R_{\text{sh}} \frac{L_{\text{ch}}}{W_g} + R_{\text{contact}} \frac{2}{W_g(d-a)} \tag{6.57}$$

  where R_{sh} is the channel sheet resistance and R_{contact} is the contact resistance between the SSPP metal and FET source-drain regions. The numerical analysis of Section 2.2 shows that the SSPP attenuation constant is most sensitive to a modulation impedance of, $Z_M \approx 2.5$ kΩ. This value supports the practicality of the proposed approach, as the ON-state resistance of micro- or nanometer scale enhancement mode transistors is much lower. For nanoscale FETs, the scale mismatch between the SSPP metal teeth (in the order of sub-millimeter) and the FET (in the order of nanometer) will supposedly make the contact resistance critical. An approach to reduce contact resistance would be to choose a large SSPP period or a small SSPP groove width, as suggested by Eq. (6.57).

- **SSPP Stopband Range:** An important design parameter related to the SSPP stopband is the groove width, which determines the coupling between the SSPP resonant mode and the free-space mode. Using Eqs. (6.49) and (6.50), it can be shown that the SSPP BG is dependent on groove width as follows:

$$\mathrm{BG} \propto \frac{a}{d}\left(\frac{\pi}{2}\frac{\coth(\kappa_e g)}{\kappa_e h} + \cot\left(\frac{g}{h}\right)\right) \tag{6.58}$$

  where κ_e is the attenuation constant defined in Section 2.1. According to Eqs. (6.57) and (6.58), enhancement of modulation bandwidth will come at the cost of increased ON-state resistance. Hence, to design the FET-enabled SSPP modulator, an optimum value of groove width needs to be reached so that appreciable modulation bandwidth and speed can be achieved.
- **Optimal Biasing Point of FET:** In order to maximize the sensitivity of the transport properties of SSPP waveguide on the state of FET, we take help from Figure 6.8c, where we show the attenuation constant for a range of variation of Z_M. It is observed that the sensitivity of the attenuation constant, $\frac{\partial \beta''}{\partial Z_M}$, is maximum when Z_M varies from 1 to 3 kΩ. Hence, to implement an ASK-modulated SSPP output with a reasonably low bit error rate, Z_M should be significantly less than 1 kΩ in the passband and significantly greater than 3 kΩ in the stopband. Thus, the DC biasing voltage V_g on the gate of the FET should be such that $(Z_{M,\mathrm{bias}} = (V_D/I_D)|_{V_{g,\mathrm{bias}}} \approx (1+3)\,\mathrm{k}\Omega/2 = 2\,\mathrm{k}\Omega)$. In our numerical analysis resulted from our developed theoretical framework, we showed that a value of $Z_M = 5\Omega$ and a value of $Z_M = 5000\ \Omega$ would closely correspond to the state of short-circuited and open-circulated states of FET respectively, as evident from the dispersion relation of the SSPP waveguide presented in Figure 6.9.

In the present study, we aim to evaluate the modulation performance of our proposed design in the regime of geometric dispersion limited modulation speed rather than the FET's transconductance-property limited speed. Hence, for simplicity, the FET has been modeled as a variable resistor throughout this work.

6.8.2 Transmission Characteristics

In order to match the modes between the feeder and the SSPP waveguide, we designed an adapter utilizing the structure of a Vivaldi antenna [27, 28]. The geometry of the converter is designed in a way to achieve approximately 50 Ω input impedance. The matching with the input impedance was confirmed by varying the impedance of a lump element at the output port and monitoring the minima of reflection coefficient at frequencies close to the spoof plasmon resonance. Figure 6.10a,b demonstrate the electric field distribution in log scale for the ON- and OFF-states of the modulator, respectively. The ON- and OFF-states of the

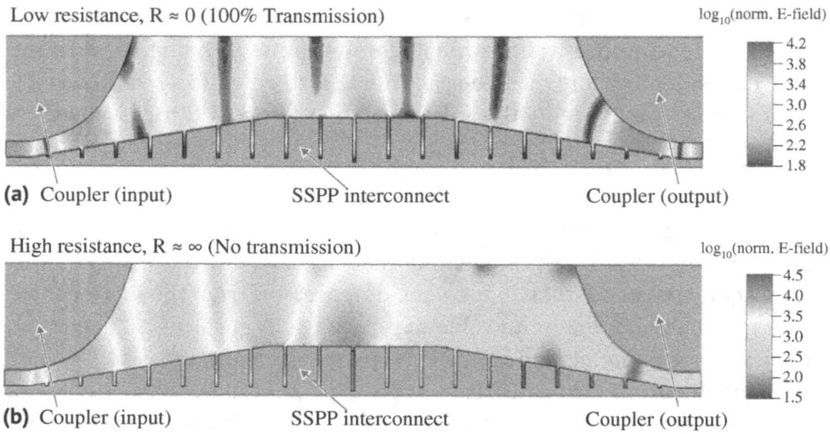

Figure 6.10 Field distribution in SSPP waveguide during ON- and OFF-states. The center groove is elongated and incorporates a variable resistor (such as, a gate-controlled FET) inside it. (a) High transmission coefficient is observed at carrier frequency when the modulator is switched ON. (b) Suppressed signal propagation through SSPP waveguide during the OFF state of the modulator.

modulator are mimicked by setting $Z_M << Z_C$, and $Z_M >> Z_C$, respectively, where Z_C is the characteristic impedance of the medium inside the groove for TEM mode propagation. It can be readily seen that the ON-state of the modulator corresponds to high transmission of surface waves, while the OFF-state corresponds to signal propagation up to the elongated groove (where the variable resistor is located), followed by dominant back-scattering.

Based on the formalism discussed in Section 6.2, three types of SSPP interconnects have been considered for finite element method-based numerical analysis. These structures have been labeled and shown as Types I, II, and III in Figure 6.11. Here, Type-I is a reference static structure with two elongated grooves, Type II is a dynamic structure with one elongated groove and a variable resistor inside it, and Type-III is an extension of Type II structure with multiple modulating grooves. All the structures, unless otherwise specified, have a regular groove length of 20 mm, elongated groove length of 24 mm, groove width of 2 mm, and periodicity of 20 mm. The objective is to show how the transmission bandwidth and roll-off of band edge can be impacted by the in situ variation of the resistance Z_M of implanted electronic modulators and also by the design choice of total number of modulating cells.

The bandwidth ($\mathrm{BW_{sp}}$) of the Type-I interconnect is given by the following relation [24].

$$\mathrm{BW_{sp}} = \frac{c}{2\pi n_{\mathrm{geff}}}\left[\frac{\pi}{2} - \frac{a}{d}\frac{d}{\sqrt{4h^2 - d^2}}\right] \tag{6.59}$$

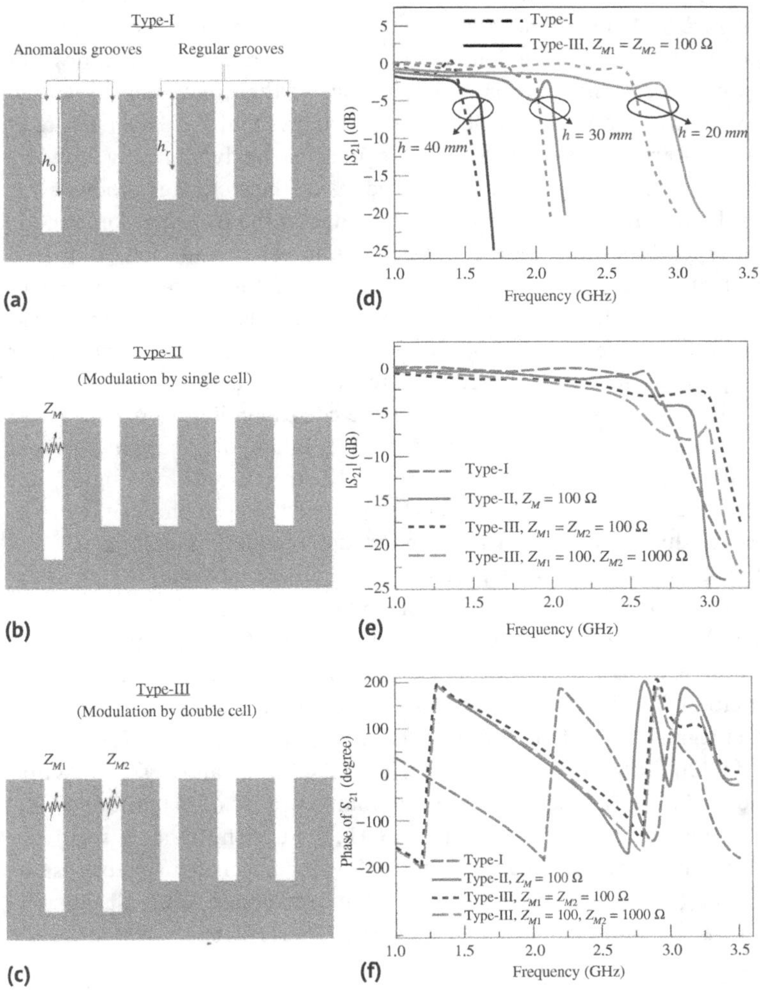

Figure 6.11 Design optimization of SSPP interconnects for signal modulation. (a–c) Schematic illustration of the three types of SSPP interconnects, labeled as Type-I, Type-II, and Type-III, depending on the presence and number of modulating resistors. (d) Comparison of S_{21} parameters of Type-I and Type-III interconnects, considering identical values of modulating resistance Z_{M1} and Z_{M2}. (e) Comparison of S_{21} parameters of Type-I, Type-II, and Type-III interconnects; both equal and non-equal values of Z_{M1} and Z_{M2} have been considered for the Type-III interconnect. (f) Phase response of S_{21} parameters corresponding to the amplitude response shown in (e).

where n_{geff} is the effective refractive index of the dielectric material in between the grooves. The numerically obtained transmittance (S_{21}) of Type-I interconnect (shown in Figure 6.11d) is found to be in accordance with this theoretical estimate of bandwidth. Figure 6.11e demonstrates that the transmission bandwidth increases with the reduction of channel resistance of the modulator. This is in accordance with the theoretical prediction, which shows that the dispersion relation and the cutoff frequency of the SSPP waveguide can be dynamically tuned by changing the modulating impedance implanted inside the elongated grooves. It is also noteworthy that the modulator-induced change in the transmission band is greater for the Type-III structure, which involves multi-cell modulation, compared to the Type-II structure with single-cell modulation. As the degree of change of transmission band is directly related to the maximum available modulation speed and also to modulation depth, we expect that Type-III structure to support higher modulation speed compared to the Type-II structure. In Section 6.8.3, we show that this is indeed the case. The phase response corresponding to the S_{21} parameters of Figure 6.11e is also shown in Figure 6.11f. Taking the phase response of Type-I structure as a reference, both Types II and III structures show a significant change of phase as Z_M is varied. This offers the prospect of both phase and amplitude modulation during signal transmission through the SSPP interconnect.

To obtain further insight into the prospect of amplitude modulation, transmission characteristics of Type II interconnect are analyzed (Figure 6.12). The −3 dB cutoff changes by 0.4 GHz as Z_M is changed from 5 Ω to 5 kΩ. The shaded region of Figure 6.12a, extending from ~2.4 to 2.8 GHz, posits the region of interest for signal modulation. To understand how a continuous change of channel resistance would impact signal transmission, the S_{21} parameter is plotted as a function of Z_M in Figure 6.12b for SSPP carrier frequencies of 2.45, 2.75, and 2.9 GHz. While transmittance at 2.75 GHz decreases by more than an order of magnitude as Z_M increases from low to high resistance, at 2.45 GHz the transmittance is relatively less sensitive to the change of Z_M. According to the prediction of our theory established in Section 6.2, for a choice of 2.75 GHz carrier frequency, the dispersion-limited maximum modulation speed would be (2.75–2.45) Gbps = 300 Mbps. In the next section, we will verify this prediction by time domain analysis.

6.8.3 Modulation Characteristics

In this section, based on time-domain analysis, we demonstrate how the proposed design can be utilized to implement the ASK modulation. The corresponding SSPP structure with adapter at input and output port and the modulating impedance halfway through the structure is shown in Figure 6.13a. For a carrier signal having frequency f_c, the message signal shown in Figure 6.13b is given as input to the control terminal of the electronic modulator. The ON- and OFF-states of the

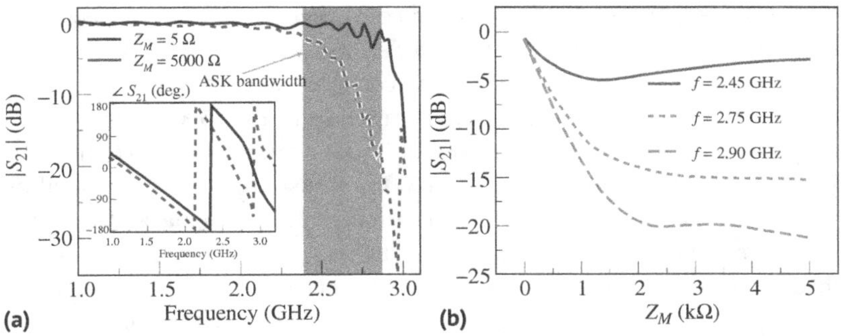

Figure 6.12 Transmission characteristics for amplitude modulation. (a) Transmission characteristics of the Type-III SSPP interconnect for $Z_M = 5$ and $5000\ \Omega$ (inset shows corresponding phase response). (b) S_{21} parameters obtained for various frequencies around the resistor-induced BG regime while Z_M is varied up to $5000\ \Omega$.

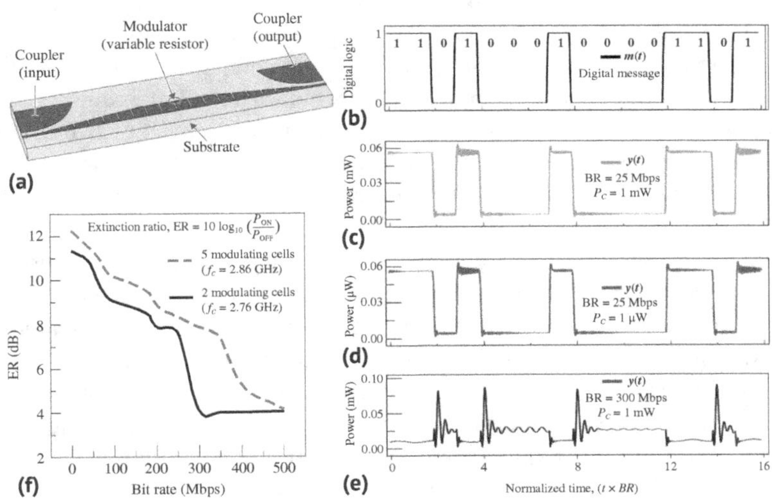

Figure 6.13 Amplitude modulation and time-domain analysis. (a) 3D schematic of the SSPP communication system comprising the SSPP interconnect, input–output couplers or adapters, and modulating variable resistor. (b) The message signal ($m(t)$) and (c–e) corresponding output signal ($y(t)$), obtained with a SSPP carrier frequency of f_c = 2.76 GHz and two-cell modulation. Here (c, d) show modulated ASK output for carrier power (P_c) of 1 and 0.01 mW, respectively, at a bit rate of BR = 25 Mbps. In (e), the case of P_c = 0.01 mW and BR = 300 Mbps is shown. (f) Extinction ratio (ER), which is defined as the ratio of output power during ON- and OFF-states of the electronic modulator (denoted as P_{ON} and P_{OFF} respectively), has been plotted as a function of modulation speed for two cases: two-cell modulation with a carrier frequency of 2.76 GHz and five-cell modulation with a carrier frequency of 2.86 GHz.

modulator correspond to modulating resistance $Z_M = 5$ and 5000 Ω, respectively. Pseudo-random bit streams of OOK-NRZ encoded message signal ($m(t)$) have been considered in the time domain analysis. The carrier frequency is maintained at 2.76 GHz and the modulating signal speed is varied from 25 to 500 Mbps.

The output signal ($y(t)$), obtained for two different power levels (P_c) of the carrier is shown in Figure 6.13. For $P_c = 1$ mW, the transmitted message signal of 25 Mbps speed remains nearly identical to the original message signal after demodulation at the receiving end (Figure 6.13c). At such a speed, which is significantly lower than the electronic modulator-switching induced BG, the transmitted signal remains quite similar to the original signal even after a reduction of carrier power by 1000 times (Figure 6.13d). However, when the bit rate is ramped up to 300 Mbps, the distortion and attenuation of the transmitted signal becomes significant (Figure 6.13e). To estimate the maximum bit rate for the considered design parameters, we calculate the signal ER as the ratio of output amplitude levels during the ON- and OFF-states of the modulator. The ER decreases at high bit rates, with a drastic decline within the 200–300 Mbps range. These characteristics can be correlated with the frequency response of the SSPP interconnect shown earlier in Figure 6.12a. According to the frequency response, the −3 dB cutoff frequency of the modulator during its OFF state is 2.45 GHz. Therefore, for a choice of 2.76 GHz carrier signal, the lower band edge of the OFF-state of the modulator is approximately 300 MHz away from the carrier frequency, limiting the modulation speed to within 300 Mbps.

However, the precise value of modulation bandwidth and degree of modulation depth (i.e., the ER) for a given carrier frequency additionally depends on the number of modulated grooves. To demonstrate this, in Figure 6.13f, we have compared the ER of two different cases of the Type-III system discussed in Section 3.2. It is observed that at any given bit rate, the ER increases as the number of modulating cells is increased. For an interconnect having five modulating cells—each comprising a variable resistor as modulator—the highest enhancement of ER is achieved around the modulation speed of 300 Mbps. Nonetheless, it is important to note that a power penalty is associated with the incorporation of electronic modulators in additional grooves of the interconnect. Each electronic modulator within the SSPP cell introduces its own I^2R loss, thereby contributing to the system's overall power consumption. Furthermore, achieving synchronization among multiple electronic modulators during high-speed modulation necessitates the design of suitable signal delay lines for each modulator. This would invariably increase the design complexity and footprint of the system.

To further evaluate the ASK modulation scheme, a standard communication system comprising a transmitter, a modulated SSPP interconnect, and a receiver

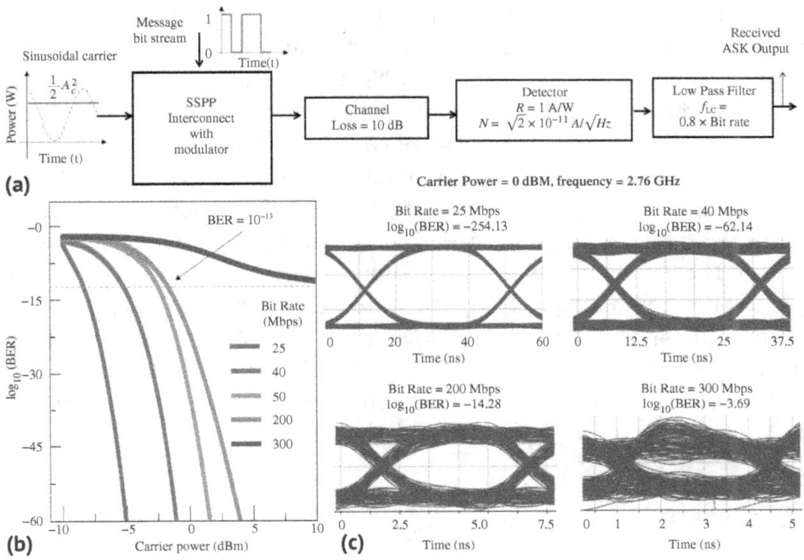

Figure 6.14 SSPP-based communication system and its figure of merit. (a) Schematic of an SSPP interconnect-based communication system comprising channel loss of 10 dB, detector responsivity $R = 1$ A/W, noise spectral current density $N = \sqrt{2} \times 10^{-11} \frac{A}{\sqrt{Hz}}$; cutoff frequency ($f_{LC}$) of the LPF is 80% of bit rate. (b) BER– obtained for different bit rates of the message signal–plotted as a function of carrier power (the dashed line indicates a threshold of 10^{-13} BER); and (c) corresponding eye diagrams considering a fixed carrier power and frequency of 0 dBm and 2.76 GHz, respectively.

is considered (Figure 6.14a). For simplicity, the channel is considered distortion-free with an attenuation of 10 dB. The detector is modeled with shot noise spectral current density of $N = \sqrt{2} \times 10^{-11} A/\sqrt{\text{Hz}}$, and responsivity of $R = 1$ A/W. The detector is followed by a low-pass filter (LPF), the cutoff frequency (f_{LC}) of which is approximately 80% of BR. The bit error rate (BER), calculated for different bit rates and carrier powers, is shown in Figure 6.14b. In accordance with previous studies [34, 35], $\text{BER} \leq 10^{-13}$ is taken as the threshold for error-free data transmission. For BR = 25 Mbps, the minimum carrier launch power required to maintain $\text{BER} \geq 10^{-13}$ is −8.46 dBm (0.14 mW). As the BR is increased to 200 Mbps, this power requirement increases to −1.63 dBm (0.7 mW). For signal frequencies at 300 Mbps or higher, BER for a given carrier power drastically increases beyond $> 10^{-13}$ prohibiting the modulation at such speed. Thus, we validate our theoretical prediction that the maximum modulation speed for an SSPP waveguide with groove length 20 mm would be ≤ 300 Mbps.

The eye diagram of Figure 6.14c is obtained using a virtual test bench of commercial software. The eye diagram shows the transmission characteristics for a carrier power of 0 dBm (1 mW) at different bit rates of the message signal. The eye-opening decreases by 26.8% as the BR increases from 25 to 300 Mbps. Also, jitter—normalized by corresponding symbol period—increases from 4% to about 17.5% as the BR is increased. With 1 mW power of carrier signal and 300 Mbps bit rate of the message signal, BER is $\sim 10^{-3}$. The BER at such high bit rates can be reduced by increasing the carrier power, as has been demonstrated in Figure 6.14b. The modulation speed of SSPP interconnects can be extended from hundreds of Mbps to hundreds of Gbps by scaling down the groove length from millimeter to micrometer dimensions. However, at the sub-THz regime of data transfer rate, the low energy efficiency of the modulator due to ohmic loss in metal would become a bottleneck for the system, as explicated by our theoretical analysis.

6.9 Summary

In this chapter, we discussed how to harness the resonating property of spoof plasmon waveguide in order to make a high-speed modulator for spoof plasmonic signal. We demonstrated that the length of the spoof plasmonic modulator can be as short as that containing a single unit cell, and described how we can design a spoof plasmon waveguide with high Q-factor resonant tunneling mode by cascading grooves of different lengths. Then we proposed an idea of dynamically altering the effective resonant frequency of the waveguide by attaching a transistor across each of the grooves of the modulator; so that the transmission property of the waveguide can be drastically changed upon switching the transistor ON and OFF. The parameters of the modulating waveguide that can be tailored in order to increase the modulation speed and depth of modulation are also discussed with analytical expressions.

References

1 Chihhui Wu, Alessandro Salandrino, Xingjie Ni, and Xiang Zhang. Electrodynamical light trapping using whispering-gallery resonances in hyperbolic cavities. *Physical Review X*, 4(2):021015, 2014.

2 Susumu Noda, Katsuhiro Tomoda, Noritsugu Yamamoto, and Alongkarn Chutinan. Full three-dimensional photonic bandgap crystals at near-infrared wavelengths. *Science*, 289(5479):604–606, 2000.

3 Wenshan Cai and Vladimir M Shalaev. *Optical metamaterials*. New York, Springer-Verlag, 2010.

4 Nader Engheta and Richard W Ziolkowski. *Metamaterials: Physics and engineering explorations*. John Wiley & Sons, 2006.

5 Eli Yablonovitch. Photonic band-gap structures. *Journal of the Optical Society of America B*, 10(2):283–295, 1993.

6 Andreas Wallraff, David I Schuster, Alexandre Blais, L Frunzio, R-S Huang, J Majer, S Kumar, Steven M Girvin, and Robert J Schoelkopf. Strong coupling of a single photon to a superconducting qubit using circuit quantum electrodynamics. *Nature*, 431(7005):162–167, 2004.

7 Yu-Jung Lu, Jisun Kim, Hung-Ying Chen, Chihhui Wu, Nima Dabidian, Charlotte E Sanders, Chun-Yuan Wang, Ming-Yen Lu, Bo-Hong Li, Xianggang Qiu, et al. Plasmonic nanolaser using epitaxially grown silver film. *Science*, 337(6093):450–453, 2012.

8 Sanfeng Wu, Sonia Buckley, John R Schaibley, Liefeng Feng, Jiaqiang Yan, David G Mandrus, Fariba Hatami, Wang Yao, Jelena Vučković, Arka Majumdar, et al. Monolayer semiconductor nanocavity lasers with ultralow thresholds. *Nature*, 520(7545):69–72, 2015.

9 Jeffrey N Anker, W Paige Hall, Olga Lyandres, Nilam C Shah, Jing Zhao, and Richard P Van Duyne. Biosensing with plasmonic nanosensors. *Nature Materials*, 7(6):442–453, 2008.

10 AV Akimov, A Mukherjee, CL Yu, DE Chang, AS Zibrov, PR Hemmer, H Park, and MD Lukin. Generation of single optical plasmons in metallic nanowires coupled to quantum dots. *Nature*, 450(7168):402–406, 2007.

11 Seunghoon Han, Yi Xiong, Dentcho Genov, Zhaowei Liu, Guy Bartal, and Xiang Zhang. Ray optics at a deep-subwavelength scale: A transformation optics approach. *Nano Letters*, 8(12):4243–4247, 2008.

12 David R Smith, John B Pendry, and Mike CK Wiltshire. Metamaterials and negative refractive index. *Science*, 305(5685):788–792, 2004.

13 Momchil Minkov and Vincenzo Savona. Automated optimization of photonic crystal slab cavities. *Scientific Reports*, 4:5124, 2014.

14 Xuetao Gan, Ren-Jye Shiue, Yuanda Gao, Kin Fai Mak, Xinwen Yao, Luozhou Li, Attila Szep, Dennis Walker Jr, James Hone, Tony F Heinz, et al. High-contrast electrooptic modulation of a photonic crystal nanocavity by electrical gating of graphene. *Nano Letters*, 13(2):691–696, 2013.

15 Fabrice Lemoult, Nadège Kaina, Mathias Fink, and Geoffroy Lerosey. Wave propagation control at the deep subwavelength scale in metamaterials. *Nature Physics*, 9(1):55–60, 2013.

16 Nadege Kaina, Fabrice Lemoult, Mathias Fink, and Geoffroy Lerosey. Ultra small mode volume defect cavities in spatially ordered and disordered metamaterials. *Applied Physics Letters*, 102(14):144104, 2013.

17 KM Ho, Che Ting Chan, and Costas M Soukoulis. Existence of a photonic gap in periodic dielectric structures. *Physical Review Letters*, 65(25):3152, 1990.

18 John D Joannopoulos, Pierre R Villeneuve, Shanhui Fan. Photonic crystals: Putting a new twist on light. *Nature*, 386(6621):143–149, 1997.

19 VN Astratov, DM Whittaker, IS Culshaw, RM Stevenson, MS Skolnick, TF Krauss, and RM De La Rue. Photonic band-structure effects in the reflectivity of periodically patterned waveguides. *Physical Review B*, 60(24):R16255, 1999.

20 William L Barnes, Alain Dereux, and Thomas W Ebbesen. Surface plasmon subwavelength optics. *Nature*, 424(6950):824–830, 2003.

21 Xian-Shi Lin and Xu-Guang Huang. Tooth-shaped plasmonic waveguide filters with nanometeric sizes. *Optics Letters*, 33(23):2874–2876, 2008.

22 Shanhui Fan. Sharp asymmetric line shapes in side-coupled waveguide-cavity systems. *Applied Physics Letters*, 80(6):908–910, 2002.

23 Zhao Xu, Kyungjun Song, and Pinaki Mazumder. Analysis of doubly corrugated spoof surface plasmon polariton (dc-sspp) structure with sub-wavelength transmission at THz frequencies. *IEEE Transactions on Terahertz Science and Technology*, 2(3):345–354, 2012.

24 Mikhail Erementchouk, Soumitra Roy Joy, and Pinaki Mazumder. Electrodynamics of spoof plasmons in periodically corrugated waveguides. *Proceedings of the Royal Society A: Mathematical, Physical and Engineering Sciences*, 472(2195):20160616, 2016.

25 Suzit Hasan Nayem, Nafiz Imtiaz, Soumitra Roy Joy, and Md Zunaid Baten. Theoretical limit and framework of dynamic modulation in spoof surface plasmon polariton interconnects. *Optics Express*, 31(18):29536–29557, 2023.

26 Soumitra Roy Joy, Mikhail Erementchouk, Hao Yu, and Pinaki Mazumder. Spoof plasmon interconnects—communications beyond RC limit. *IEEE Transactions on Communications*, 67(1):599–610, 2018.

27 Soumitra Roy Joy, Md Faizul Bari, Md Zunaid Baten, Feng Lan, and Pinaki Mazumder. A reconfigurable interconnect technology based on spoof plasmon. In *2019 IEEE 19th International Conference on Nanotechnology (IEEE-NANO)*, pages 135–140, 2019. https://doi.org/10.1109/NANO46743.2019.8993934.

28 Md Faizul Bari, Soumitra Roy Joy, Md Zunaid Baten, and Pinaki Mazumder. Process variation in spoof plasmon interconnect: Consequences and compensations. In *2020 IEEE Radio and Wireless Symposium (RWS)*, pages 283–286. IEEE, 2020.

29 Soumitra Roy Joy, Mikhail Erementchouk, and Pinaki Mazumder. Spoof surface plasmon resonant tunneling mode with high quality and purcell factors. *Physical Review B*, 95:075435, 2017. https://doi.org/10.1103/PhysRevB.95.075435.

30 Anastasia Rusina, Maxim Durach, and Mark I Stockman. Theory of spoof plasmons in real metals. In *SPIE NanoScience+ Engineering*, pages 77572R. International Society for Optics and Photonics, 2010.

31 Pei Hang He, Yi Fan, Ling Yun Niu, Hao Chi Zhang, Cun Yue Wei, Junwei Zhang, Hongfei Gao, Yifei Huang, Jie Xu, Wenxuan Tang, and Tie Jun Cui. Analysis, reduction, and utilization of loss in reconfigurable spoof surface plasmon polaritons. *IEEE Transactions on Microwave Theory and Techniques*, 71(3):945–955, 2022.

32 B Sell, S An, J Armstrong, D. Bahr, B. Bains, R. Bambery, K. Bang, D. Basu, S. Bendapudi, D. Bergstrom, et al. Intel 4 CMOS technology featuring advanced FinFET transistors optimized for high density and high-performance computing. In *2022 IEEE Symposium on VLSI Technology and Circuits (VLSI Technology and Circuits)*, pages 282–283, 2022. https://doi.org/10.1109/VLSITechnologyandCir46769.2022.9830194.

33 Han Wui Then, M Radosavljevic, P Koirala, N Thomas, N Nair, I Ban, T Talukdar, P Nordeen, S Ghosh, S Bader, et al. Advanced scaling of enhancement mode high-K gallium nitride-on-300mm-Si (111) transistor and 3D layer transfer GaN-silicon finfet CMOS integration. In *2021 IEEE International Electron Devices Meeting (IEDM)*, pages 11.1.1–11.1.4. IEEE, 2021.

34 David Fritsche, Paul Stärke, Corrado Carta, and Frank Ellinger. A low-power SiGe BiCMOS 190-GHz transceiver chipset with demonstrated data rates up to 50 Gbit/s using on-chip antennas. *IEEE Transactions on Microwave Theory and Techniques*, 65(9):3312–3323, 2017.

35 Pedro Rodríguez-Vázquez, Janusz Grzyb, Bernd Heinemann, and Ullrich R Pfeiffer. A 16-QAM 100-Gb/s 1-M wireless link with an EVM of 17% at 230 GHz in an SiGe technology. *IEEE Microwave and Wireless Components Letters*, 29(4):297–299, 2019.

7

Process Variation Effect on Spoof Plasmonic Interconnect: Compensations

7.1 Introduction

The issue of interconnect designing has so far been dealt with by intelligent designing, such as multiple-layer interconnects, crossbar data bus, etc. [1], but we are gradually running out of design space, and the circumstances call for finding a new type of interconnect obeying different physics. The present thesis proposes a technique of information transfer over short distances, spanning $0.1 - 10\,\text{cm}$ by spoof surface plasmon polariton (SSPP), an entirely different class of quasi-particle accommodated by a metallic surface with a deliberate design pattern [2]. Spoof plasmon interconnect is a novel communication system that leverages the unique electromagnetic properties of metasurface for high-speed data transfer while maintaining high signal integrity by suppressing cross talk between adjacent wires. Unlike conventional interconnects, which incur aggravated signal fidelity at elevated frequencies owing to cross talk, spoof plasmon channels demonstrate quite the reverse trend: the cross talk is suppressed further at the high frequency end of their frequency band, allowing the possibility of faster data transfer with signal integrity.

7.2 Background Studies: Process Variation in Interconnects

The effect of process variation in deep submicron interconnects on-chip is a well-studied field. However, the standard interconnects usually have a very simple and plain geometry, such as that of an electrical interconnect with a rectangular cross-section, and therefore the nominal variation of process parameters resulting in the variation of the geometry of the interconnect does not pose a major issue in

Artificial Plasmonics for VLSI Interconnects: Bridging the Gap between Electronics and Optics, First Edition. Soumitra R. Joy and Pinaki Mazumder.

the electronic conduction mechanism of the standard channels. However, pattern irregularities in spoof plasmon interconnect may warrant us taking the effect of process variation more seriously than ever before in order to deal with potential new issues while testing its feasibility. To elaborate, the unique ability of the metasurface to carry information with high fidelity is attributed to its carefully crafted geometric pattern [3]; and hence it might be critical to brood over the possibility of inadvertent pattern irregularity in the metasurface due to fabrication imperfection, and about the effect of this undesired irregularity on signal integrity. To elaborate, there are finite probabilities that the patterns will not be identical to the design due to variation in the fabrication process, such as mask misalignment, diffraction in lithography, nonuniform metal deposition, polishing, etc. As the electromagnetic properties of a metasurface are governed by the geometry, the inadvertent process variation may lead to aberration of the data transmission behavior of the channel.

It might be argued that the state-of-the-art fabrication technologies have become mature enough to provide sub-nanometer scale resolution [4], and therefore the study of the impact of process variation in SSPP structures with micron-scale feature sizes is now little more than an academic interest. While the argument is valid for a planar or horizontal interconnect structure, an act of metal patterning on a vertical direction, for instance, if an SSPP interconnect has to replace conventional through-silicon-vias (TSV) in a three-dimensional large-scale integration (3D-LSI) system [5], would definitely confront far more challenges and could easily be resulted into significant pattern variation.

7.3 Author's Contribution in the Field of Spoof Plasmon Signal Restoration

In this chapter, we delve into some practical fabrication nonidealities that can impact the performance of spoof plasmon channels. We investigated the frequency responses of an ideal SSPP channel with a regular pattern, and analyzed how these responses get affected when the metal patterning becomes irregular owing to fabrication imperfection. We then elaborated on the consequences of the geometric pattern irregularity of SSPP on the integrity of the digital signal transmitted through the channel; and explained how one can build a compensating analog circuit to recuperate the loss in signal quality in real time.

7.4 Frequency Response of SSPP Channel

7.4.1 SSPP Channel with Ideal Pattern

In order to make point-to-point electrical signal transmission via SSPP interconnect, we designed an appropriate adapter at the transmitting end to convert the transverse electromagnetic (TEM) mode into a surface-bound mode. The adapter

Figure 7.1 Geometric features of the SSPP channel, including the electronic-SSPP converter on both end.

comprises a Vivaldi antenna, and the length of the grooves of the adapter has been gradually increased to match the TEM mode with SSPP mode over broadband. On the receiving side, we employed the mirrored structure of the transmitting adapter. The geometry of an isolated SSPP interconnect with a converter is shown in Figure 7.1.

SSPP metasurface is characterized by a few features, including length (h) and width (a) of groove, thickness (t) of the metal film, and refractive index (n_s) of the substrate. Among them, the most critical parameter happens to be h, since it determines the resonant frequency (ω_r). Near resonance, the SSPP bandgap arises; the detuning of the band-edge frequency (ω_B) from ω_r is proportional to the coupling between the modes in the resonating groove and that in free space. The fundamental band of SSPP spans from zero frequency up to ω_B. The field confinement gradually increases as one proceeds along the band to the highest frequency ω_B. Because of the nominal field confinement in the lower part of the band, we have to forsake some lower fraction of the band to avoid cross talk between neighboring channels. Joy et al. [6] showed that when cross talk remains below −3 dB at any transmitted frequency component in a parallel data bus, the maximum bandwidth density corresponds to the utilization of an upper one-third $\left(\frac{1}{3}\right)$ of the full fundamental band of SSPP for data transfer.

In the present study, we designed the groove length (h) such that the resonant frequency (ω_r) appears to be close to 300 GHz frequency. The ratio a/d is taken to be 0.3. For a given period d in the metasurface, an increase of $\frac{a}{d}$, at the cost of a nominal decrease in bandwidth, benefits us in two ways: it increases field confinement for a broader range of frequencies on the fundamental band, which helps combat cross talk, and it also reduces ohmic loss in the metal, as the loss is proportional to a/l_s, where l_s is the skin depth of the metal at the operating frequency.

7.4.2 SSPP with Pattern Irregularity

We assumed that the process variation would result in a Gaussian distribution of the feature size of the SSPP metasurface, as such sorts of distributions are also reported in the features of standard VLSI interconnects [7]. If needed, our method could be easily adapted for other types of statistics, such as Poisson distribution analysis, by redefining the rate parameters. In our model, the number of unit cells having groove length h is defined as follows:

$$N(h) = \frac{N_T}{\sigma_h \sqrt{2\pi}} \exp\left(-\frac{(h - h_m)^2}{2\sigma_h^2}\right) \tag{7.1}$$

where N_T is the total number of unit cells in the SSPP channel, h_m is the mean value of groove length in the structure, and $\sigma_h^2 = 1/N_T \sum_i (h_i - h_m)^2$ is the variance of the distribution for h, which depends on the particular fabrication process. Since we are dealing with discrete numbers, we ensure that the length of grooves of all the unit cells in a given structure is accounted for within the range of $\pm 3\sigma_h$. As per the suggestion of Joy et al. [6], the signal propagation length through the SSPP channel with a groove length of 250 μm can be up to 5 cm. We have taken the period of the structure to be 375 μm, and the groove width is taken as 125 μm. The rationale for choosing these values of h parameter follows from the fact that the corresponding resonant frequency of the unit cell is 300 GHz, and this is the spectral ballpark where the state-of-the-art sub-terahertz complementary metal-oxide semiconductor (CMOS) oscillator can generate carrier frequency efficiently [8]. The value of period d is chosen in a way so that we meet the condition ($2h > d$) of yielding high field confinement zone of SSPP [2], but at the same time evade being in too deep SSPP region, where the advantage of strong field confinement is smeared by the drawback of dispersion-induced pulse distortion and heightened ohmic loss. With the chosen value of period d in a 5 cm-long interconnect, there will be more than 120 unit cells in the channel, which is a reasonably large number to test the impact of inhomogeneity in their geometric shape.

In order to study the impact of geometric shape variation of SSPP structure on the information capacity of the channel, we took different values of σ_h. For instance, for a 3% variation, the length of the groove h among all the unit cells in the 5 cm-long channel will vary within the range of (250 ± 7.5) μm around the designed length of 250 μm, which corresponds to $3\sigma_h = 7.5$ μm. Figure 7.2 shows

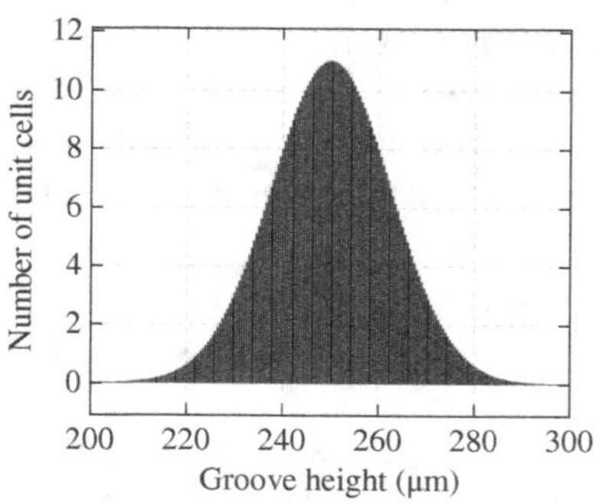

Figure 7.2 Probability distribution of groove length in a 5 cm SSPP channel, where the length varies around the designed mean value by 3%.

the Gaussian distribution of the groove lengths, centered around the mean value of 250 μm, that we have taken to simulate the behavior of the SSPP channel with pattern irregularity.

7.5 Performance Loss for Structural Imperfections

7.5.1 Bandwidth Degradations

Because of the random variation of the length of the groove in the SSPP structure, the transmission characteristics over the frequency range are affected. Figure 7.3a,b shows the magnitude and phase of the transmission coefficient, respectively, for different degrees of coefficient of variance of SSPP pattern irregularity. We defined the bandwidth of a perfect SSPP channel by the difference

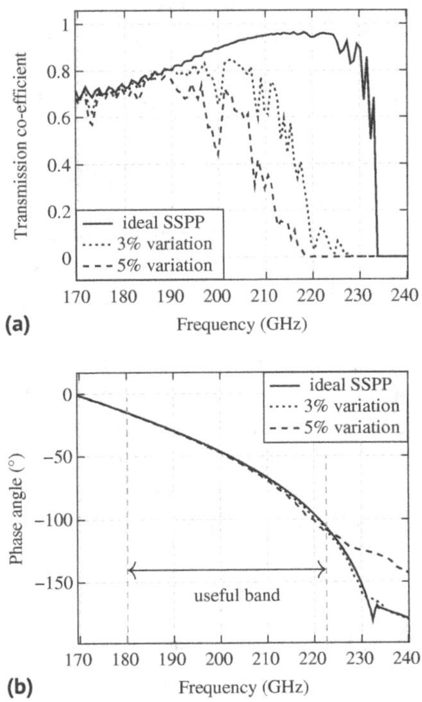

Figure 7.3 (a) Magnitude and (b) phase response of ideal SSPP, SSPP with 3 and 5% coefficients of variance of the irregular metasurface pattern. The relatively linear segment of the band that can be utilized for data transfer is also shown with a double arrow sign.

between the upper (f_U) and lower cutoff frequencies (f_L). The upper cutoff frequency is marked by the drop of power gain from the maxima by −3 dB, while the lower cutoff frequency is marked by the frequency below which the transmission coefficient drops below 0.7. We observed that the upper cutoff frequency is quite sensitive to the variation of groove length, and shows a red shift with the increase of σ_h. On the contrary, the lower cutoff frequency is relatively less sensitive to the variation of groove length. In addition, with the increase in statistical variation of groove length, the roll-off of the transmission power with frequencies around the upper cutoff frequencies reduces.

7.5.2 Loss of Signal Integrity

In order to evaluate the integrity of the transmitted signal, we must carefully consider both the magnitude and phase responses of the SSPP structure. In the absence of any geometric variation, the magnitude of the transmission coefficient shows a value of ≥ 0.8 from 180 to 230 GHz. However, near 230 GHz, we observed a sudden spike in the gain magnitude. This spike is likely to be a result of the Fano antiresonance. At the same time, the phase response also shows a sudden change around Fano antiresonance near 230 GHz. As the strong nonlinearity in phase response can result in significant distortion of the pulse, a fraction of the bandwidth in the upper extreme of the SSPP spectrum has to be sacrificed for the sake of maintaining signal integrity. Therefore, while transmitting data through the SSPP channel, we restrict the bandwidth of the data pulse to be within $(225 - 180)$ GHz $= 45$ GHz, the spectrum over which the gain remains roughly constant and the phase remains roughly linear, as indicated in Figure 7.3b. We transmitted the double-sideband data; hence, the carrier frequency is selected to be the mid-frequency $f_{\text{carrier}} = (225 + 180)/2$ GHz ≈ 203 GHz of the available transmission window of the SSPP channel. We sampled random binary data at a sampling rate of F_s. Since a rectangular function in the time domain spreads up to infinite frequency in the spectral domain, we truncated the spectrum of the pulse by a low-pass filter with a cutoff frequency at $(225 - 203)$ GHz $= 22$ GHz. F_s is chosen in a way so that the main lobe, in addition to eight side lobes on each side of the main lobe of the spectrum of a rectangular pulse, can fit within the transmission window of SSPP.

$$8F_s = 22 \text{ GHz}$$

Theoretically, it could have been possible to make the data transfer rate F_s equal to half of the available bandwidth in the channel. However, in a real set-up, doing so could have resulted in a worse bit error rate on the receiving side due to the presence of noise and the jitter in signal recovery timing.

Figure 7.4 depicts the input signal, taken as "1010" bit pattern, sampled at a frequency of $F_s = 2.75$ GHz and modulated by frequency $f_{\text{carrier}} = 203$ GHz, that is fed to the SSPP channel after appropriate filtering and modulation. Depending on the degree of patterning irregularity in the SSPP metasurface, the fidelity of the signal emerging from the other end of the channel aggravates. Figure 7.5 shows the time domain "1010" signals that emerge at the receiving side of the SSPP channel and compares the signal integrity for different extents of pattern imperfection in

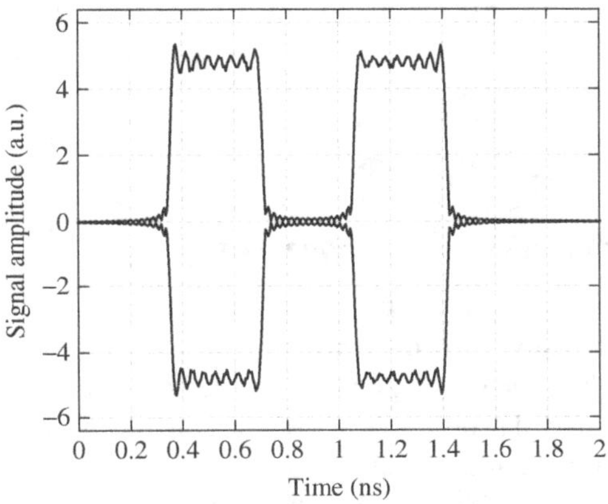

Figure 7.4 Sub-terahertz carrier frequency modulated random digital data, after being filtered, is provided as input to the SSPP interconnect.

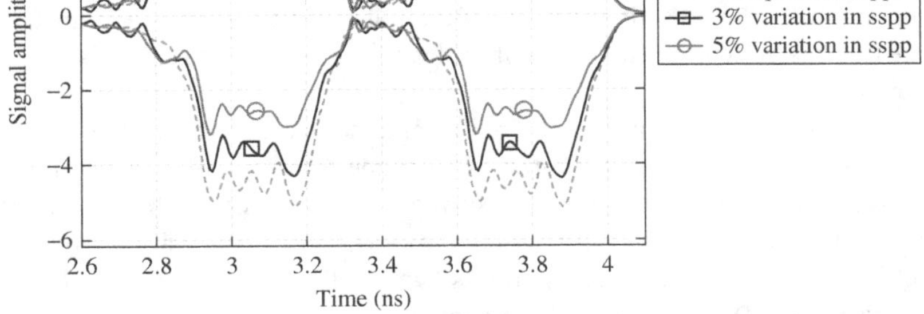

Figure 7.5 Received signal after transmitting through ideal SSPP (solid line), SSPP with 3% variation (square symboled line), and SSPP with 5% variation (circle symboled line). Besides signal attenuation, the signal also suffers from shape distortion.

the channel. The comparison vividly shows that, with the pattern irregularity getting worse, the amplitude of the signal deteriorates. However, if it were only the amplitude would aggravate with pattern irregularity, we could have recuperated the signal by simply deploying an appropriate amplifier. Unfortunately, due to the apparent nonuniformity in the frequency response of the imperfect SSPP channel over the chosen 45 GHz bandwidth, the signal also undergoes pulse-shape distortion, shown in Figure 7.5, which complicates the design process of the compensation circuit for regaining signal quality.

7.6 Mitigation of Performance Degradation

7.6.1 Mathematical Functions to Compensate for Signal Loss

We analyzed the mathematical function of the amplitude response of the compensation circuits so that they can restore the distorted signal to its original shape. In order to do so, we have taken the envelope of a modulated signal of a "...000101000..." data stream. This particular bit pattern was chosen since it represents the most critical transition of bits, and hence is expected to undergo the most severe effect of dispersion and bandwidth loss. The chosen bit stream is filtered and supplied to the input of two different kinds of SSPP channels: one whose geometry is periodic and one whose geometry is not. Thus, we obtained a pair of envelope functions of the output data streams as a function of time, and performed appropriate signal processing in order to express the output signal amplitude of an imperfectly shaped SSPP channel (with variance in groove length) as a function of that of a perfect SSPP channel. The resultant transfer function is then smoothed in order to approximate it with a piecewise linear function. This is the function that must have to be the amplitude transfer characteristics of the nonlinear compensation circuit in order to recuperate the signal from the distortion incurred from the irregularity in the metasurface channel. Figure 7.6 shows the approximated transfer function of the compensating circuit. Different linear segments of the transfer curve are designated with encircled numbers from ① to ⑦ for the ease of subsequent referencing.

Readers should note that, in actual circuit implementation, the transfer curve depicted in Figure 7.6 should retain the same shape over the entire bandwidth of data around the carrier frequency. Ensuring the wide-bandwidth amplitude response of the compensating circuit around the carrier frequency is a challenging task and merits a separate manuscript on its own. In the present work, we limit our analysis to the f DC characteristics only of the compensating circuit.

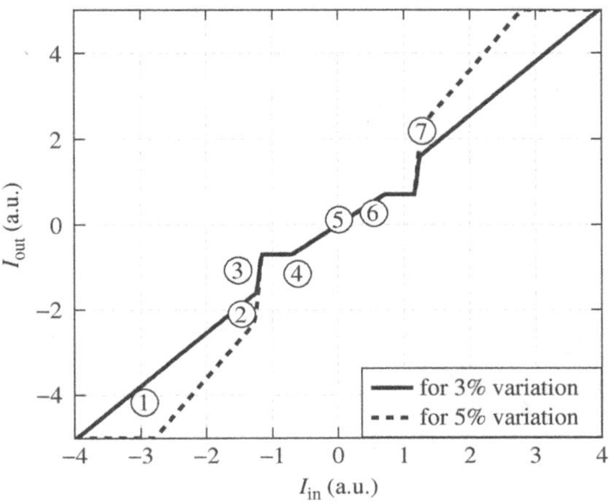

Figure 7.6 Transfer function of nonlinear compensation circuit for different degrees of structural irregularities in SSPP channel. The different linear pieces on a particular compensation function are designated with numbers from 1 to 7.

7.6.2 Nonlinear Circuit Design for Real-time Compensations

There can be multiple ways of implementing a nonlinear circuit with the desired amplitude response function. We have chosen the technique of "summation of current" in a circuit node to generate the particular nonlinear response, since the addition of current can be as simple as joining wires to a common output node. The basic building blocks of the designed circuit are current mirrors (CMs). The CM blocks are supplied with two bias currents, I_1 and I_2. One pair of the field-effect transistor (FET) has a width ratio of $\frac{1}{\alpha}$. The I/V characteristics of each of the CM blocks are a straight line with a slope α, which is clipped at two different coordinates: the bottom coordinate corresponds to $C_1 \equiv (I_1, 0)$, the top one to $C_2 \equiv \left(I_1 + \frac{I_2}{1+\alpha}, \frac{I_2\alpha}{1+\alpha}\right)$. Transition of output current from C_1 to C_2 follows a straight line with slope $= \alpha$. Figure 7.7 shows the basic CM block and its amplitude transfer function. It has been shown in Wilamowski et al. [9] that the above-described CM blocks can be combined to generate an arbitrary nonlinear transfer function.

For a concise representation of the final design of the full compensation circuit, we shall represent the CM block as a black box, characterized by the ratios of transistors' widths. In order to acquire the desired transfer characteristics (shown in Figure 7.6) of the compensating circuit, we added the output nodes of five such CM blocks, where the ratios of transistors' widths of each of the CM blocks have been chosen carefully to generate the full transfer curve, as shown in Figure 7.8. As the individual CM blocks cannot provide bidirectional current, we have to

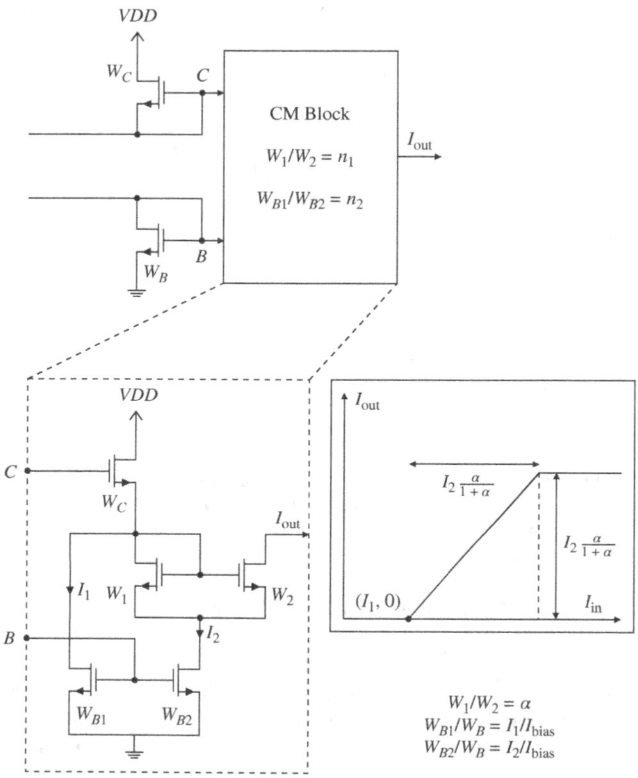

Figure 7.7 Basic CM block of the compensation circuit with its piecewise linear amplitude response shown in the inset. Such sorts of blocks will be combined to construct the full compensation circuit.

also add an offset current source at the input terminal of the compensation circuit to shift the entire transfer curve either to the right or to the left of the x-axis (i.e., to a quadrant of the Cartesian coordinate system where input current is unidirectional). The output of the compensation circuit will be fed to the next SSPP channel through a transformer to truncate the DC component from the output signal.

7.6.3 Dynamic Tunability of Compensation Circuit

Despite a standard fabrication process characterized by minimum resolution features and expectation value of errors, it is unlikely that we would have prior exact knowledge of the degree of variations in a specific sample of the SSPP channel. Hence, the compensation circuit ought to have in situ tunability, as the fine details of the input-output characteristics of the required compensating circuit depend

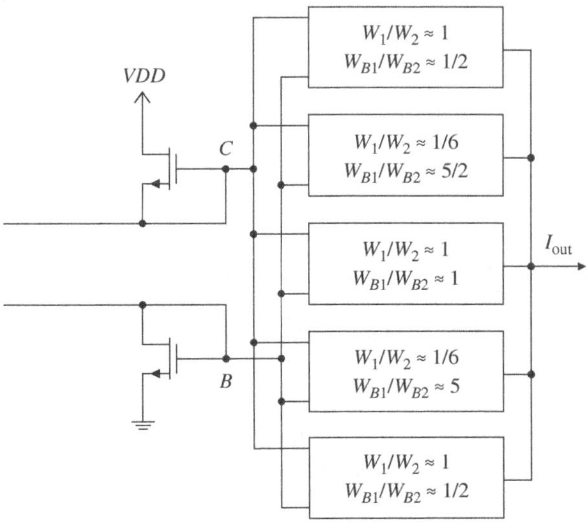

Figure 7.8 Full circuit design for compensation of the performance loss of the SSPP channel for structural irregularity. The CMs are drawn in the form of black boxes, characterized by the ratios of transistor widths for concise representation.

upon the exact coefficient of variance of the geometric pattern of the particular sample interconnect chip.

As suggested by our analysis, regardless of the degree of pattern irregularity, the required input-output amplitude transfer behavior of the compensation circuit remains mostly identical for regions ③, ④, ⑤, designated in Figure 7.6. The parts of the compensation function designated as regions ①, ②, ⑥, ⑦ vary with the degree of geometric irregularity in terms of their slopes, where the slopes have to become steeper with increased degree of pattern irregularity. The tunability of the compensation circuit can be readily obtained by the inclusion of additional FETs parallel to that designated with transistor width W_2 in Figure 7.7. Each of the additional FETs will be controlled by a series-connected transistor. Further tunability, if needed, can also be obtained by the provision of extra bias current sources.

7.7 Summary

With the beginning of the era of "tyranny of interconnect," the quest for a novel interconnect technology can potentially be satiated by spoof surface plasmon-based channels, which get rid of conventional parasitic effects such as channel capacitance and inductance. In this chapter, we endeavor to analyze the collateral fabrication issues that may emerge with the introduction of the new technology, the ramifications of the same on channel performance, and the methodology for compensation for degraded performance by proper circuit design. This may constitute an important step toward the realization of the novel technology and integrating it with the standard CMOS fabrication process.

References

1 Dongkook Park, Soumya Eachempati, Reetuparna Das, Asit K Mishra, Yuan Xie, Narayanan Vijaykrishnan, and Chita R Das. Mira: A multi-layered on-chip interconnect router architecture. In *2008 International Symposium on Computer Architecture*, pages 251–261. IEEE, 2008.

2 Mikhail Erementchouk, Soumitra Roy Joy, and Pinaki Mazumder. Electrodynamics of spoof plasmons in periodically corrugated waveguides. *Proceedings of the Royal Society A: Mathematical, Physical and Engineering Sciences*, 472(2195):20160616, 2016.

3 Paloma Arroyo Huidobro, Antonio I Fernández-Domínguez, John B Pendry, Luis Martín-Moreno, and Francisco J García-Vidal. Spoof surface plasmon metamaterials. *MRS Bulletin*, 45(4):318–318, 2020.

4 Yu Kyoung Ryu Cho, Colin D Rawlings, Heiko Wolf, Martin Spieser, Samuel Bisig, Steffen Reidt, Marilyne Sousa, Subarna R Khanal, Tevis DB Jacobs, and Armin W Knoll. Sub-10 nanometer feature size in silicon using thermal scanning probe lithography. *ACS Nano*, 11(12):11890–11897, 2017.

5 Makoto Motoyoshi. Through-silicon via (TSV). *Proceedings of the IEEE*, 97(1):43–48, 2009.

6 SR Joy, M Erementchouk, H Yu, and P Mazumder. Spoof plasmon interconnects—communications beyond RC limit. *IEEE Transactions on Communications*, 67(1):599–610, 2019. https://doi.org/10.1109/TCOMM.2018.2874242.

7 KG Verma, BK Kaushik, and R Singh. Effects of process variation in VLSI interconnects – a technical review. *Microelectronics International*, 26(3):49–55, 2009.

8 Yahya M Tousi, Omeed Momeni, and Ehsan Afshari. A 283-to-296GHz VCO with 0.76 mW peak output power in 65nm CMOS. In *2012 IEEE International Solid-State Circuits Conference*, pages 258–260. IEEE, 2012.

9 BM Wilamowski, ES Ferre-Pikal, and O Kaynak. Low power, current mode CMOS circuits for synthesis of arbitrary nonlinear functions. In *9th NASA Symposium on VLSI Design*, pages 7–3. IEEE, 2000.

8

Future Research Avenues for Spoof Plasmonic Interconnects

8.1 Introduction

The book, so far, has endeavored to build a coherent theory for data transfer through the bus of spoof plasmon interconnect for point-to-point parallel communication. It began with reviewing the state-of-the-art interconnect technologies and identifying the potential technology gap for energy-efficient information transfer at a short scale (few millimeters to centimeters), a critical length scale for interchip communication (Chapter 1). At the end of Chapter 1, we, the authors, introduced the concept of a new class of interconnect based on spoof plasmonics and discussed how it may contribute to bridging the technology gap. Then, in Chapter 2, we discussed how the idea of surface plasmon polariton emerged, and how the limiting factors of surface plasmon polariton-based devices led the scientists to innovate a new structure to artificially mimic the behavior of surface plasmon polariton at a designer's choice of frequency. Then, we gave an overview of how, along with other researchers around the world, the Mazumder Lab at the University of Michigan has contributed to the field of spoof plasmonics by developing the physics of spoof plasmonics and based on the theory, they proposed two new kinds of very large-scale integration (VLSI) circuit components (such as spoof plasmonic switches, spoof plasmonic Mach–Zehnder Interferer, spoof plasmonic ADC, etc.). In Chapter 3, we took a technical deep dive to understand the mathematical framework for explaining spoof plasmonic interaction with materials, particularly metals and dielectrics. We showed the process of determining the electromagnetic dispersion relation in a periodically corrugated film of metal residing on a substrate – the most common form of in-plane spoof plasmon interconnect. In Chapter 4, based on the developed dispersion relation, we analyzed various performance metrics of spoof plasmon interconnect, such as bandwidth density, propagation loss, bit-error-rate, and their trade-off relations, which will

Artificial Plasmonics for VLSI Interconnects: Bridging the Gap between Electronics and Optics, First Edition. Soumitra R. Joy and Pinaki Mazumder.
 Companion Website: https://www.wiley.com/go/artificialplasmonics

help to identify the potential area of application of spoof plasmonic interconnect based on quantifiable figures of merit. In Chapter 5, we introduce a new concept of hybridized spoof plasmonic mode, which shows several unique advantages over pure, standard spoof plasmonic mode: the hybridized mode can support baseband data transfer and therefore provides an energy-scalable solution for interchip communication. While cross-coupling among wires naturally increases with frequency (as per Faraday's law of electromagnetic induction), a hybridized spoof plasmonic mode can reverse this trend, and therefore can increase the bandwidth of a channel in systems by at least a factor of $\sqrt{2}$. This may prove immensely beneficial for devices with high wiring density, for example, a 200 Gbps bandwidth can readily be enhanced by approximately another 100 Gbps with the hybridized spoof plasmonic mode to accommodate a total bandwidth of almost 300 Gbps without any additional circuitry for driving the signal, and without any additional footprint on the substrate. We also showed how the hybridized spoof plasmonic mode can reduce the latency of data transfer, which may contribute to the long-standing memory-wall problem in VLSI. We also showed the proof-of-concept demonstration of the existence of hybridized spoof plasmonic mode and through experimental realization, and demonstrated the mechanism of bandwidth augmentation. In Chapter 6, we proposed the technique of high-speed modulation of surface waves in spoof plasmonic interconnect, and analyzed how to tailor the speed of modulation, and quantified the fundamental trade-off relation between signal modulation speed and energy efficiency. We showed that the spoof plasmonic mode can propagate up to 100 mm distance at 10 Gbps data transfer speed with >80% energy efficiency.

In this concluding chapter, we shall discuss what the path forward is in realizing a spoof plasmonics-based interconnect system for integration into VLSI circuits. We shall identify what the major roadblocks are in the overarching goal, and how we can systematically address the challenges, so as to bring the spoof plasmonic interconnect technology from the research lab into industrial implementation.

8.2 New Research Frontiers for Spoof Plasmonic Interconnect

8.2.1 New Material Integration with Spoof Plasmonics

One of the primary challenges of spoof plasmonic interconnects is material loss. The spoof plasmonic effect relies on the resonance phenomenon within grooves, which means that to support the propagation of surface waves along the skin of the interconnect, an auxiliary wave must repeatedly bounce back and forth inside the metallic grooves. This process significantly increases ohmic losses in the channel. To address this issue, it is essential to explore alternative methods for supporting spoof plasmonic modes, particularly by identifying materials that are resistant to electron scattering losses. Readers can refer to Chapters 4 and 6 of the book for an in-depth discussion on how energy consumption in spoof plasmonic interconnects escalates as data transfer rates increase.

While the feasibility of achieving spoof plasmonic behavior in all-dielectric media remains uncertain, such a breakthrough could drastically reduce material losses at high frequencies. Superconductors and topological insulators, in this regard, can be investigated as promising materials to be integrated with spoof plasmonic technology, so as to harness the advantages of minimized energy loss from the new materials, as well as the advantages of reduced electromagnetic coupling for high signal integrity.

Superconductors and topological insulators are emerging as promising materials for reducing energy loss and enabling high energy efficiency in ultra-broadband communication systems. Superconductors, with their zero-resistance property below critical temperatures, eliminate ohmic losses, making them ideal for power transmission and signal propagation in high-speed communication systems. Meanwhile, topological insulators, which exhibit surface states that conduct electricity with minimal dissipation while their bulk remains insulating, offer robust performance against scattering and defects, providing high energy efficiency and stable signal transmission.

However, these materials face limitations. Superconductors require extremely low temperatures to operate, necessitating expensive and complex cryogenic systems, which limit their practical deployment. Topological insulators, while promising, often suffer from challenges in material fabrication, such as achieving high-quality thin films and maintaining stability under operational conditions. Additionally, their performance can be compromised by unwanted bulk conduction or disorder in the crystal structure. Overcoming these challenges in these advanced material systems and integrating them with spoof plasmonic technology will be crucial for realizing their full potential in next-generation communication technologies. Figure 8.1 illustrates the purpose and concepts for implementing spoof plasmonic interconnect with advanced materials with low loss.

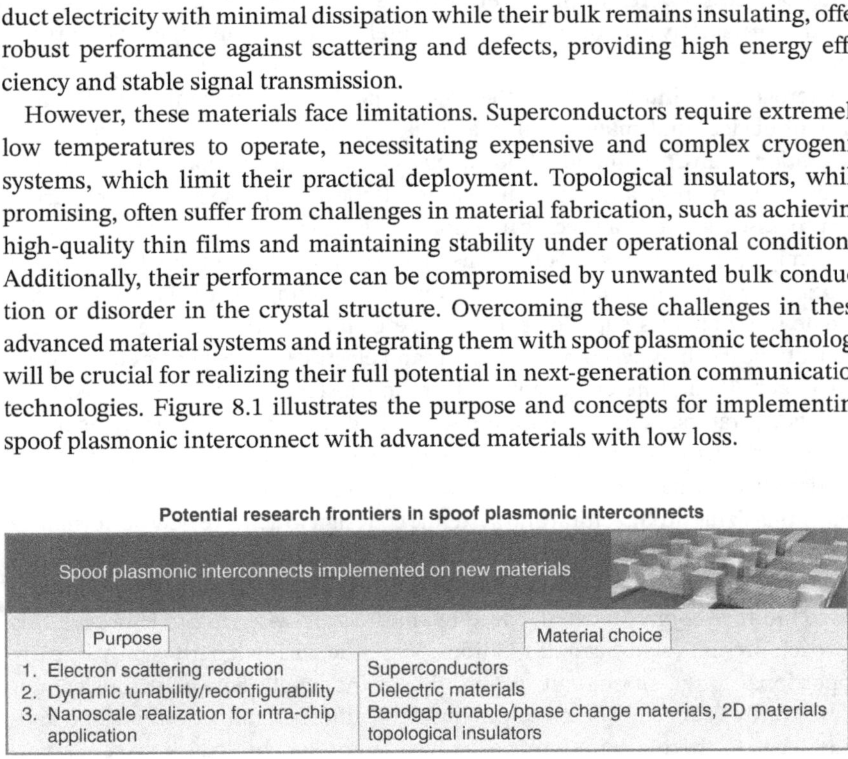

Figure 8.1 Illustration of the idea of realizing spoof plasmonic mode in new materials with better properties, such as low-loss medium, in situ tunable materials, etc.

8.2.2 Spoof Plasmonics for Higher Device Integration

In the post-Moore era, as traditional transistor scaling approaches its physical and economic limits, the focus has shifted toward advanced packaging techniques like three-dimensional (3D) integration and heterogeneous integration to sustain performance improvements in electronic devices. These methodologies enable the continuation of performance scaling by addressing challenges in power efficiency, interconnect density, and functionality enhancement, paving the way for next-generation computing and communication systems.

3D integration involves stacking and vertically interconnecting multiple layers of devices or circuits using through-silicon vias (TSVs), hybrid bonding, or similar technologies. This vertical stacking significantly reduces the physical distance between layers, enabling faster interconnects, and lower latency compared to traditional planar designs. Furthermore, 3D architectures provide higher bandwidth and energy efficiency by reducing signal propagation delays and interconnect power dissipation. For instance, memory modules can be directly stacked on logic processors, creating high-performance compute-in-memory architectures that are particularly beneficial for artificial intelligence (AI) and machine learning workloads.

Heterogeneous integration, on the other hand, focuses on combining different types of devices and materials into a single package. This approach allows the integration of diverse functionalities, such as logic, memory, analog, RF, and photonics, in a compact and efficient manner. By leveraging specialized materials and processes tailored for specific functions, heterogeneous integration maximizes performance for targeted applications. For example, integrating silicon-based processors with compound semiconductors like gallium nitride (GaN) or silicon carbide (SiC) enables superior performance in high-power and high-frequency applications, such as 5G communications and electric vehicle power electronics.

Together, 3D and heterogeneous integration foster innovation in areas, such as AI accelerators, edge computing devices, and quantum systems by enabling higher performance-per-watt, reduced form factors, and enhanced system-level functionality. These technologies represent a paradigm shift from traditional scaling, emphasizing architectural and systemic codesign to achieve new performance benchmarks.

The current packaging techniques for heterogeneous integration predominantly aim to improve communication speed by minimizing wire length. However, this approach presents significant limitations. First, the wiring length is only linearly proportional to the speed, and second, the wiring length between chips cannot be arbitrarily reduced. The close proximity of different chips on the same substrate leads to thermal interference, as heat from one chip can impact the performance of adjacent chips. Furthermore, integrating chips manufactured using disparate process nodes onto a single substrate is challenging due to the lack of a substrate that simultaneously provides robust mechanical, thermal, and signal

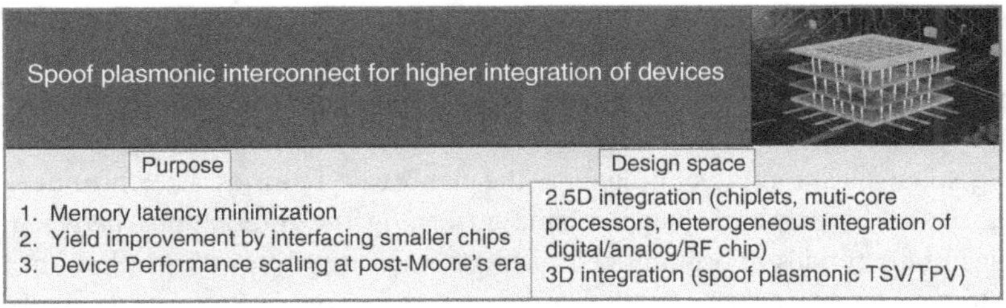

Figure 8.2 Illustration of the idea of realizing spoof plasmonic mode of communication for 2.5D and 3D integration of devices for higher bandwidth of interchip communication.

integrity while supporting high wiring density. These constraints highlight the need for alternative solutions for interchip communication that do not solely rely on wire length reduction. Exploring novel methods, such as spoof plasmonic interconnects, offers a promising paradigm for achieving high-bandwidth communication with reduced delay and energy consumption. In this book, we primarily discussed spoof plasmonic interconnect on planar technologies, we eventually need to invest in research on developing 2.5D and 3D spoof plasmonic interconnect, such as the development of spoof plasmonic interposers, and spoof plasmonic through-silicon-via/through-package-via, spoof plasmonic redistribution layers, etc. Figure 8.2 illustrates the potential research avenues for spoof plasmonic interconnect in designing high integration of devices.

8.2.3 Nanoscale Realization of Spoof Plasmonic Interconnect

The nanoscale realization of spoof plasmonic technology could represent a significant advancement in the field of metamaterials and nanophotonics. Spoof plasmonics, which mimic the behavior of surface plasmons at lower frequencies, are engineered by structuring metallic surfaces with periodic grooves or patterns. While this technology has been extensively explored in the microwave and GHz ranges of frequency, and at sub-THz frequency, scaling this technology to the nanoscale would open a whole new vista by enabling tighter confinement of electromagnetic waves and improved performance in applications like sensing, communication, and energy transfer.

At the nanoscale, spoof plasmons can support highly localized surface wave propagation, enabling the development of compact, high-speed interconnects and devices. This is particularly advantageous for next-generation communication systems where bandwidth and energy efficiency are critical. The key challenge lies in the precise fabrication of nanoscale features, which requires advanced lithographic and etching techniques to achieve the necessary resolution and uniformity. Moreover, nanoscale spoof plasmonic devices must address increased material losses due to electron scattering and resistive heating, which are more pronounced at smaller dimensions.

The successful realization of nanoscale spoof plasmonics can hold promise for a wide range of applications, including ultrafast optical computing and energy-efficient interconnects, paving the way for breakthroughs in nanotechnology and beyond.

Scaling spoof plasmonic technology to the nanoscale requires a deep understanding of the fundamental mechanisms governing the interactions between spoof plasmonic modes and quantized particles such as photons and phonons. These interactions are crucial for determining how energy is transferred, dissipated, and confined at the nanoscale dimensions, which directly impacts the efficiency and functionality of spoof plasmonic devices.

Spoof plasmonic modes are engineered surface waves that arise from structured metallic surfaces, and their behavior at the nanoscale introduces complexities not seen at larger scales. For instance, the interaction of spoof plasmons with photons can lead to unique scattering phenomena and altered dispersion properties, affecting how electromagnetic waves propagate. Similarly, the coupling of spoof plasmons with phonons, the quantized vibrations of a material's lattice, can result in energy loss through processes like heat generation, or, conversely, open pathways for novel hybridized states that could enhance performance.

Despite the potential of these interactions to unlock new functionalities, the theoretical and experimental models describing these mechanisms are still underdeveloped. Challenges include accounting for quantum mechanical effects, nonlocal responses of materials, and nanoscale fabrication imperfections. Furthermore, accurately modeling the coupling strength and energy exchange between spoof plasmonic modes and quantized particles requires advancements in computational techniques and experimental methods.

Developing a robust understanding of these interactions is essential for optimizing nanoscale spoof plasmonic devices. It could lead to breakthroughs in areas like energy-efficient signal processing, quantum information technologies, and highly sensitive nanosensors. Future research must focus on creating comprehensive models and experimental frameworks to explore and harness these fundamental interactions effectively. Figure 8.3 shows the research direction involving advanced modeling of spoof plasmonic interconnect at nanoscale to mesoscale.

8.2.4 Development of Multi-level Computation Model for THz Surface Wave Network

The limitations of conventional simulation techniques of electromagnetic systems are that they cannot recognize the different levels or hierarchies of a system. Instead, they take each part of the geometry of the system on equal footing and generate a large amount of extraneous information that is not of principal interest. In order to make a systematic analysis of THz spoof surface plasmon polariton (SSPP) network with greater computational speed and accuracy, the future

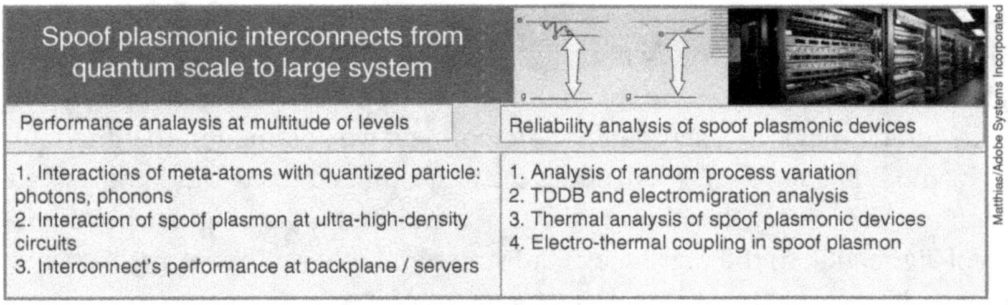

Figure 8.3 Illustration of the research path for advanced modeling of spoof plasmonic interaction at various scales: from the quantum scale to large-scale VLSI.

research strategy may employ a multi-level modeling approach. Under this strategy of computation, the system can be divided into four main levels, and computation mechanisms with closed-form analytical solutions customized for each individual level may be deployed. A brief description of the levels and associated computational physics is described as follows.

8.2.4.1 Meta-atom Level

The fundamental level of the analysis can be the meta-atom level. The subwavelength unit-cells that appear in a periodic manner in the SSPP structures create effective permittivity and permeability of the structures to the electromagnetic mode that is different from that of the materials themselves. These unit-cells can be considered as the constituent meta-atoms governing the dispersion properties of the periodic structure. In finite and thin-film structures, the dispersion relation will be computed by partitioning the meta-atom in two modular spaces: (i) along the long axis of the pattern, where the scalar wave equation will be employed, and (ii) along a short axis of the pattern, where electrostatic formulas will be applied. Both analyses are significantly simpler to solve than the full-wave Maxwell's equations.

8.2.4.2 Channel Level

This can be the second-order level in the envisioned hierarchical modeling. The solution of the wave vectors obtained as a function of operating frequencies and geometric shape of the meta-atoms calculated in the meta-atomistic level can be deployed to analyze the scattering at the heterogeneous interface of the channel. The necessary boundary conditions for channel-level analysis will follow from the principle of energy flux continuity and the continuity of the average electric field over the attenuation length of the evanescent mode. In addition, the electrothermal analysis of the channels resulting from the estimation of effective channel resistance can be employed to incorporate the stochastic behavior of the channels' instantaneous response to incoming signal pulses.

8.2.4.3 Data-bus Level

The third hierarchy of the proposed multi-level analysis can be the data-bus level. In this level, the cross-interference of channels will be modeled so as to modify the terminal characteristics of a standalone interconnect we obtained in the previous levels. The modeling technique will adopt the first-order coupled-wave analysis so as to compute the energy flow into and out of a channel to its neighborhood and any loss of signal due to the presence of substrate and other adjacent media, such as heat-absorbing layers for thermal management of the interconnects.

8.2.4.4 System Level

The fourth and the highest level of the envisioned modeling can be the system level. This level will analyze the complex system comprising electronic transmitters and receivers that exchange packets of data through THz surface-wave. Standard system analysis method, namely Markov chain process, will be employed to characterize the performance of the system to model the stochasticity, and corresponding transition probability matrices describing the propagation among different states of the system will be applied to obtain desired performance metrics such as speed of convergence to stationary states and time-domain response.

Figure 8.4 illustrates the concept of a multi-level modeling approach of a complex SSPP network.

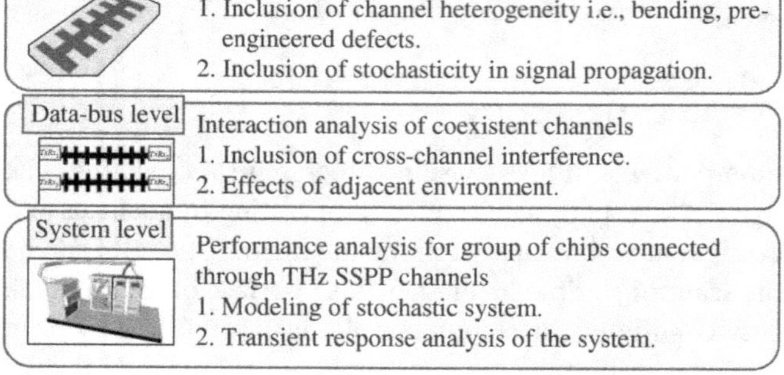

Figure 8.4 Multi-level approach for comprehensive modeling of THz surface wave network, starting from the fundamental meta-atomic level to the system level. The proposed stratified modeling can compute large-scale network with accuracy and time efficiency.

8.3 Summary of the Chapter

The development of spoof plasmonic interconnects holds immense promise for advancing energy-efficient and high-bandwidth communication technologies. However, realizing their full potential requires addressing key challenges, such as minimizing material losses, exploring alternative wave-propagation methods, and scaling the technology to the nanoscale. Achieving this demands a deeper understanding of the interplay between spoof plasmonic modes and quantized particles, to optimize energy transfer and reduce dissipation. The integration of advanced materials, including superconductors and topological insulators, offers a pathway to overcoming these limitations by mitigating ohmic losses and enhancing performance under high-frequency operation. Furthermore, the potential realization of spoof plasmonic behavior in all-dielectric media and the exploration of hybrid nanostructures could open new frontiers in minimizing material loss and improving wave confinement. These challenges and opportunities collectively point to an exciting research trajectory, where the fundamental physics of spoof plasmonic systems and the development of novel materials and fabrication techniques will drive breakthroughs in energy-efficient, high-bandwidth interconnects for exascale computing and beyond.

Index

Note: Page numbers in *italics* and **bold** refers to figures and tables respectively.

a

analog-to-digital converters (ADCs) 77
ADCs. *see* analog-to-digital converters (ADCs)
AI. *see* artificial intelligence (AI)
artificial intelligence (AI) 268
ASK modulation scheme 235, 244, 246–247

b

back-end-of-line (BEOL) 7, 143
backward wave oscillators (BWOs) 57, 58
  baleen whales 103–104, *104*
bandgap (BG) 115, 215
bandwidth 11
  in crosstalk-mediated SSPP channels 147–152, *148, 152*
  degradations 257–258, *257*
  mechanism for bandwidth augmentation 198–209
  modulation of confined mode 131–132
baseband communication, optimal design technique for hybrid-SSPP waveguide for 193–194
BEOL. *see* back-end-of-line (BEOL)
BER. *see* bit error rate (BER)
BG. *see* bandgap (BG)
biosensors 63–64, 81, *91*
  MZI-SSPP-based biosensor 91–92
  THz biosensor design 91–93
bit error rate (BER) 156, 176, 201, 247, 265
  probabilistic 157
  for two sets of interconnect pair *207*
"brute-force" method 138
BWOs. *see* backward wave oscillators (BWOs)

c

carbon nanotubes (CNTs) 8
  interconnect 32–33, *33*
characteristic impedance 186, 234, 242
chip package signaling 18–20, *18–19*
chip-scale interconnect technologies 14–15, *15*
  electrical interconnects 15–16, *16*
Clausius–Mossotti equation 174
CMOS. *see* complementary metal-oxide semiconductor (CMOS)
CNTs. *see* carbon nanotubes (CNTs)

Artificial Plasmonics for VLSI Interconnects: Bridging the Gap between Electronics and Optics, First Edition. Soumitra R. Joy and Pinaki Mazumder.

communications challenges 4–5, *4*
compensation circuit 174, 260
 dynamic tunability of 262–263, *263*
 nonlinear 260, *261*
complementary metal-oxide semiconductor (CMOS) 3, 11, 35, 105, 256
confined modes 129
 bandwidth modulation of 131–132, *132*
 degree of confinement 129–131, *130*
conventional strategies for crosstalk reduction 173–175, *175*
copper interconnects 15, 28
coupled cavity structures 59
critical length 34–35, 145, 160, 190, 265
crosstalk 1–2
 bandwidth in crosstalk-mediated SSPP channels 147–152, *148, 152*
 conventional strategies for crosstalk reduction 173–175, *175*
 in data bus 175
 interconnect, advantages of proposed 175–176, *176*
 length 159
 lossless electrical bus, crosstalk limited bandwidth in 175–179, *178*
 lossless SSPP Bus, crosstalk limited bandwidth in 179–183, *181, 183*
 severity of 171–172, *172*
C-SSPP. *see* cylindrical spoof surface plasmon polariton (C-SSPP)
cylindrical geometries, SSPP in 72–73, *72, 73*
cylindrical spoof surface plasmon polariton (C-SSPP) 88, 89

d

data-bus level in multi-level computation model 272
data transfer 33–34
 alternative interconnect technology 35, *36*
 last centimeter barrier 34–35, *34*
 spoof plasmon interconnect 36–42
 time variation in data traffic 35

DC. *see* direct current (DC)
"defect" 220
demultiplexer (DEMUX) 24
DEMUX. *see* demultiplexer (DEMUX)
differential signaling 173
direct current (DC) 12
dispersion law 109
 double-sided SSPP structure *115*, 115–116
 half-closed SSPP structure *110*, 111–113, *114*
 open SSPP structure 113–115, *114*
 of SSPP waveguides 111–115
dispersion-limited modulation speed, prediction of 232–234
dispersion relation 67–68, 70, 72, 191, 271
 electromagnetic 265
 of spoof SPPs *71, 73*
domino plasmons (DPs) 76
domino structure, SSPP in 75–76, *75*
doped semiconductor 85–86
double-sided SSPP structure, dispersion in 115–116, *115*
doubly corrugated spoof surface plasmon polariton (DC-SSPP) 77–79, 84, 91–92
DPs. *see* domino plasmons (DPs)
Drude–Lorentz model 66–69
"dynamical theory of grating" 62

e

EDM. *see* electric dipole moment (EDM)
eigen-vector 189
electrical channel of length 161
electrical interconnects 5, 15–16, 23, 199
 crosstalk in 160–162, *162*
 optical interconnect *vs.* *25*, 28–29, *29*
electric dipole moment (EDM) 106–107
electrodynamics
 Maxwell's 40
 of spoof plasmon in finite structure 116–117

electro/electronic-SSPP mode 194–195
  experimental verification *196*
  in spoof plasmon waveguide 165
electromagnetic interference (EMI) 27, 210
electromagnetic waves 5, 37, 57, 103, 269–270
  phase velocity 59
  surface-coupled 66
electron beams 57, 59
electro-SSPP interconnect-pair 199, 201, 204, 207
embedded multi-die interconnect bridge (EMIB) technology 142–143
EMI. *see* electromagnetic interference (EMI)
energy-efficiency 35
  analysis of 234–235
  trade-off between modulation speed and 237–238, *238*
enhanced radiation rate 225–226, *226*
ER. *see* extinction ratio (ER)
extinction ratio (ER) 218

f

Fairchild semiconductor 14
far-end crosstalk (FEXT) 20, 172, 177, 195, 197
FD. *see* finite difference (FD)
FDM. *see* frequency division multiplexing (FDM)
FDTD. *see* finite time-domain method (FDTD)
FEXT. *see* far-end crosstalk (FEXT)
FF. *see* form factor (*FF*)
field-effect transistor (FET) 227, 261
  ON-resistance 240
  optimal biasing point of 241
figure-of-merit (FoM) 12, 23, 150
finite difference (FD) 108
finite time-domain method (FDTD) 108
FoM. *see* figure-of-merit (FoM)
form factor (*FF*) 180, 182
Foveros 142
frequency division multiplexing (FDM) 21
front-end (FE) nodes 143

g

gallium nitride (GaN) 268
gigahertz carrier 32
graphene nanoribbons 8

h

half-closed waveguide 109, *110*, 111–113, *114*
helix structures 59
HEMTs. *see* high electron mobility transistors (HEMTs)
heterogeneous integration 2, 141, 268
heterogeneous structures
  SSPP scattering in 221–225
  transmission spectra of 218–221, *218–219, 221*
high electron mobility transistors (HEMTs) 240
high frequency modulated data transmission 192–193
homogeneous structures, transmission spectra of 218–221, *218–219, 221*
human *vs.* machine-level communications challenges 4–5, *4*
hybrid-SSPP interconnect 199, 203
hybrid-SSPP mode 183, *184, 187*
  coupling among large number of parallel hybrid-SSPP channels 190–192
  coupling between two channels of 188–190
  design technique of 192–193
hybrid-SSPP waveguide, optimal design technique for 193–194

i

ICs. *see* integrated circuits (ICs)
ILD. *see* interlayer dielectric material (ILD)

information capacity of spoof plasmonic interconnect
  bandwidth in crosstalk-mediated SSPP channels 147–152, *148, 152*
  data transfer 138–140, *139*
  dual mode in spoof plasmon waveguide 165–166, *165*
  field of interconnect design 145–147
  fundamentally different propagation mode for millimeter-scale packaging 144
  in limit of thermal noise 154–157, *158*
  limitations of standard interconnect technologies 145
  millimeter-scale chip packaging, industry investment in 142–144, *143*
  millimeter-scale communication 140–142, *142*
  SSPP interconnect 157–165
  SSPP mode 152–154
  Transatlantic Telegraph (1858) 137–138, *138*
innovations in interconnect frontier 6, *7*
  chip-scale interconnect technologies 14
  material level 6–8, *9*
  network level 8–10, *10*
  waveguide level 10–14
integrated circuits (ICs) 16–17, 32–33, 35, 36
integrated photonic circuits on silicon 24–27, *25*
Intel's Teraflop Research chip 16
interchip communications 33–34
interconnecting silicon circuit components 14
interconnect research at waveguide level
  plasmonic interconnects 12–13, *13*
  terahertz interconnects 13–14, *14*
  wireless communication at chip scale 10–12, *11*
interconnects 5
  advantages of proposed 175–176, *176*
  field of interconnect design 145–147
  frontier 6–16
  at material level 6–8, *9*
  at network level 8–10, *10*
  modes 5–6, *6*
  at waveguide level 10–16
interlayer dielectric material (ILD) 174
International Technology Roadmap for Semiconductors (ITRS) 18, 23
internet of things network 13
ITRS. *see* International Technology Roadmap for Semiconductors (ITRS)

l

last centimeter interconnect barrier 34–35, *34*
localized surface plasmons (LSPs) 66, 95–97
LSPs. *see* localized surface plasmons (LSPs)

m

Mach–Zehnder interferometers (MZIs) 81
maximum channel density 192
Mazumder Laboratory 40, 42, 76–93, 104, 106
MDM. *see* mode division multiplexing (MDM)
mean free path (MFP) 8
metallic nanoparticle (MNP) 93, 106–107
metallic nanowires (MNWs) 40, 106–107
metal-oxide-semiconductor field-effect transistors (MOSFETs) 240
metal strip 126
MFP. *see* mean free path (MFP)
microfabrication process 198
microwave technology 57, 59
milled wires, SSPP in 73–74, *74*

millimeter-scale packaging
  fundamentally different propagation mode for 144
  industry investment in 142–144, *143*
millimeters-scale communication 141–142, *142*
  challenges in 140
millimeter-wave (mmWave) 29
mmWave. *see* millimeter-wave (mmWave)
MNP. *see* metallic nanoparticle (MNP)
MNWs. *see* metallic nanowires (MNWs)
mode division multiplexing (MDM) 24
modern interconnect technologies 1, 5
  communications challenges 4–5, *4*
  data transfer 33–42
  innovations 6–14, *7*
  modes of 5–6, *6*
  optical interconnect 23–33, *23*
  overview and motivation 1–2, *2*
  problem specifics 2–3, *3*
  research on spoof plasmon wave 42–47, *43*
  scaling issue of 16–23, *17*
modulation characteristics in signal modulation 244–248, *245*, *247*
modulation speed 218, 248
  prediction of dispersion-limited 232–234
  trade-off between energy-efficiency and 237–238, *238*
Moore's Law 15, 141–142, 144
MOSFETs. *see* metal-oxide-semiconductor field-effect transistors (MOSFETs)
MPSoC. *see* multiprocessor-SoC (MPSoC)
multi-level computation model for THz surface wave network 270–272
multiprocessor-SoC (MPSoC) 9
MUX. *see* three-mode MDM multiplexer (MUX)
MZIs. *see* Mach–Zehnder interferometers (MZIs)
MZI-SSPP-based biosensor 91–92, *91*

n

nanophotonics 12, 57
nanoscale realization of spoof plasmonic interconnect 269–270
near-end crosstalk (NEXT) 20
nematic liquid crystal (N-LC) 79–80
network-on-chip (NoC) topologies 9, *10*
NEXT. *see* near-end crosstalk (NEXT)
N-LC. *see* nematic liquid crystal (N-LC)
nonlinear bistable SSPP device design 82–83, *83*
nonlinear circuit design for real-time compensations 261–262, *262*
numerical analysis 235–237, *236*

o

on-chip planar interconnect technology 14
one-dimensional (1D)
  periodic grooves of quarter-wavelength height 119
  plasmonic waveguides 66
  SSPP grating structure 216
open SSPP structure 113–115, *114*
optical carrier 32
optical grating 59–60, *60*
optical interconnect 5, 23–24, *23*
  carbon nanotube interconnect 32–33, *33*
  crosstalk in 158–160, *160*
  electrical *vs.* optical cost comparison *25*, 28–29, *29*
  integrated photonic circuits on silicon 24–27, *25*
  short range 27–29
  wireless network on chip 29–31, *30*
optical-SSPP mode, verification of 195–198, *197*
optimal Shannon information capacity 157
opto-SSPP mode in spoof plasmon waveguide 165, 194–195

p

parallel hybrid-SSPP channels, coupling among large number of 190–192
particle-motion control 93–94, *94*
patterned metallic structures 59–60, *60*
PCBs. *see* printed circuit boards (PCBs)
PCI-Express data bus 21
PEC. *see* perfect electric conductor (PEC)
perfect electric conductor (PEC) 69, 94, 121, 199
performance degradation, mitigation of 260–263
phase-accumulation in SSPP for DAC 87–88, *88*
photonic crystals 13, 93, *222*
  BG formation in 222
  wavelength-scale 217
photonic integrated circuits (PICs) 25
photonic topological insulators (PTIs) 14
photon scanning tunneling microscope (PSTM) 37–38, *38*
PICs. *see* photonic integrated circuits (PICs)
PIST. *see* Plasmonic Interconnect Simulator Tool (PIST)
Plasmonic Interconnect Simulator Tool (PIST) 106
plasmonic interconnects 12–13, *13*, 28
plasmonic waveguides 37, 45, 93, 106, 133
power spectral density 155
printed circuit boards (PCBs) 95
process variation effect on spoof plasmonic interconnect
  bandwidth degradations 257–258, *257*
  dynamic tunability of compensation circuit 262–263, *263*
  frequency response of SSPP channel 254–257
  interconnects, process variation in 253–254
  loss of signal integrity 258–260, *259*
  mathematical functions to compensate for signal loss 260–261, *261*
  mitigation of performance degradation 260–263
  nonlinear circuit design 261–262, *262*
  performance loss for structural imperfections 257–260
  spoof plasmon signal restoration 254
propagating SPPs 66, 106–107
PSTM. *see* photon scanning tunneling microscope (PSTM)
PTIs. *see* photonic topological insulators (PTIs)

q

Q-factor 225–226, *226*

r

radar systems 57, *58*
radio frequency to terahertz (RF–THz) 105
Rayleigh–Schrodinger perturbation theory 109
Rayleigh's theory 62
RDL. *see* redistribution layers (RDL)
real-time compensations, nonlinear circuit design for 261–262, *262*
redistribution layers (RDL) 142
resistor-controlled modulation scheme 239
resonant tunneling 217, 220, 224
RFID tags 97
RF–THz. *see* radio frequency to terahertz (RF–THz)
RLCG model 107

s

scaling issue of system level interconnect 16–17, *17*
  chip package signaling 18–20, *18–19*
  crosstalk, issue of 20–22, *21*
  high power consumption in system-level interconnect 22–23, *22*

scanning tunneling microscope (STM) 37
Schottky effect 86
semiconductor technology 141
Shannon information capacity 147, 154, 156–157
Shannon's paradigm 140
short range optical interconnect 27–29
signal fidelity 32, 161, 253
signal filtering 59–60
signal integrity, loss of 258–260, *259*
signal loss, mathematical functions to compensate for 260–261, *261*
signal modulation by spoof plasmonics 215, *216*
  design considerations 240–241
  design of modulator 215–217
  dynamic switching of SSPP transmission property 226–239
  experimental considerations and 240–248
  modulation characteristics 244–248, *245, 247*
  Q-factor and enhanced radiation rate 225–226, *226*
  spoof plasmon, controlling 217–218
  SSPP scattering in heterogeneous structures 221–225
  transmission characteristics 241–244, *242–243, 245*
  transmission spectra 218–221, *218–219, 221*
signal-to-noise ratio (SNR) 154
silicon carbide (SiC) 268
silicon nitride (SiN) 8
silicon-on-insulator (SOI) 8, 25
silicon photonic integrated circuits 24–26, *25*
  high propagation losses 27
  integration complexity 26
  limited wavelength range 27
  on-chip laser source, absence of 26
  packaging and fiber coupling challenges 27
  thermal management 27
SiN. *see* silicon nitride (SiN)
single sided spoof surface plasmon polariton (SS-SSPP) 85
single-walled CNTs (SWCNTs) 33
SiP. *see* system-in-package (SiP)
slow wave structures 57–61, *58*
  coupled cavity structures 59
  helix structures 59
  modern microwave technology 59
  optical grating 59
  patterned metallic structures 59–60, *60*
  Pendry's discovery 60
  signal filtering 59–60
  surface-plasmon polariton 61
  variants on 59–62, *59, 60*
SNR. *see* signal-to-noise ratio (SNR)
SoC. *see* system-on-a-chip (SoC)
SOI. *see* silicon-on-insulator (SOI)
Sound Fixing and Ranging (SOFAR) channel 103
SP. *see* surface plasmon (SP)
SPICE simulation tool 106–108, *107–108*
spoof plasmon
  controlling 217–218
  researchers technology 43–47, *44*
  signal restoration 254
  toward CMOS compatibility 42–47, *43*
  wave 42–43, *43*
spoof plasmonic interconnect 36–42, *37*, 210–212, *212*
  channel level 271
  data-bus level 272
  for higher device integration 268–269, *269*
  meta-atom level 271
  multi-level computation model 270–272
  nanoscale realization of 269–270
  new material integration with 266–267, *267*
  new research frontiers for 266
  plasmonics 36–42, *38–40, 42*
  system level 272, *272*

spoof plasmonic mode 117–124
  absence of Ez field 121–122
  aided high speed VLSI communication 104–108
  analysis of 117–124
  baleen whales 103–104, *104*
  fundamental electrodynamics of 117–124
  impact of dielectric half-space on dispersion 119–120, *120*
  dispersion relation in finite thickness structures 120–123
  dispersion relation in wide structures 117–119, *118*
  effective refractive index, planar SSPP 124
  electrodynamics in finite structure 116–117
  properties of confined modes 129–132
  effect of substrate 123
  thin-film of SSPP 124–129
  TM mode prevalence 122–123
  universal theoretical framework 109–116, *110*
spoof plasmonics 210–212, *212*
  advancement in 94–97, *96*
  background studies 171–172, *172*
  conventional strategies for crosstalk reduction 173–175, *175*
  crosstalk in data bus 175–183
  experimental characterizations of SSPP data bus 194–198
  experimental validation of bandwidth enhancement 208–209, *208–209*, **209**
  hybrid-SSPP mode 183–193, *184*, *187*
  mechanism for 198–209, *200*, *202–208*
  optimal design technique for hybrid-SSPP waveguide 193–194
  particle-motion control, SSPP waveguide 93–94, *94*
  slow wave structure 57–59, *58*, 59–62, *59*, *60*
  SSPP-based THz circuits, Mazumder laboratory 76–93, *77*
  surface plasmon polariton 62–64, *62*, *63*, *64*, 64–65, 66–69, *67*, *68*
  technology 36–41
spoof surface plasmon polariton (SSPP) 36, 57, 105, 109, 253
  analysis of energy efficiency 234–235
  based ADC 85–87, *85*, *86*
  biosensor design 91–93
  channel 151, 154, 215
  circuit component design 77, 82–91
  crosstalk in 162–165, *164*
  in cylindrical geometries 72–73, *72*, *73*
  for DAC, phase-accumulation 87–88
  dispersion in presence of modulator *227–232*, *227*
  in domino structure 75–76, *75*
  dual mode in 165–166, *165*
  dynamic switching of 226–239
  electrical interconnects, crosstalk in 160–162, *162*
  electromagnetic analysis 77–79, *79*
  electronic-SSPP mode, verification of 195, *196*
  experimental characterizations of 194–198
  frequency response of 254–257
  fundamental physics and electrodynamic analysis 76–77
  gap structure from high sensitivity biosensing 92–93, *93*
  with ideal pattern 254–255, *255*
  interconnects 145–146, 157, 194
  metasurface 255
  microfabrication process 198, *198*
  in milled wires 73–74, *74*
  mode in lossy metal 152–154
  modulators 238–239
  network 270
  numerical analysis 235–237, *236*
  on-chip ultrafast communication systems 77

optical interconnect, crosstalk in 158–160, *160*
optical-SSPP mode, verification of 195–198, *197*
with pattern irregularity 255–257, *256*
in planar geometry 69–76
prediction of dispersion-limited modulation speed 232–234
scattering in heterogeneous structures 221–225
spoof plasmonic biosensor design 77
stopband range 241
structures 216
switch design 80–82
textured surfaces 69–70, *69*
THz beam steering by structure 83–84, *84–85*
THz circuits 76–77, *77*
trade-off between modulation speed and energy-efficiency 237–238, *238*
in 2D dimple arrays 70–72, *71*
waveguide, particle-motion control 93–94

spoof surface plasmons (SSPs) 95
Georgia State University 45
Harvard University 44
Imperial College London 44
Purdue University 46
Southeast University China 44–45
technology 43–47
University of Michigan 46
University of Pennsylvania 45
University of Texas at Austin 45–46

SPP. *see* surface plasmon polariton (SPP)

SPR. *see* surface plasmon resonance (SPR)

SSPP. *see* spoof surface plasmon polariton (SSPP)

SSPs. *see* spoof surface plasmons (SSPs)

SS-SSPP. *see* single sided spoof surface plasmon polariton (SS-SSPP)

standard interconnect technologies 145

STM. *see* scanning tunneling microscope (STM)

superconductors 267

surface plasmon (SP) 63, 106–107

surface plasmon polariton (SPP) 12, 36–37, 61, 105
Drude's model, explanation through 66–69, *67, 68*
extraordinary transmission 64–65, *64*
in metal 62–65, *62, 63*

surface plasmon resonance (SPR) 62, 63

surface wave *61*, 61–62

SWCNTs. *see* single-walled CNTs (SWCNTs)

switching 9
with liquid crystal 79–80, *80*
resonance and absorption 80–81, *81*

system-in-package (SiP) 10

system-on-a-chip (SoC) 9, 20, 29, 142

t

terahertz (THz) 57
biosensor design 91–93
circuit components design 77
interconnects 13–14, *14*
multi-level computation model for 270–272

terahertz SSPP beam splitter 90–91, *91*
beam steering by structure 83–84, *84–85*
circuit component design 82–91
with liquid crystal 79–80, *80*
nonlinear bistable device 82–83, *83*
phase-accumulation for DAC 87–88, *88*
polarizer 88–89, *90*
resonance and absorption 80–81, *81*
SSPP-based ADC 85–87, *85, 86*
switch design 80–82
WCW-based all-optical switch 81–82, *82*

textured surfaces 69–70, *69*

thermal noise power 156
 information capacity in limit of 154–157, *158*
thin-film of SSPP
 capacitance between edges of neighboring conductor plates 126–127
 effective thickness 124–126, *125*
 effect of substrate 127–129, *128*
three-dimensional (3D)
 chip stacking technology 142
 integration 77, 268
three-dimensional large-scale integration (3D-LSI) system 254
three-mode MDM multiplexer (MUX) 24
through silicon vias (TSVs) 16, 268
THz. *see* terahertz (THz)
TL. *see* transmission line (TL)
topological insulators 267
Transatlantic Telegraph (1858) 137–138, *138*
transmission line (TL) 108
transmission resonance 220, 224
transmission spectra of homogeneous and heterogeneous structures 218–221, *218–219, 221*
transverse electromagnetic (TEM) mode 165, 175, 254
traveling wave tubes (TWTs) 57, *58*
TSMC 141–143
TSVs. *see* through silicon vias (TSVs)
2D dimple arrays, SSPP in 70–72, *71*
TWTs. *see* traveling wave tubes (TWTs)

u

ultra high-density fan-out (UHD FO) solutions 142
ultraviolet (UV) 66
universal theoretical framework for spoof plasmonics 109–111, *110*
 dispersion in double-sided SSPP structure 115–116, *115*
 dispersion law of SSPP waveguides 111–115
 half-closed waveguide *110*, 111–113, *114*
 open SSPP structure 113–115, *114*
UV. *see* ultraviolet (UV)

v

varactor-controlled modulation scheme 239
VCVS. *see* voltage-controlled voltage source (VCVS)
very large scale integration (VLSI) 18, 104–106, 141
 circuit components 265
 SPICE simulation tool 106–108, *107–108*
virtual grounding technique 173
VLSI. *see* very large scale integration (VLSI)
voltage-controlled voltage source (VCVS) 107

w

waveguide-cavity-waveguide (WCW) 80
 based all-optical switch 81–82, *82*
wavelength division multiplexing (WDM) 8, 24
WCW. *see* waveguide-cavity-waveguide (WCW)
WDM. *see* wavelength division multiplexing (WDM)
WiNoC. *see* wireless network on chip (WiNoC)
wireless communication at chip scale 10–12, *11*
wireless network on chip (WiNoC) 10–12, *11*, 29–31, *30, 31*
 intrachip/interchip wireless channels 31–32, *32*
 signal fidelity 32